ADHESIVES – RECENT DEVELOPMENTS

ADHESIVES

Recent Developments

Bernard S. Herman

NOYES DATA CORPORATION

Park Ridge, New Jersey London, England

1976

Library of Congress Catalog Card Number: 76-2195
ISBN: 0-8155-0613-9
Printed in the United States

Published in the United States of America by
Noyes Data Corporation
Noyes Building, Park Ridge, New Jersey 07656

FOREWORD

The detailed, descriptive information in this book is based on U.S. patents issued since 1974 and relating to the technology of adhesives.

This book serves a double purpose in that it supplies detailed technical information and can be used as a guide to the U.S. patent literature in this field. By indicating all the information that is significant, and eliminating legal jargon and juristic phraseology, this book presents an advanced, technically oriented review of recent developments in the adhesives industry.

The U.S. patent literature is the largest and most comprehensive collection of technical information in the world. There is more practical, commercial, timely process information assembled here than is available from any other source. The technical information obtained from a patent is extremely reliable and comprehensive; sufficient information must be included to avoid rejection for "insufficient disclosure." These patents include practically all of those issued on the subject in the United States during the period under review; there has been no bias in the selection of patents for inclusion.

The patent literature covers a substantial amount of information not available in the journal literature. The patent literature is a prime source of basic commercially useful information. This information is overlooked by those who rely primarily on the periodical journal literature. It is realized that there is a lag between a patent application on a new process development and the granting of a patent, but it is felt that this may roughly parallel or even anticipate the lag in putting that development into commercial practice.

Many of these patents are being utilized commercially. Whether used or not, they offer opportunities for technological transfer. Also, a major purpose of this book is to describe the number of technical possibilities available, which may open up profitable areas to research and development. The information contained in this book will allow you to establish a sound background before launching into research in the field.

Advanced composition and production methods developed by Noyes Data are employed to bring our new durably bound books to you in a minimum of time. Special techniques are used to close the gap between "manuscript" and "completed book." Industrial technology is progressing so rapidly that time-honored, conventional typesetting, binding and shipping methods are no longer suitable. We have bypassed the delays in the conventional book publishing cycle and provide the user with an effective and convenient means of reviewing up-to-date information in depth.

The Table of Contents is organized in such a way as to serve as a subject index. Other indexes by company, inventor and patent number help in providing easy access to the information contained in this book.

15 Reasons Why the U.S. Patent Office Literature Is Important to You –

1. The U.S. patent literature is the largest and most comprehensive collection of technical information in the world. There is more practical commercial process information assembled here than is available from any other source.

2. The technical information obtained from the patent literature is extremely comprehensive; sufficient information must be included to avoid rejection for "insufficient disclosure."

3. The patent literature is a prime source of basic commercially utilizable information. This information is overlooked by those who rely primarily on the periodical journal literature.

4. An important feature of the patent literature is that it can serve to avoid duplication of research and development.

5. Patents, unlike periodical literature, are bound by definition to contain new information, data and ideas.

6. It can serve as a source of new ideas in a different but related field, and may be outside the patent protection offered the original invention.

7. Since claims are narrowly defined, much valuable information is included that may be outside the legal protection afforded by the claims.

8. Patents discuss the difficulties associated with previous research, development or production techniques, and offer a specific method of overcoming problems. This gives clues to current process information that has not been published in periodicals or books.

9. Can aid in process design by providing a selection of alternate techniques. A powerful research and engineering tool.

10. Obtain licenses – many U.S. chemical patents have not been developed commercially.

11. Patents provide an excellent starting point for the next investigator.

12. Frequently, innovations derived from research are first disclosed in the patent literature, prior to coverage in the periodical literature.

13. Patents offer a most valuable method of keeping abreast of latest technologies, serving an individual's own "current awareness" program.

14. Copies of U.S. patents are easily obtained from the U.S. Patent Office at 50¢ a copy.

15. It is a creative source of ideas for those with imagination.

CONTENTS AND SUBJECT INDEX

INTRODUCTION

While solvent-based adhesives will continue to be used in some parts of the world, particularly in the developing nations, it is expected that antisolvent legislation will have a major impact in the major adhesive consuming countries. With the advent of the EPA Clean Air Act of 1970, OSHA, worker safety concerns and the Arab oil embargo, the need for solvent-free adhesives has been dramatically accelerated.

In 1974, it was estimated that Europe and North America consumed some 300 million gallons of solvent-based adhesives, with over fifty percent of the total being consumed for the manufacture of pressure sensitive labels and tapes.

Major replacements for solvent-based adhesives over the next decade are expected to be two-component reactive adhesives, hot melt, photoreactive and water-based systems.

An adhesive bonding agent is generally characterized as a material capable of bonding separate articles together by adhering to the surfaces of both articles. The selection of an adhesive bonding agent for a particular use is governed by the chemical and physical properties of the adhesive bonding agent composition and the chemical and physical properties of the surfaces of the articles to be bonded. Some of the chemical and physical properties to be considered when an adhesive bonding agent is selected are:

- —the affinity of the surfaces to be bonded to the adhesive bonding agent
- —the ability of the adhesive bonding agent to adequately wet the surface to be bonded
- —the ability of the adhesive bonding agent to be placed in intimate contact with the surface to be bonded
- —the cohesive strength of the adhesive bonding agent itself
- —the chemical reactivity of the adhesive agent with the surface to be bonded.

Hence, the degree of difficulty in selecting or producing an adhesive bonding agent for adhering a plurality of surfaces to each other is directly proportional to the diversity of compositions of surfaces to be bonded and the bonding characteristics of the surfaces themselves.

Generally, adhesive bonding agents are organic polymers having functionality which is adapted to the surfaces to be bonded. Organic polymers are usually employed due to their wide degree of variation in chemical and physical properties.

In addition to the bonding characteristics of adhesive bonding agents, many uses for such agents require sealing characteristics. Thus, the bonding agent must form a continuous mass between the bonded articles which is minimally permeable, if not impermeable to the flow of gases and liquids through the adhesive bonding agent portion of the bonded article. Therefore, adhesive bonding agents having sealing characteristics are generally designated as adhesive-sealants.

In recent years, synthetic polymers have been developed for a wide variety of adhesive applications ranging from the tires we ride on to the many highly demanding uses in space vehicles. In the 1970s particular attention has been devoted to thermoplastic polymer research and formulation techniques for pressure sensitive and hot melt adhesives. Additionally, adhesive systems for plywood impregnation, textile bonding and a number of rubber-to-metal and metal-to-metal structural adhesives, have been optimized to provide environmentally safe products which afford handling ease, rapid cure and processing and, of course, durable composites.

Specialty adhesives and sealants have been developed for dental and prosthetic use, underwater application, electronic components and high temperature composites for a wide variety of modern fabrication techniques.

This book describes over one hundred different bonding processes and provides hundreds of adhesive formulations as noted in the U.S. Patent literature since 1974.

PRESSURE SENSITIVE ADHESIVES

Pressure sensitive adhesive products encompass typically webs of paper, plastic film or sheets, or metal foils carrying on one side a coating or layer of a solventless, solid adhesive composition that possesses sufficient tackiness or tack, as it is referred to in the art, to permit such webs of paper, plastic or metal to be adhered to the surface of an object by merely contacting and momentarily pressing the adhesive layer-carrying side of the web directly against the object surface.

In formulating pressure sensitive adhesive compositions, the skilled adhesive artisan is concerned with producing a composition that will have a certain minimum balance of physical and chemical properties so as to be able to withstand the stresses and environment that the product will encounter and be expected to survive in use.

Generally, in the technology of pressure sensitive adhesives, the physical properties known in the art as tack, peel adhesion, and creep resistance are paramount in determining the suitability of any given pressure sensitive adhesive composition to provide the necessary application properties and in-service durability required of the pressure sensitive adhesive product.

The first property, tack, is in effect the instantaneous adhesion of the adhesive layer of the web to the surface of an object that is developed immediately after contacting the web with the surface. One widely accepted method of measuring tack, endorsed by the Pressure Sensitive Tape Council, is test method PSTC-6 evolved and published as a standard test by the council. It involves rolling a stainless steel ball down an inclined plane unto the surface of a pressure sensitive adhesive layer deposited on a horizontal plane, and measuring the distance in inches, the ball rolls across the adhesive. In this test, tack is expressed in inches of ball travel, and this is an inverse function of the distance the ball rolls, high tack adhesive stopping the ball within 0.5 to 3 inches, low tack adhesives allowing the ball to roll more than 12 inches.

The second important property, peel adhesion, is the adhesion of the pressure sensitive adhesive product after the adhesive layer has been in contact with an adherent surface for a sufficient period of time to reach a steady state or equilibrium with the surface. It is normally measured by American Standards Test Method (ASTM) D-903-49, which involves pressing a one-inch wide strip of an adhesive-coated web unto a test surface, such as a metal or glass plate, allowing the laminate to age under specified conditions of time, temperature and humidity, and then peeling the web from the plate at a 180° angle at a peeling rate of six inches per minute.

For pressure sensitive adhesive products that are meant to be only temporarily adhered, such as masking tapes and descriptive labels, peel adhesion of two pounds or less is usual. Conversely, products that are to be permanently adhered, such as product, trademark or instruction labels, must exhibit peel adhesion of at least three pounds with levels of five pounds or more being preferred.

The third physical property utilized to define the suitability of a pressure sensitive product for a given use is its creep resistance. This test measures in essence the durability and permanence of the adhesive bond of the adhered product when subjected to a constant delaminating force, such as would be present when a pressure sensitive adhesive product is adhered to a vertical or convex surface. It is measured by the Pressure Sensitive Tape Council's test method PSTC-7 that consists of overlapping one inch of a one-inch wide web unto a test surface, such as a plate of metal or glass, conditioning the test specimen for a specified period under standard temperature and humidity conditions, and then imposing a two pound continuous separation force on the specimen in a direction coaxial to the length of the web in a $90°\pm0.5°F$ air environment, and measuring the time for the web to pull away from the plate.

Ideally, pressure sensitive adhesive products, whether removable or permanent, are sought that exhibit no separation. In reality, however, products withstanding fifty hours or more of stress are considered to be superior, with the usual normal range being twenty to forty hours.

Besides these physical property parameters, the chemical attributes of the pressure sensitive adhesive composition must also be considered in order to provide a pressure sensitive adhesive product having the requisite durability in the environment in which it is to be used. Thus, for example, the pressure sensitive adhesive must be formulated so as to exhibit resistance to adhesive-degrading environmental conditions such as elevated temperatures, sunlight, organic solvents, moisture, acids, and bases to which the pressure sensitive adhesive product will be exposed in service.

ACRYLIC POLYMERS

Alkyl Acrylates, Unsaturated Acids and Hydroxylated Amides

According to a process described by *A.M. Coffman; U.S. Patent 3,900,674; August 19, 1975; assigned to The B.F. Goodrich Company* pressure sensitive adhesive polymers have been discovered that have an improved balance of instantaneous adhesion and wettability when applied to a surface and of high cohesive strength and low peel strength. The polymers consist essentially of: (1) from about 76 to 98% by weight of an alkyl acrylate wherein the alkyl group contains at least 6 carbons; (2) from about 2 to 4% by weight of an α,β-olefinically unsaturated carboxylic acid; (3) about 0.04 to 0.15% by weight of an N-alkylol amide of an α,β-olefinically unsaturated carboxylic acid; and (4) up to 20% by weight of an alkyl acrylate wherein the alkyl group contains 4 to 5 carbons. The polymers are particularly useful in the preparation of transparent pressure sensitive adhesive tapes, decorative tapes, and like articles.

The polymers are characterized by having a solution viscosity ranging from about 25,000 to 200,000 cp as measured at 25°C with a Brookfield model LVT viscometer using spindle No. 7 at 10 rpm on a 50% by weight solution of polymer in a mixture of 20 parts by weight of methyl ethyl ketone and 80 parts by weight of cyclohexane. The polymer can also be characterized by recovering it, heating the polymer for 15 minutes at 140°C, and testing for the amount of crosslinking. The polymers have a Williams Plasticity (WP) value ranging from about 1.7 to 2.8 mm, and more preferably from about 2.0 to 2.6 mm as measured using a Williams parallel plate plastometer.

The value is the height in millimeters of a 2.0 gram ball of polymer after being pressed under a 5 kilogram load for 14 minutes at 100°F. Weight percent gel and volume percent swell of the gel can also be used to characterize the crosslinked polymers. The weight percent gel range is from about 20 to 60% by weight of the polymer. The percent swell range is from about 20 to 150% by volume, while a more preferred range is from about 40 to 100% by volume.

Weight percent gel and volume percent swell are measured in the same test. 0.5 gram of polymer is admixed with 100 ml of methyl ethyl ketone in a Pyrex bottle which contains a screen rack of five 50 mesh screens. The mixture sets at room temperature for 16 hours and then the screen rack is withdrawn. Weight percent gel is determined by running a total solids on the solution to measure percent soluble polymer, and the difference calculated. Volume percent swell is determined by weighing the solvent swollen insoluble gel on the screens and comparing this weight to the dry gel weight.

The particular novelty of the polymers of this process over known pressure sensitive adhesive polymers is in their excellent balance between instantaneous adhesion and cohesive strength as measured in the Rolling Ball Tack Test and the 20° angle Shear Test. The polymers also have excellent wettability when applied to a surface, spreading out under no pressure other than the weight of the backing. Although they have good cohesive strength, they have peel strengths around two pounds per inch so they can be easily removed from a surface, and, when removed, they leave no deposit or residue on the surface.

Examples 1 through 12: A series of solution polymerizations were made using a mixture of 80 parts by weight of cyclohexane and 20 parts by weight of methyl ethyl ketone as the solvent. Examples 6, 11 and 12 employed 97.5 parts by weight of cyclohexane and 24.5 parts by weight of methyl ethyl ketone as the solvent. Examples 1 through 7 and 10 through 12 were batch charged into a reactor vessel and heated and agitated for the indicated times. Examples 8 and 9 were handled similarly except that the benzoyl peroxide catalyst was proportioned throughout the run. Examples 10 through 12 were prepared using formulations outside of the claims of this process. Percent conversions were determined from percent total solids measurements. All of the examples reached essentially 100% conversion of monomers to polymer. The formulations and polymerization data are shown in the table below in which the components are in parts by weight.

	1	2	3	4	5	6
2-Ethylhexyl acrylate	97	97	87	87	97	97
n-Butyl acrylate	—	—	10	10	—	—
Acrylic Acid	2.85	2.90	2.92	2.94	2.95	2.96
N-methylol acrylamide	0.15	0.10	0.08	0.06	0.05	0.04
Benzoyl peroxide	0.15	0.15	0.15	0.15	0.15	0.15
Polymerization time, hours	66	64	64	64	64	—
Temperature, °C.	55	50	45	45	50	40
Percent Conversion	100	100	100	100	100	100
Solution viscosity, cp	43000	61000	18000	152000	58000	30000
	7	**8**	**9**	**10**	**11**	**12**
2-Ethylhexy acrylate	86	87	87	97	97	57
n-Butyl acrylate	10	10	10	—	—	40
Acrylic Acid	3.90	2.94	2.945	2.97	2.8	2.8
N-methylol acrylamide	0.10	0.06	0.055	0.025	0.2	0.2
Benzoyl peroxide	0.15	0.15	0.15	0.15	0.15	0.15
Polymerization time, hours	64	—	—	64	70	64
Temperature, °C.	50	45	45	45	55	60
Percent Conversion	100	100	100	100	100	100
Solution viscosity, cp	140000	61000	62400	45800	8400	16600

All of the samples were evaluated as transparent tape pressure sensitive adhesives. The polymer solutions were applied to a Mylar (polyethylene terephthalate) film using a 3 mil draw bar. The solution coated films were heated to 140°C for 15 minutes. The resultant polymer films were about 1 to 1.5 mils in thickness. The adhesive tapes were then evaluated on the Rolling Ball Tack Test, the 20° angle Shear Test, and checked for peel strength and wettability. A small portion of each polymer solution was dried down to obtain a sample for property analysis. After being cured at 140°C for 15 minutes, each polymer sample was analyzed for weight percent gel and volume percent swell, and for its Williams Plasticity. Property data and pressure sensitive adhesive test results are listed in the following table.

	1	2	3	4	5	6	7	8	9	10	11	12
Weight Percent Gel	53	42	39	31	30	36	44	52	39	0	54	—[6]
Volume Percent Swell	36	46	57	95	87	63	48	39	67	0	29	—[6]
Williams Plasticity, mm	2.8	2.1	2.3	2.1	1.7	1.8	2.4	2.5	2.2	1.5	2.4	2.1
20° Shear, minutes	45[2]	98[5]	70[4]	197[5]	44[3]	40[2]	53[1]	240[1]	240[1]	6[1]	11[1]	28[1]
Rolling Ball Tack, inches stopping distance	0.3	0.3	0.2	0.4	—[6]	—[6]	—[6]	0.2	0.2	—[6]	—[6]	—[6]

[1]One Test
[2]Average of 2 Tests
[3]Average of 3 Tests
[4]Average of 4 Tests
[5]Average of 5 Tests
[6]Not Tested

Examples 1 through 9 demonstrate an excellent balance of instantaneous adhesion and cohesive strength, and show the criticality of the polymer composition on the performance of the adhesives. For instance, Example 10, having only 0.025% by weight of the unsaturated hydroxylated amide exhibited a poor 20° shear value (6 minutes); while Example 6, having 0.04% by weight of the unsaturated hydroxylated amide, exhibited a shear value more than six times as great (40 minutes).

Similarly, Example 11, having a high of 0.2% by weight of the unsaturated hydroxylated amide, exhibited a poor 20° shear value (11 minutes); while Example 1, having 0.15% by weight of the unsaturated hydroxylated amide, exhibited a value more than four times as great (45 minutes). Hence, the critical range of the unsaturated hydroxylated amide is amply demonstrated.

All of the adhesives had peel strengths of about two pounds per inch, and all showed good wettability on a smooth, clean glass surface. The adhesives are useful at low temperatures, the compositions having glass transition temperatures (Tg) down to -60°C (Example 2), as measured using a Differential Thermal Analysis.

Ionomeric Copolymers

R.D. Bohme and P.A. Ilagan; U.S. Patent 3,890,292; June 17, 1975; assigned to Daubert Chemical Company describe water dispersible pressure sensitive adhesive compositions comprising ionomeric copolymers of hydrophilic and hydrophobic monomers. In their preferred form the compositions comprise a copolymer formed of an ionic monomer such as a water-soluble alkali metal salt of an α,β-unsaturated monocarboxylic acid, a nonionic monomer such as a water-soluble polyoxyalkylene ester of an α,β-unsaturated monocarboxylic acid, and copolymerizable monomers comprising an α,β-unsaturated monocarboxylic acid, and an insoluble alkyl ester of such an acid. The compositions in dry form are normally tacky and substantially moisture stable. The compositions are useful in the making of various types of tapes, including transfer tapes.

The water-soluble nonionic monomers used in the compositions are polyalkylene oxide condensation products with an α,β-unsaturated monocarboxylic acid exemplified by acrylic acid and methacrylic acid. These esters advantageously contain at least 2, usually from about 10 to 15, and upwards of 100, or more, alkylene oxide groups such as ethylene oxide, propylene oxide, butylene oxide, and the like. Typical of such products are the polyethylene oxide acrylates known as Tergitol 15-S-12 and Carbowax 550. Each of these products contains 12 ethylene oxide units, and differs from each other in the number of carbons in the alkyl group. The following examples are illustrative of the water dispersible pressure sensitive adhesive compositions of this process.

Example 1: Into a one liter flask equipped with a reflux condenser are added 140.6 grams of 2-ethylhexyl acrylate, 27.6 grams of a 50% water solution of polyethylene oxide acrylate (Tergitol 15-S-12), 31.8 grams of acrylic acid, and 250 grams of methyl ethyl ketone. The mixture is purged with nitrogen to expel oxygen, and is then blanketed with nitrogen. To this mixture, 0.4 gram of n-butyl mercaptan and 0.4 gram of a α,β-azobisisobutyronitrile are added, and the mixture is polymerized at 65°C for about 18 hours. To 80 parts of the resulting polymer in solution, 5.8 grams of a 50% solution of potassium hydroxide are added. The final copolymer comprises 75.4% 2-ethylhexyl acrylate, 5.25% potassium acrylate, 11.8% acrylic acid, and 7.4% polyethylene oxide. acrylate. When coated and dried, the polymer is tacky and water soluble.

Example 2: Several copolymers are prepared in accordance with the procedure outlined in Example 1, and evaluated with the copolymer of Example 1 with respect to solubility and tack. The results are tabulated in the table below.

	2-ETHYLHEXYL ACRYLATE GRAMS	POTASSIUM ACRYLATE GRAMS	ACRYLIC ACID GRAMS	TERGITOL 15-S-12 ACRY-LATE GRAMS	TACK LEVEL	SOLUBILITY IN WATER
Comp. of Ex. 1	75.4	5.25	11.8	7.4	Tacky	Soluble
Comp. 2	75.4	0	17.0	7.4	Tacky	Insoluble
Comp. 3	75.4	0	0	24.4	High	Insoluble
Comp. 4	75.4	0	24.4	0	Low	Insoluble
Comp. 5	75.4	14	10	0	Very Low	(Formed Gel)

Example 3: Following the procedure outlined in Example 1, a copolymer is prepared in 100 grams of ethyl acetate from 30 grams of 2-ethylhexyl acrylate, 5 grams of methylacrylate, 40 grams of ethyl acrylate, 10 grams of acrylic acid, and 15 grams of polyoxyethylene acrylate (Tergitol 15-S-12). To one portion (3a) of the copolymer, sufficient potassium hydroxide is added to neutralize 50% of the acrylic acid. To another portion (3b) of the copolymer, sufficient potassium hydroxide is added to neutralize all of the acrylic acid present. The results are set forth in the table below.

	NON-FUNCTION-AL MONOMER, g	ACRY-LIC ACID, g	POTAS-SIUM ACRY-LATE, g	TERGITOL 15-S-12 ACRYLATE, g	SOLUBILITY	% PLASTI-CIZER	ADHESIVE PERFOR-MANCE
Comp. of Ex. 3	75	5	0	15	Insoluble	3	—
Comp. 3a	75	2.5	3.9	15	Insoluble	3	825 g/in.peel
Comp. 3b	75	0	7.8	15	Soluble	3	800 g/in.peel

Example 4: To each of the copolymers of Example 2, equal amounts of a water-soluble plasticizer and a water-insoluble plasticizer are admixed. The water-soluble plasticizer is a polyethylene glycol monophenyl ether having an average molecular weight of 270 which is known as Pycal 94. The water-insoluble plasticizer is di-isodecylphthalate (DIDP). The results of the compounding with respect to tack, tack sensitivity to humidity, and water dispersibility are set forth in the table below.

	g PYCAL 94/PER 100g POLYMER	g DIDP PER 100g POLYMER	ADHESIVE APPEAR-ANCE	SURFACE TACK 90% rH	SURFACE TACK 50% rH	SURFACE TACK 10% rH	DISPERSA-BILITY IN WATER
Comp. of Ex. 1	10	6.8	Soft	125	195	153	Soluble
Comp. 2	10	6.8	Soft	113	76	87	Insoluble
Comp. 3	10	6.8	Very Weak	10	10	34	Insoluble
Comp. 4	10	6.8	Hard	500	450	183	Insoluble
Comp. 5	10	6.8	Not suitable for adhesive due to gellation				

"Soft" indicates that surface tack is sufficiently high to cause transfer of the adhesive to a test loop under the conditions of the test.

To determine the adhesive surface tack as a function of relative humidity of coated samples, the adhesive is coated 1 mil thick onto the surface of a glass plate and conditioned for 24 hours in a dessicator partially filled with a moisture controlling salt solution.

The loop method is used for determining the tack wherein a strip (½ inch wide, 6 inches long) of polyester film is looped and allowed to contact the adhesive under its own weight. After a 10 second residence time the loop is lifted and the force required to remove the loop is measured. The readings obtained, averaged over the number of measurements are recorded as tack in the table above.

Interpolymers Containing Enolizable Keto Group and Orthotitanic Acid Chelate

A process described by *L.W. McKenna, Jr.; U.S. Patent 3,900,610; August 19, 1975; assigned to Monsanto Company* provides pressure sensitive resin solutions with improved viscosity and flow. These resin solutions are made of two components.

The first component is an interpolymer having a weight average molecular weight in the range of 10,000 to 500,000 and a glass transition temperature in the range of -15° to -75°C. The interpolymer comprises between 0.5 and 20 weight percent of at least one monomer containing a hydroxyl, carboxyl or enolizable keto group; at least one monomer selected from the group consisting of esters of acrylic acid and methacrylic acid containing from 6 to 20 carbons; and optionally, a monomer selected from the group consisting of α-olefins containing from 2 to 10 carbons, vinyl esters of alkanoic acids containing from 3 to 10 carbons, ethyl and methyl esters of acrylic and methacrylic acids, acrylonitrile, methacrylonitrile, styrene and vinyl chloride.

The second component of the resin solution is a chelate ester of orthotitanic acid having the formula:

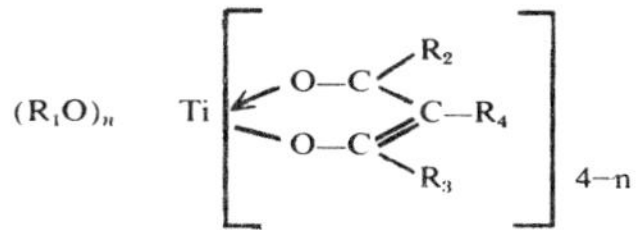

where n is an integer of 2 or 3, R_1 is a C_2 to C_{10} alkyl, alkenyl, substituted alkyl or substituted alkenyl group, R_2 is a C_1 to C_6 alkyl, alkoxy, alkenyl or alkenoxy group, R_3 is a C_1 to C_6 alkyl or alkenyl group or a C_6 to C_{10} aryl group and R_4 is hydrogen or a C_1 to C_6 alkyl or alkenyl group, and R_2 and R_3 may be combined as an ethylene or a trimethylene group.

The process involves the preparation of the interpolymer which is then formulated with the chelate ester of orthotitanic acid in a suitable organic solvent. The solution is cast on a substrate and the resulting film is dried and cured to a creep-resistant permanently tacky composition.

The following examples illustrate the preparation formulation and testing of the special interpolymers which are to be used in the process. All parts and percentages are by weight unless otherwise specified and the expressions polymer and interpolymer are used interchangeably.

Example 1: This example illustrates the preparation of an interpolymer comprising 50.5 parts 2-ethylhexyl acrylate, 45 parts methyl acrylate and 4.5 parts 2-hydroxyethyl acrylate. The polymer is prepared under reflux conditions in a kettle equipped with a stirrer, condenser, holding tanks and pumps.

Ingredients	Initial Charge*	Subsequent Charges* 1	2	3
2-ethylhexyl acrylate	14.1	-	6.97	-
Methyl acrylate	12.6	-	6.20	-
2-hydroxyethyl acrylate	1.26	-	0.62	-
Ethyl acetate	19.69	12.3	-	9.0
Hexane	5.14	-	2.74	9.26
α,α'-azo-diisobutyronitrile	0.076	0.038	-	-
Dodecyl mercaptan	0.0027	0.0013	-	-

*Parts by weight

Charges 1 and 2 are made at uniform rates over the 1¼ hour period after the initial charge reaches reflux. After 6½ hours of reflux, the batch is cooled and charge 3 is added. The solids content is 41.0%. The Brookfield viscosity is 3,000 cp. The relative viscosity (2 g resin in 100 ml benzene) equals 4.1.

Examples 2 through 11: The general procedure of Example 1 is followed except that different monomers and monomer ratios are used in order to illustrate a variety of interpolymers used in the practice of this process.

Compositions of Interpolymers

Example	Monomeric Components	Weight Ratio
1	EHA/MA/HEA	50.5/45/4.5
2	EHA/VAc/AA	59/39.5/1.5
3	EHA/VAc/AA	52/47/1.0
4	EHA/MA/AA	61/33/6.2
5	EHA/MA/HEA	71/24/5
6	iBA/MA/HEA	71/24/5
7	VAc/EHA/HPA	45/50/5
8	VAc/EHA/HEMA	35/60/5
9	VAc/EHA/HEF	40/55/5
10	AN/EHA/HEMA	25/70/5
11	VCl/EHA/HEMA	30/65/5

Note: EHA - 2-ethylhexyl acrylate
MA - methyl acrylate
HEA - 2-hydroxyethyl acrylate
VAc - vinyl acetate
AA - acrylic acid
HPA - 3-hydroxypropyl acrylate
HEMA - 2-hydroxyethyl methacrylate
HEF - bis-(2-hydroxyethyl)fumarate
AN - acrylonitrile
VCl - vinyl chloride
iBA - isobutyl acrylate

Solution Properties of Interpolymers

Ex.	SOLVENT Weight % Ethyl Acetate	Toluene	Hexane	iPrOH	EtOH	Solids, %	Viscosity, cp	Relative Viscosity
1	70	—	30	—	—	41	3,000	4.1
2	50	—	50	—	—	50	15,000	2.7
3	10	19	71	—	—	56	30,000	3.2
4	75	—	15	10	—	39	15,000	4.3
5	70	—	20	10	—	41	5,200	4.7
6	60	20	—	—	20	41	12,000	—
7	74	—	18	—	8	42	8,800	5.7
8	74	—	—	—	26	40	8,500	—
9	74	—	—	—	26	43	9,200	—
10	20	80	—	—	—	37	—	—
11	100	—	—	—	—	41	—	—

After the preparation of the interpolymer, the interpolymer solution is formulated with at least one chelate ester of orthotitanic acid of the general formula:

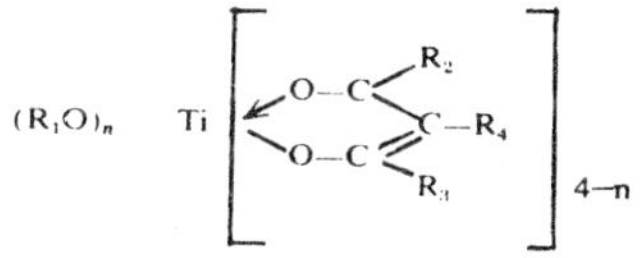

where n is an integer of 2 or 3, R_1 is a C_2 to C_{10} alkyl, alkenyl, substituted alkyl or substituted alkenyl group, R_2 is a C_1 to C_6 alkyl, alkoxy, alkenyl or alkenoxy group, R_3 is a C_1 to C_6 alkyl or alkenyl group or a C_6 to C_{10} aryl group and R_4 is hydrogen or a C_1 to C_6 alkyl or alkenyl group, and R_2 and R_3 may be combined as an ethylene or a trimethylene group.

The chelate esters are prepared by conventional methods such as the addition of the appropriate chelating agent to an alkyl or alkenyl titanate where the alkyl or alkenyl group contains 2 to 10 carbons. The preferred chelate esters are dialkoxy titanium diacetylacetonates, dialkyl titanium di(1,3-cyclopentandionates) and dialkoxy titanium di(1,3-cyclohexandionates).

In the evaluation of pressure sensitive resin solutions containing esters of orthotitanic acid, the solutions are formulated to 32 weight percent solids. The esters of orthotitanic acid are added in ethanol solution gradually with stirring to the pressure sensitive resin solution, the amount of ethanol being 25 weight percent of the total solvent in the final solution.

The solutions are subjected to determination of viscosity and stringiness index. Viscosity is determined at 20°C by the conventional method using a Brookfield Model LVF Viscometer and the appropriate spindle and speed. Viscosity stability is determined from viscosity measurements over a 7 day period.

The stringiness index (SI) is determined by immersing a platinum surface tension ring of 3.6 cm diameter, 1 inch below the surface of the pressure sensitive resin solution contained in a 3 inch diameter jar. The ring is mounted in the jaw of an Instron Test Machine and is withdrawn from the solution by allowing the crosshead upon which the jar rests to descend at a rate of 5 inches per minute. The tensile force exerted on the ring is recorded by means of the A cell on the chart set at 10 grams full scale and moving at 5 inches per minute. The stringiness index is measured from the peak of maximum extensional force to the failure point of the elongating curtain of solution.

Five measurements are made in rapid succession. The arithmetic average expressed in inches is the stringiness index. In general, a stringiness index of less than 0.30 inch indicates that a resin solution has good flow properties and that the solution will not form strings on conventional high-speed roll coaters at normal speeds of operation. The tendency of a pressure sensitive resin solution to develop stringiness is determined by measurement of the stringiness index over a 7 day period.

The data for viscosity and stringiness index of the interpolymer solutions of Examples 1 through 4 formulated with various esters of orthotitanic acid are presented in the tables below. The solids content of the solutions is 32%. Ethanol forms 25 percent of the solvent. In the tables, TBT signifies tetrabutyl titanate and TAA signifies diisopropoxy titanium diacetylacetonate which is formed by reaction of 2 mols of acetylacetone with 1 mol of tetraisopropyl titanate. In the tables below the viscosities are given in centiposes and the stringiness index in inches.

Effect of Titanate Type on Solution Viscosities

Titanate Concentration, grams per 100 g. resin	Type	RESIN: Example 1 Initial	Example 1 Aged 1 week	Example 2 Initial	Example 2 Aged 1 week	Example 3 Initial	Example 3 Aged 1 week	Example 4 Initial	Example 4 Aged 1 week
0	none	800	850	195	180	375	380	—	—
0.2	TBT	1,170	1,140	270	280	535	560	—	—
	TAA	810	840	235	240	415	410	—	—
0.4	TBT	1,920	1,800	470	500	1,110	1,200	Gel	Gel
	TAA	870	900	300	345	450	460	7,700	14,000
0.6	TBT	Gel	Gel	1,130	1,350	2,550*	3,000*	Gel	Gel
	TAA	940	1,000	250	250	455	480	9,000	18,000

*Very ropy, stringy; not coatable.

Effect on Titanate Type on Stringiness Index

Titanate Concentration, grams per 100 g. resin	Type	RESIN: Example 1 Initial	Example 1 Aged 1 week	Example 2 Initial	Example 2 Aged 1 week	Example 3 Initial	Example 3 Aged 1 week
0	None	0.18	0.21	0.23	0.19	0.13	0.14
0.2	TBT	0.22	0.25	0.21	0.30	0.14	0.17
	TAA	0.18	0.21	0.20	0.29	0.18	0.18
0.4	TBT	0.35	0.27	0.37	0.41	0.21	0.20
	TAA	0.20	0.19	0.18	0.26	0.18	0.18
0.6	TBT	—	—	0.39	0.53	0.64	0.72
	TAA	0.20	0.22	0.22	0.27	0.18	0.18

The data show the higher viscosities, the tendency to gel and the higher stringiness indices of pressure sensitive resin solutions containing tetrabutyl titanate in comparison with solutions containing diisopropoxy titanium diacetylacetonate especially at higher concentrations of the titanate ester. Data for stringiness index of interpolymer solutions of Examples 3 and 6 through 11 formulated with various chelate esters of orthotitanic acid are shown in the table below. The solids content of the solutions is 30 percent. Ethanol forms 25 percent of the solvent. The concentration of chelate ester of titanium is 0.4 part per 100 parts by weight of resin.

Resin Ex.	TITANIUM CHELATE ESTER: n	R_1	Chelating Agent	Stringiness Index(in.)
3	2	iso-propyl	acetylacetone	<0.3
6	3	iso-propyl	acetylacetone	<0.3
7	2	octyl	benzoylacetone	<0.3
8	2	butyl	1,3-cyclohexanedione	<0.3
9	2	octyl	1,3-cyclopentanedione	<0.3
10	2	butyl	ethyl acetoacetate	<0.3
11	2	iso-propyl	acetylacetone	<0.3

The pressure sensitive resin solutions are cast on silicon release paper, and the cast films are dried at room temperature, then at 90°C for 2 minutes. The (0.8 mil thickness) films are cooled to room temperature and applied to polyvinyl chloride film (Ultron Polyvinyl Chloride UL-58 Film, 3 mils thick). The laminate is cut into one inch strips. The strips are conditioned at 70°F and 50% relative humidity for 24 hours. The release paper is removed and the strips are applied to steel panels (ASTM 1000-65) with a Pressure Sensitive Tape Council roller. Peel strength is determined on an Instron Test Machine by peeling the strip at a 180° angle and at a rate of 6 inches per minute from the steel panel.

Green strength is gauged by comparing the peel strength of the bond 20 minutes after formation and 24 hours after formation. Data are presented in the table below. The peel strength of the pressure sensitive resins is given in pounds per inch width.

Titanate Concentration, grams per 100 g. resin	Type	RESIN Example 1 20 min.	Example 1 24 hr.	Example 2 20 min.	Example 2 24 hr.	Example 3 20 min.	Example 3 24 hr.
0	none	3.8	4.0	5.0	8.6_c*	4.2	4.7
0.2	TBT	3.3	4.0	4.0	7.0_c*	3.9	4.4
	TAA	3.4	3.8	3.7	7.5_c*	3.6	4.6
0.4	TBT	3.0	3.9	3.1	4.8	3.7	4.2
	TAA	3.0	4.2	3.0	5.5	3.2	4.2
0.6	TBT	—	—	2.5	4.4	—	—
	TAA	3.0	3.5	2.7	4.5	2.9	4.0

*Subscript "c" denotes cohesive splitting of the film.

The data show that titanate ester causes a decrease in peel strength in direct relation to the concentration of titanate ester. However, the effects of tetrabutyl titanate and diisopropoxy titanium diacetylacetonate are essentially equivalent and in no case is the peel strength reduced below a useful value.

The cohesive strength of the pressure sensitive adhesive is gauged from the creep resistance of a 0.25 square inch 0.8 mil thick bond formed by the film of adhesive between a strip of Mylar Polyester film and a polished stainless steel bar. The Mylar strip is loaded with a one-pound weight. The bond is held in a vertical plane at 70°F and 50% relative humidity and the time in hours for failure after application of the load is determined. The data are presented in the table below. The creep resistance of the pressure sensitive resins is given in hours to failure.

Titanate Concentration, g. per 100g. resin	Type	RESIN Ex. 1	Ex. 2	Ex. 3	Ex. 4
0	none	0.2	0.3	0.7	0.5
0.2	TBT	3	3	4.6	—
	TAA	3	2	4	10
0.4	TBT	10	55	40	—
	TAA	25	15	55	25
0.6	TBT	—	52	—	—
	TAA	55	145	155	—

Note that at low concentrations of titanate ester, the initial creep resistance is roughly the same for resins containing tetrabutyl titanate and those containing diisopropoxy titanium diacetylacetonate; but at higher concentrations, the creep resistance improves to a higher level for TAA formulations than for the TBT systems. An exception is the Example 2 system for which TBT may be a more efficient crosslinker.

Interpolymers Containing Acid-Epoxy Groups and 1,3-Bis(Dimethylamino)-2-Hydroxypropane

D.M. Gardner and L.W. McKenna; U.S. Patents 3,893,982; July 8, 1975; and 3,903,057; September 2, 1975; both assigned to Monsanto Company describe pressure sensitive resin solutions containing an acid-epoxy self-curing interpolymer and 1,3-bis(dimethylamino)-2-hydroxypropane. The solutions exhibit improved viscosity stability and provide pressure sensitive resin films with cohesive strength and low release values from silicon release paper. The interpolymers are conveniently prepared by the process set forth in U.S. Patent 3,284,423 to E.C. Knapp.

The following resins were made by the process described in U.S. Patent 3,284,423. All parts and percentages are by weight unless otherwise specified. The terms interpolymer and resin are used interchangeably.

Examples 1 through 4:

Resin Solution Compositions, Parts by Weight

	Ex. 1	Ex. 2	Ex. 3	Ex. 4
Acrylic Acid	6.6	6.9	7.4	5.0
Glycidyl Methacrylate	0.07	0.2	0.35	—
Allyl Glycidyl Ether	—	—	—	1.0
2-Ethylhexyl Acrylate	60.0	60.0	60.0	74.0
Methyl Acrylate	33.0	32.5	32.0	—
Vinyl Acetate	—	—	—	20
Ethyl Acetate	95.0	95.0	95.0	80.0
Hexane	19.0	19.0	19.0	—
Toluene	—	—	—	24.0

Formulation of the acid-epoxy pressure sensitive resin solutions with the amine catalyst is conveniently carried out by preparing a solution of the amine in an organic solvent at a concentration between 5 and 50% and adding the solution to the pressure sensitive resin solution slowly with stirring to disperse the amine uniformly throughout. The amount of 1,3-bis(dimethylamino)-2-hydroxypropane added to the resin solution is between 0.01 and 1.0 part by weight per 100 parts by weight of resin and is preferably between 0.05 and 0.4 part by weight for adequate viscosity stability and curing rate.

The formulated solutions are subjected to determination of viscosity over a period of at least 24 hours at 25°C to determine the pot life. The viscosity is determined at 25°C by the conventional method using a Brookfield Model LVF Viscometer and the appropriate spindle and speed.

Data for Resin Examples 1 and 2 diluted to 30 percent solids are presented in the table below. In the tables which follow, triethylenediamine is designated as TDA and 1,3-bis(dimethylamino)-2-hydroxypropane is designated as BDMAHP. The data show that resin solutions containing 1,3-bis(dimethylamino)-2-hydroxypropane are vastly superior to resin solutions containing triethylenediamine in viscosity stability and pot life. The effect of catalyst on the curing rate is determined by comparing the creep resistance of the catalyzed and uncatalyzed pressure sensitive resins and by comparing the degree of cure of the catalyzed and uncatalyzed resins.

Resin Ex. 1	Parts Catalyst per 100 Parts Resin	Brookfield Viscosity (centipoises) Initial	315 Hrs.
Uncatalyzed	—	3,150	3,350
TDA	0.2	3,550	>50,000
BDMAHP	0.15	3,750	3,800
	0.30	4,740	5,000
Resin Ex. 2			
Uncatalyzed	—	2,300	2,400
TDA	0.2	2,300	Gel
BDMAHP	0.2	2,300	2,600

The pressure sensitive resin solutions are cast on silicon release paper and the cast films are dried for 15 minutes at room temperature and 2 minutes at 90°F. The films (0.8 mil thickness) are cooled to room temperature and transferred to a film of polyester known as Mylar. The Mylar film is cut into ½-inch strips which are applied to polished stainless steel to form 0.25 square inch bonds. The bonds are held in a vertical plane at 70°F and 50% relative humidity and are loaded with one pound weights. The time in hours for failure of the bond after application of the load is determined. Data for the resin of Example 1 cured for 2 minutes and for 60 minutes at 100°C are shown in the table below.

Resin Ex. 1	Parts Catalyst per 100 Parts Resin	Creep Time to Failure, Hrs. 2 min.	60 min.
Uncatalyzed	—	1.0	13
TDA	0.2	1.5	22
BDMAHP	0.15	1.3	29
	0.30	1.6	41

The data show that cured films containing 1,3-bis(dimethylamino)-2-hydroxypropane as the cure catalyst are superior in cohesive strength to films of the same polymer without catalyst or with triethylenediamine catalyst.

The degree of cure of the pressure sensitive resin is determined by heating a dry film of the resin for five minutes at 110°C. A one gram sample of the resin is stirred in a 100 ml graduated glass cylinder with 99 ml toluene. The resin swells and dissolves at least partially. After twenty-four hours of stirring, the swollen resin is allowed to settle in the graduated glass cylinder for twenty-four hours. The volume of swollen resin is noted and the swelling index is calculated as the ratio of the volume of swollen resin to the volume of the unswollen resin.

A 20 ml sample of the supernatant solvent is taken out through a plug of cotton into a pipette. The resin content of the sample is determined, and, hence, the fractions of soluble resin and insoluble resin in the cured film are determined. In the table below, data for resin Example 3 catalyzed with triethylamine, triethylenediamine and 1,3-bis(dimethylamino)-2-hydroxypropane are compared with an uncatalyzed sample of resin. The tertiary amine catalyst is present in a concentration of 1.78 mmol per 100 grams resin.

Resin Example 3	Insoluble Fraction	Swelling Index
Uncatalyzed	0	—
Triethylamine	0	—
TDA	0.36	50
BDMAHP	0.10	80

The data show that triethylenediamine and 1,3-bis(dimethylamino)-2-hydroxypropane are very effective cure catalysts in comparison with triethylamine. In general, an insoluble fraction of 0.02 or greater is associated with appreciable cohesive strength in the pressure sensitive resin.

Keil release values for the pressure sensitive resins are determined by coating a silcon release paper with pressure sensitive resin solution and allowing the solution to dry for 15 minutes, at 70°F and for 2 minutes at 90°C. The dry film thickness is 2 mils. Mylar polyester film of 1 mil thickness is applied with the standard 4-pound roller specified by the Pressure Sensitive Tape Council to the pressure sensitive film.

Strips of the Mylar release paper laminate 1 inch in width are cut. The strips are subjected to a T-peel test on an Instron Tensile Tester Machine, with a crosshead speed of 12 inches per minute. The peel value or Keil release value expressed as grams per inch width is obtained. The release values, initial, after 24 hours at 70°C and after 88 hours at 70°C are determined. The data for resin Example 2 are presented in the table below. The catalyst is present in a concentration of 1.78 mmol per 100 grams resin.

	Keil Release Value, grams per inch width		
Catalyst	Initial	24 Hrs. 70°C.	88 Hrs. 70°C.
TDA	22	38	77
BDMAHP	19	25	29

Keil release values less than 30 grams per inch width are generally acceptable. The data show the rapid increase in Keil release value of the resin containing triethylenediamine to an unacceptable value upon accelerated aging at 70°C.

The compositions may be used as the adhesive component in pressure sensitive tapes, films and foams. They adhere well to resin surfaces such as plasticized polyvinyl chloride, Mylar, cellulose acetate, nylon, polyethylene and polypropylene, as well as to paper, metal and painted surfaces.

These compositions are especially useful as the adhesive component of decorative vinyl sheets and decals, conferring excellent shrink resistance to vinyl film. Their outstanding tack, wetting and holding power may be used to advantage in transfer adhesive applications.

Electron Radiation Technique

The forming of pressure sensitive adhesives by polymerizing monomers or mixtures of monomers is well known in the art. The monomers employed often include functional monomers which give the resulting adhesives greater strength. In producing adhesives by such methods, the monomers are put into a solution along with a free-radical initiator and heated at an elevated temperature to polymerize the monomers. The product is utilized by applying a layer of the polymer solution on to a substrate whereupon the solvent evaporates to form the adhesive film.

The above method of forming pressure sensitive adhesives, however, involves considerable expense. Because the resultant polymer is of extremely high molecular weight, thus resulting in a high viscous solution, the polymer solution used to apply the pressure sensitive adhesive must be maintained at a rather low nonvolatile solids content, which is conventionally about 25 to 35 percent solids. The solvents used are quite expensive and cannot be economically recovered after the adhesive solution is dried. Among other problems inherent in the prior art method of forming pressure sensitive adhesives is that a great many of the solvents used are highly inflammable and explosive, and extra care must be exercised when using them.

R. Dowbenko and R.M. Christenson; U.S. Patent 3,897,295; July 29, 1975; assigned to PPG Industries, Inc. have discovered that pressure sensitive adhesives having greater tack and adhesive strength are obtained using a method which eliminates the use of nonrecoverable solvents and expensive manufacturing and application processes.

In this method, any of a number of acrylates or acrylamides is mixed with a homopolymer of any of a number of acrylates or acrylamides or a copolymer of any of a number of acrylates or acrylamides to form an essentially solventless composition and subjected to ionizing irradiation to form the pressure sensitive adhesive material. Using this process, no solvents need be used as the process may be carried out at 100 percent solids. The resulting polymer has greatly improved tack and adhesive strength over the conventional solution polymerized polymers.

The pressure sensitive products of this process are useful as pressure sensitive adhesive coatings on many substrates, including aluminum foil, steel or other metal panels; asbestos sheets, plastic substrates, such as polyvinyl chloride, Mylar and the like; cellulosic substrates such as paper, wood, etc.; fabrics such as cotton, nylon, wool and the like. Especially useful are products such as ahdesive tapes and the like wherein the adhesive is formed on a thin, flexible backing sheet material. In the following examples all parts and percentages are by weight unless otherwise indicated.

Example 1: A pressure sensitive adhesive was prepared by precipitating with methanol, 2.4 grams of poly(2-ethylhexyl acrylate) out of a 12.1 percent solution of poly(2-ethylhexyl acrylate) and drying the polymer and dissolving it in 5 grams of N-(2-ethylhexoxymethyl)acrylamide monomer. A layer of 3 mil thickness of the mixture of monomer and polymer was coated onto a vinyl sheet and subjected to electron beam impingement at a rate of 22 feet per minute at an accelerating potential of 375 kv and a tube current of 14 ma. The total dosage given the coating was 4.4 megarads and the resulting product was a completely cured and tacky pressure sensitive adhesive. This pressure sensitive adhesive was compared to pressure sensitive adhesives formed by polymerizing only monomers of ethylhexyl acrylate and N-(2-ethylhexoxymethyl)acrylamide. Under the same ionizing irradiation conditions, the latter two products were found to be either too soft or too hard to be utilized as pressure sensitive adhesives and both polymers were insufficiently tacky to be useful for most adhesive purposes.

Examples 2 through 18: A series of pressure sensitive adhesives were prepared from the following mixtures:

Example	Monomer	Amount (grams)	Polymer	Amount (grams)
2	2-ethylhexyl acrylate	10	2-ethylhexyl acrylate	2.4
3	3-methoxybutyl acrylate	5	2-ethylhexyl acrylate	2.4
	2-ethylhexyl acrylate	5		
4	N-(2-ethylhexoxymethyl)-acrylamide	5	2-ethylhexyl acrylate	2.4
	2-ethylhexyl acrylate	5		
5	N-(2-ethylhexoxymethyl)-acrylamide	5	N-(2-ethylhexoxymethyl)-acrylamide	2.5
6	2-ethylhexyl acrylate	5	N-(2-ethylhexoxymethyl)-acrylamide	2.5
7	3-methoxybutyl acrylate	5	N-(2-ethylhexoxymethyl)-acrylamide	2.5
8	2-ethylhexyl acrylate	5	2-ethylhexyl acrylate	1.25
			N-(2-ethylhexoxymethyl)-acrylamide	
			2-ethylhexyl acrylate	1.25
9	N-(2-ethylhexoxymethyl)-acrylamide	5	N-(2-ethylhexoxymethyl)-acrylamide	1.25
10	3-methoxybutyl acrylate	5	2-ethylhexyl acrylate	1.25
			N-(2-ethylhexoxymethyl)-acrylamide	1.25
11	2-ethylhexyl acrylate	5	2-ethylhexyl acrylate	1.25
			N-isooctylacrylamide	1.25
12	N-(2-ethylhexoxymethyl)-acrylamide	5	2-ethylhexyl acrylate	1.25
			N-isooctylacrylamide	1.25
13	2-ethylhexyl acrylate	5	N-isooctylacrylamide	2.5
			2-ethylhexyl acrylate	2.5
14	2-ethylhexyl acrylate	5	2-ethylhexyl acrylate	2
			N-isooctylacrylamide	2
15	2-ethylhexyl acrylate	5	2-ethylhexyl acrylate	1.25
			N-isooctylacrylamide	3.75
16	2-ethylhexyl acrylate	5	2-ethylhexyl acrylate	1.25
			N-isooctylacrylamide	2.50
17	2-ethylhexyl acrylate	5	2-ethylhexyl acrylate	1.25
			N-isooctylacrylamide	1.25
18	N-(2-ethylhexoxymethyl)-acrylamide	5	N-(2-ethylhexoxymethyl)-acrylamide	2.5

The above mixtures were coated on vinyl substrates and subjected to electron beam impingement in an atmosphere of nitrogen gas at a line speed of 30 feet per minute at an accelerating potential of 365 kv and a tube current of 6 ma. The samples were polymerized under total dosages ranging from 2.15 megarads to 6.5 megarads and the resultant products were tacky and had excellent pressure sensitive adhesive properties.

RUBBER BASE

Polypentenamers

R. Hombach and D. Theisen; U.S. Patents 3,843,590; October 22, 1974; and 3,862,917; January 28, 1975; both assigned to Bayer AG, Germany describe contact adhesives which are composed of at least one polyalkenamer dissolved in an inert organic solvent, 10 to 250 percent by weight, based on the polyalkenamer, of a natural or synthetic resin, 0 to 5 percent by weight, based on the polyalkenamer, of an antiageing composition and 0 to 100 percent by weight of a filler and/or pigment, based on the polyalkenamer.

Suitable polyalkenamers for the pressure sensitive adhesives are any polyalkenamers and their mixtures which are obtained by ring opening polymerization of cycloalkenes having 5 or 7 to 12 carbon atoms and an olefinic double bond in the ring. Polymers in which a high proportion of the carbon-carbon double bonds are in trans configuration are particularly suitable. It is preferred to use polymers having a Mooney plasticity ML-4 (100°C) (DIN 53 523) of 20 to 200.

The polyalkenamer is preferably dissolved in an inert organic solvent at a concentration of 10 to 20 percent by weight, based on the solution, most preferably 10 to 15 percent by weight. Most suitable solvents are, e.g., liquid aliphatic hydrocarbons such as n-hexane or heptane or liquid mixtures of aliphatic hydrocarbons such as gasoline fractions boiling in the range of 65° to 95°C, cycloaliphatic hydrocarbons such as cyclohexane, aromatic hydrocarbons such as benzene, toluene or xylene, the corresponding chlorinated compounds such as chlorobenzene, trichloroethylene or methylene chloride or mixtures thereof.

Especially suitable natural resins are the colophony resins, e.g., esters or unmodified or modified colophony and among these particularly the esters of hydrogenated, disproportionated or dimerized colophony, or polyterpene resins. Any synthetic resins which increase the bond strength of the mixture are suitable as additives. The additive may also consist of a mixture of synthetic resins, a mixture of natural or synthetic resins or a mixture of natural resins. Among the synthetic resins, terpene-phenol resins, coumarone resins and phenol-formaldehyde resins are especially suitable.

Fillers and pigments such as zinc oxide, titanium oxide, aluminum oxide, lithopones, chalk, siliceous chalk, heavy spar, finely dispersed silicic acid and carbon black are added in quantities of 0 to 100 percent by weight, based on the polyalkenamer, to reduce the plasticity of the adhesives. The fillers are preferably added in the form of a triturate obtained on a ball mill. Any reduction in the bond strength of the adhesives during storage is prevented by the addition of 0 to 5 percent by weight based on the polyalkenamer of an antiageing compound. The known stabilizers such as sterically hindered amines and phenols are used for this purpose.

The adhesives can be coated onto the materials to be bonded by means of continuously or intermittently operating coating apparatus in known manner. Suitable materials which can be bonded by the adhesives according to the process are, in particular, paper, flexible synthetic resin foils (from plasticized polyvinyl chloride, cellophane, polyethylene, terephthalic acid polyester, polystyrene, etc.) and textile fabrics or fleeces made of natural or synthetic fibers of the kind commonly used for the manufacture of adhesive tapes. Typical formulations include: Adhesive A, 100 grams of trans-polypentenamer having a Mooney plasticity ML-4' (100°C) of 46.75 of terpene-phenol resin (softening range 76° to 88°C) and 2 grams of 2,6-di-tert-butyl-p-cresol (antiageing compound) were dissolved in 825 grams of a 9:1 mixture of petroleum hydrocarbons and benzene with slow stirring. Adhesive B, has the same composition as Adhesive A except that a trans-polypentenamer having a Mooney plasticity ML-4' (100°C) of 70 was used. Adhesive C, has the same composition as Adhesive A except that a trans-polypentenamer having a Mooney plasticity ML-4' (100°C) of 94 was used. Adhesive D, has the same composition as Adhesive A except that a trans-polypentenamer having a Mooney plasticity ML-4' (100°C) of 122 was used. Adhesive E, 75 grams of zinc oxide was incorporated in 100 grams of a trans-polypentenamer having a Mooney plasticity ML-4' (100°C) of 77 on rollers. This mixture together with 150 grams of terpene-phenol resin (softening range 76° to 88°C) and 2 grams of the age resister 2,6-di-tert-butyl-p-cresol was dissolved in 725 grams of 9:1 mixture of petroleum hydrocarbons and benzene with slow stirring.

Adhesive F, 100 grams of trans-polypentenamer having a Mooney plasticity ML-4' (100°C) of 77, 150 grams of terpene-phenol resin (softening range 76° to 88°C) and 2 grams of the age resister 2,6-di-tert-butyl-p-cresol were dissolved in 750 grams of a 9:1 mixture of petroleum hydrocarbons and benzene with slow stirring. 75 grams of zinc oxide which had previously been triturated with a small quantity of benzene in a ball mill were introduced into this solution.

Adhesive G, 100 grams of trans-polypentenamer having a Mooney plasticity ML-4' (100°C) of 77, 75 grams of polyterpene resin (softening point 115°C) and 1 grams of the age resister 2,6-di-tert-butyl-p-cresol were dissolved in 825 grams of a 9:1 mixture of petroleum hydrocarbons and benzene with slow stirring.

Adhesive H (comparison to Adhesive A), 100 grams of natural rubber (pale crepe) which had been masticated at 60°C for 5 minutes, 75 grams of terpene-phenol resin (softening range 76° to 88°C) and 2 grams of the age resister 2,6-di-tert-butyl-p-cresol were dissolved in 825 grams of a 9:1 mixture of petroleum hydrocarbons and benzene with slow stirring.

Adhesive A through H were applied to a polyethylene glycol terephthalate foil 50 μ in thickness with a coating knife. The quantity of solids applied was about 100 grams. After application of the adhesives, the coated foils were stored under dust-free conditions to enable the solvent to evaporate until they were tested.

To test the bond strength of the adhesives, strips measuring 30 x 1.5 cm were cut out of the coated foils and, using a test method based on test method 4001 P 7 of AFERA (Association des Fabricants Europeens de Rubans Auto-adhesifs) they were applied to a high gloss refined steel plate in such a way that no bubbles were trapped and they were then pressed on to the plate by rolling a 3 kg aluminum roller over them twice. After 30 minutes' storage, the adhesive tapes were stripped from the steel plate at the rate of 300 mm per minute at a stripping angle of 180°. The forces required for this operation are shown in the table below.

To test the heat endurance, strips 2 cm in width were stuck over a length of 1 cm to the refined steel plate previously used for the bond strength test. A weight of 0.5 kg was attached to the free end of the adhesive tape. The length of time for which the bonds could withstand the given load of 0.5 kg at a test temperature of 50°C was recorded. The results obtained are also indicated in the table below. It is clear from the bond strength and heat endurance values shown in the table that the adhesives according to the process have very high bond strength and heat endurance.

Example	Adhesive	Bond Strength (p/cm)	Heat endurance (minutes)
1	A	1500	6
2	B	1660	7
3	C	1340	5
4	D	1260	6
5	E	1250	7
6	F	800	6
7	G	870	4
8	H	400	2

Polystyrene-Polyisoprene-Polystyrene Block Copolymers

R.E. Downey; U.S. Patent 3,880,953; April 29, 1975; assigned to The Goodyear Tire & Rubber Company describes an adhesive composition which comprises a mixture of (a) an unvulcanized elastomeric block copolymer having the general configuration A-B-A, and (b) a compatible, thermoplastic tackifying resin containing carbon-to-carbon unsaturation therein and derived from hydrocarbons having from 5 to 8 carbon atoms and at least one carbon-to-carbon double bond therein selected from aliphatic hydrocarbons and alicyclic hydrocarbons.

In the elastomeric block copolymer (a) each A is an independently selected nonelastomeric monovinyl aromatic hydrocarbon polymer block having an average molecular weight of 2,000 to 100,000 and a glass transition temperature above about 25°C, the total block A content being from about 10 to 50 percent by weight of the copolymer, and B is an elastomeric conjugated diene polymer block having an average molecular weight from about 25,000 to 1,000,000 and a glass transition temperature below about 10°C, the copolymer having a tensile strength at 25°C in excess of about 200 pounds per square inch. The following examples illustrate the process in which parts and percentages are by weight unless otherwise indicated.

Example 1: A thermoplastic tackifying resin was prepared by the following method: To a reactor was charged 200 parts of heptane and 6 parts of anhydrous aluminum chloride at a temperature of about 25°C. While continuously stirring the mixture, 400 parts of a hydrocarbon mixture containing unsaturated hydrocarbons was slowly added to the reactor over a period of about 90 minutes. The temperature of the reaction was thereafter maintained in a range of about 25° to 30°C for about 60 minutes. The hydrocarbon mixture had the following composition:

Component	Percent
2-pentene	5
2-methyl-2-butene	33
Isoprene	3
1,3-pentadiene (piperylene)	28
2,3-dimethyl-1-butene	2
4-methyl-2-pentene	19
Unsaturated hydrocarbons containing 5 to 6 carbon atoms	10

After this polymerization, 25 parts of water, 8 parts of particulate hydrated calcium hydroxide and 8 parts of Attapulgus clay were added to the polymerizate. Agitation was provided to suspend the hydrated calcium hydroxide at about 90°C and continued for about 120 minutes. The mixture was then filtered to remove the insoluble catalyst residue, excess calcium compound and clay. The filtrate was then steam distilled to a pot temperature of about 235°C. The resulting residual molten resin was poured from the reactor onto an aluminum tray and cooled to about 25°C to form 173 parts of a hard brittle resin having a pale yellow color, and having a softening point, according to ASTM Method E28-58T of 94°C.

Example 2: A polystyrene-polyisoprene-polystyrene unvulcanized elastomeric block copolymer was prepared by the following method. A solution was prepared consisting of 2 grams of styrene in a 20 milliliter solution of benzene. The solution was run through silica gel at about 25°C to remove trace amounts of water and then sparged with dry nitrogen for about 30 seconds to remove trace amounts of air. To the solution was added 0.2 millimol of secondary butyl lithium in a 0.3 molar heptane solution. The mixture was allowed to polymerize and set aside for about 16 to 18 hours at about 25°C.

A solution was prepared consisting of 16 grams of isoprene and 2 grams of styrene in 80 milliliters of a benzene solution. The solution was run through silica gel at about 25°C to remove traces of water and sparged with dry nitrogen to remove trace amounts of air. The mixture of isoprene, styrene and benzene was mixed with the original polymerized styrene mixture to the exclusion of air in a reactor. The reactor was sealed and placed in a heated liquid polymerization bath at about 50°C for about 4 hours until about a 100 percent completion of the polymerization occurred.

The resulting polymer was coagulated with methyl alcohol containing a very small amount of a phenolic antioxidant and dried. The resulting polymer had an intrinsic viscosity of 1.12 in toluene at 30°C. It contained 18 percent by weight units derived from styrene.

A portion of the polymer was molded in the form of dumbbells at 138°C. The dumbbells had a thickness of 0.6 inch, a width of 0.1 inch at its narrow portion and a length of 0.8 inch. The dumbbells were tested on an Instron tester at 25°C with a crosshead separation speed of 2 inches per minute. The tests indicated that the prepared block copolymer had an ultimate tensile strength of about 250 pounds per square inch at an elongation of 1,500 percent.

Analyses indicated that the block copolymer had a molecular weight of about 100,000. Thus, assuming that the terminal polystyrene blocks were of equal molecular weight, the block copolymer comprised terminal polystyrene blocks having molecular weights of about 9,000 and internal polyisoprene having a molecular weight of about 82,000.

Example 3: Mixtures of various tackifying resins with a polystyrene-polyisoprene-polystyrene block copolymer, obtained as Kraton 107 were prepared. The block copolymer was similar to the copolymer prepared according to Example 2. It contained about 12 percent styrene by weight, a glass transition temperature of about -70°C and a tensile strength of about 1,200 pounds per square inch with an elongation of about 1,300 percent measured at 25°C, according to the method used in Example 2. The copolymer had an intrinsic viscosity of 1.1 as measured in toluene at 30°C. The mixtures of tackifying resin and block copolymer were prepared by the following method. One part of an amine type of nonstaining antioxidant was dissolved in 300 parts of a hydrocarbon solvent at about 25°C. The hydrocarbon solvent consisted of 20 parts of toluene and 80 parts of hexane. This mixture of hydrocarbons was used for the solvent to provide an optimum viscosity for sample preparation. To the solvent mixture was added 100 parts of the block copolymer of polystyrene-polyisoprene-polystyrene and 100 parts of a tackifying resin. The mixture was agitated by rolling in a container at about 25°C for about 8 to 12 hours.

The mixture was coated onto a polyester film having a 1 mil thickness (known as Mylar) with a doctor blade to provide a 1 ml coating on the polyester film. The coating was allowed to partially dry at about 25°C for about 2 minutes. The polyester film with the coat thereon was then placed in a hot air oven having a temperature of 70°C for about 5 minutes. The sample was removed from the hot air oven and allowed to cool to about 25°C and, for protective purposes, a release paper was placed over the coating.

The release paper was a clay coated sheet of paper with a silicon release agent overcoat. The sample was cut into strips one inch wide for testing purposes. The results of 180° peel strength and shear strength tests both on freshly prepared samples and on samples allowed to age at 70°C for about 5 days are shown in the table below.

Sample No.	Tackifying Resin Origin	Iodine No. (1)	180° Peel Strength (lbs) Fresh Sample	After Aging	Shear Strength Fresh Sample	After Aging
1	Aliphatic Hydrocarbons (2)	70–110	3.5	4.0	No Failure	No Failure
2	Aliphatic Hydrocarbons (3)	145	3.7	3.3	No Failure	No Failure
3	Aliphatic Hydrocarbons (4)	44	2.5	1.5	No Failure	No. Failure
4	Styrenated Resin (5)	92	None	None	None	None
5	Styrene-Acrylonitrile (6)		None	None	None	None
6	Aromatic Hydrocarbons (7)		None	None	None	None
7	Styrene, Vinyl Toluene (8)		None	None	None	None
8	Styrene (9)		None	None	None	None

(1) According to ASTM Method D-1959-61.
(2) Prepared according to the method of Example I.
(3) Obtained as Piccopale SF-100 from The Pennsylvania Industrial Chemicals Company.
(4) Obtained as Piccopale H-2, a hydrogenated resin, from The Pennsylvania Industrial Chemicals Company.
(5) Obtained as Nevex from The Neville Chemical Company.
(6) Obtained as Piccoflex 100 from The Pennsylvania Industrial Chemicals Company.
(7) Obtained as Picco 6100 from The Pennsylvania Industrial Chemicals Company.
(8) Obtained as Piccotex 100 from The Pennsylvania Industrial Chemicals Company.
(9) Obtained as Piccolastic A75 from The Pennsylvania Industrial Chemicals Company.

Samples 9 through 12 were prepared and tested according to the method used for Sample 1, except that various amounts of the aliphatic hydrocarbon resin obtained as Wing-Tack 95 were used with 100 parts of the block copolymer. In Sample 10, 20 parts of mineral oil were also added. The adhesive of Sample 10 was suitable as a hot melt pressure sensitive adhesive.

Sample No.	Resin Parts	180° Peel Strength (lbs)	Shear Strength
9	60	2.8	No failure
10	100	2.6	48 hours
11	150	4.8	No failure
12	200	3.8	No failure

In this example the 180° peel strength tests were conducted according to Method PSTC-1 of the Pressure Sensitive Tape Council. This test measures the average pull value required to peel the adhesive from the substrate at a 180° angle in pounds per one inch width of the test sample. The shear strength was measured in this example according to the method described as PSTC-7 of the Pressure Sensitive Tape Council. In particular, it was measured by applying a 1,000 gram weight to the sample and measuring the time required for the tape to separate from the substrate. For the purposes of this example a time of about 5 days was used.

In Samples 1, 2, 3, 9, 11 and 12, all of the samples exhibited peel strength to be suitable as a pressure sensitive adhesive. In Samples 4 through 8 none of the samples indicated a measurable peel strength. In Samples 1, 2, 3 the shear strength test indicated no failure after about 5 days. In Sample 10 the shear strength test indicated a failure after 48 hours for the hot melt pressure sensitive adhesive. Such a shear strength for the adhesive is adequate and suitable for a pressure sensitive adhesive. In Samples 4 through 8 the shear strength indicated an almost immediate failure.

Tackifier for Butadiene-Styrene Copolymers

H.L. Bullard, D.R. St. Cry and R.A. Osborn; U.S. Patent 3,846,352; November 5, 1974; assigned to The Goodyear Tire & Rubber Company describe resinous material which is characterized with a softening point in the range of about 60° to 110°C and is prepared by polymerizing piperylene, 2-methyl-2-butene, dicyclopentadiene and α-methyl styrene in the presence of a catalyst selected from aluminum chloride and ethyl aluminum dichloride and a solvent selected from aliphatic and aromatic hydrocarbons. The resinous material has particular utility as a tackifier for rubbery butadiene styrene copolymers. The following examples illustrate the process in which parts and percentages are by weight unless otherwise indicated.

Example 1: Into a reactor were placed 100 parts of heptane and 3 parts of anhydrous aluminum chloride. While continuously stirring the mixture, 200 parts of a hydrocarbon mixture was slowly added to the reactor over a period of about 90 minutes. The hydrocarbon mixture had the following composition:

Component	Percent
Piperylene (1,3-pentadiene)	28.5
2-methyl-2-butene	28.0
Dicyclopentadiene	20.0
α-methyl styrene	20.0
Other unsaturated hydrocarbons having 4 to 6 carbon atoms	3.5

The temperature of the reaction was maintained in a range of about 25° to 30°C. After an hour of agitation from the time of final addition of the hydrocarbon mixture, approximately 5 parts water and 20 parts lime, hydrated calcium hydroxide, were added to neutralize and decompose the aluminum chloride. The mixture was filtered to remove particles of decomposed aluminum chloride and lime.

The filtrate was steam distilled to a pot temperature of about 235°C. The resulting residual molten resin was poured from the pot onto an aluminum tray and cooled to about 23°C to form 144 parts of a hard brittle pale yellow resin having a Gardner color (50% in toluene) of 8.5 and a softening point (Ball and Ring), according to ASTM Method E 28-58 T of 96°C. The distillate comprised 138 parts of a hydrocarbon mixture. An analysis of the hydrocarbon mixture was compared to that of the hydrocarbon mixture originally added to the reactor which indicated that the prepared resin comprised the following units:

Component	Percent
Piperylene	32.1
2-methyl-2-butene	22.9
Dicyclopentadiene	22.2
α-methyl styrene	21.4
Other unsaturated hydrocarbons containing 4 to 6 carbon atoms	1.4

Example 2: Into a reactor were placed 100 parts of toluene and 3 parts of anhydrous aluminum chloride. While continuously stirring the mixture, 200 parts of a hydrocarbon mixture were slowly added to the reactor over a period of about 90 minutes. The hydrocarbon mixture had the following composition:

Component	Percent
Piperylene	28.5
2-methyl-2-butene	28.0
Dicyclopentadiene	20.0
α-methyl styrene	20.0
Other unsaturated hydrocarbons containing 4 to 6 carbon atoms	3.5

The temperature of the reaction was maintained in a range of about 25° to 30°C. After an hour of agitation from the time of final addition of the hydrocarbon mixture, approximately 5 parts water and 20 parts lime, hydrated calcium hydroxide, were added to neutralize and decompose the aluminum chloride. The mixture was filtered to remove particles of decomposed aluminum chloride and lime.

The filtrate was steam distilled to a pot temperature of about 235°C. The resulting residual molten resin was poured from the pot onto an aluminum tray and cooled to about 23°C to form 153 parts of a hard brittle pale yellow resin having a Gardner color (50% in toluene) of 6.5 and a softening point (Ball and Ring), according to ASTM Method E 28-58 T of 82.5°C.

The distillate comprised 118 parts of a hydrocarbon mixture. An analysis of the hydrocarbon mixture was compared to that of the hydrocarbon mixture originally added to the reactor which indicated that the prepared resin comprised the following units shown below.

Component	Percent
Piperylene	30.0
2-methyl-2-butene	26.0
Dicyclopentadiene	19.5
α-methyl styrene	21.3
Other unsaturated hydrocarbons containing 4 to 6 carbon atoms	3.2

Example 3: A pressure sensitive adhesive was prepared by mixing 13.5 parts of a resin prepared in the presence of heptane according to the method of Example 1, 13.5 parts of a rubbery butadiene-styrene copolymer having a butadiene to styrene weight ratio of about 77:23 and a Mooney viscosity (ML-4) at 100°C of about 52, and 86.5 parts of toluene at about 25°C. After the rubber and resin dissolved in the toluene, a 1 mil coating was spread on a polyester (polyethylene terephthalate) film. The coating was dried and the coating film laminate cut into test strips. The adhesive properties of the coating were tested utilizing methods of the Pressure Tape Council, PSTC-6 for tack, PSTC-7 for shear strength (500 g/½ in^2), and PSTC-1 for peel. The results were as follows: tack, 9 inches; shear, 42.5 hours; and peel, 45 ounces. The resin used in this example was prepared from a monomer mixture having the following composition.

Component	Percent
Piperylene	20.0
2-methyl-2-butene	25.0
Dicyclopentadiene	19.0
α-methyl styrene	25.0
Other unsaturated hydrocarbons containing 4 to 6 carbon atoms	11.0

Pressure Sensitive Hot Melt

T.E. Russell; U.S. Patent 3,862,068; January 21, 1975; assigned to Flintkote Company describes a pressure sensitive adhesive composition having instant room temperature tack, good cold flow resistance and good shear strength. The adhesive comprises as a first component a resinous rubbery block copolymer of styrene and butadiene or isoprene, as a second component a rubber extending petroleum oil, as a third component a modified or unmodified rosin, a coumarone-indene resin, a polyterpene resin, a diene-olefin aliphatic hydrocarbon resin or a polystyrene resin, and as a fourth component a resinous atactic polypropylene.

Adhesives according to this process, which are at least essentially 100% solids, have a Brookfield Viscosity (No. 7 spindle at 300°F, 20 rpm) in the range of about 25,000 to 150,000 cp, preferably about 35,000 to 65,000 cp, a softening point (Ring and Ball, ASTM D-36-26) of at least about 200°F, preferably from about 225° to 275°F; a penetration hardness (ASTM D-5-49, room temperature) in the range of about 0.3 to 1.3 cm, preferably about 0.5 to 0.8 cm; and a loss on heating (ASTM D-6-39T, 5 hours at 300°F) of about 10% maximum, preferably below about 5%.

Example 1: A mix of 22.00 parts (by weight unless otherwise specified) of Kraton 101 polystyrene-polybutadiene-polystyrene block copolymer crumb and 22.00 parts of Shellflex 371-N, a rubber-extending petroleum oil, specific gravity

0.90, viscosity 427 SSU at 100°F, and 90% distillates at 865°F, was tumbled for about 10 hours at room temperature until all the oil had been absorbed. The crumb-oil composition was then melted at about 275° to 300°F in a heated vessel having a stirrer. An antioxidant, 0.25 part of Ionox 330 hindered phenol, was added as the preblend melted. Thereafter 55.75 parts of Piccolyte Alpha 115 polyterpene resin were added to the hot melt. Mixing of the hot melt was continued for about one-half hour until a clear, smoothly-flowing, homogeneous composition was formed. The composition was poured into a mold and cooled to a resilient block at room temperature.

A portion of the solid adhesive thus prepared was melted and coated at about 300°F onto 9 x 9 inch vinyl-asbestos floor tiles. The pressure sensitive coating, which was about 4 mils thick on the tiles, had good quick tack, good pressure-sensitivity and virtually no cold flow. Some of the tiles were placed on a cleaned concrete floor on which they had excellent shear strength. Similar results were obtained with other tiles placed on a vertical plywood wall.

The adhesive composition was coated onto various materials and tested for peel strength (Pressure Sensitive Tape Council No. 1 at 180°F). The results were greater than 160 ounces per inch on plywood, stainless steel, asbestos cement and vinyl asbestos substrates. The results were essentially the same both after 2 months under water and after accelerated shelf life at 130°F. The alkaline nature of moist cement did not affect the adhesion. Silicon-treated parchment paper was placed on the adhesive coatings during storage, and could be peeled off easily for use. A coating 2 mils thick on vinyl-asbestos tile produced a shear adhesion, or holding power, at 130°F (PSTC No. 7) of about 30 minutes.

Example 2: A pressure sensitive hot-melt adhesive was prepared substantially as described in Example 1 from a preblend of 24.90 parts of Kraton 101 and 24.90 parts of Shellflex 371-N to which preblend were added 0.40 part of Ionox 330 antioxidant and 49.80 parts of Piccolyte Alpha 115. Coatings of the adhesive on various materials produced similar satisfactory results as described in Example 1.

Urethane Base with Diene Polymer

According to a process described by *D.O. Kest; U.S. Patent 3,879,248; April 22, 1975* a pressure sensitive adhesive product is produced by casting on a substrate web, such as paper, plastic film, or metal foil, a layer of a mixture consisting of: (a) a diene polymer or copolymer having at least 1.6 terminal active-hydrogen groups and an equivalent weight of 500 or more; (b) an organic polyisocyanate; and (c) at least 75 parts per 100 parts of the diene polymer and the organic polyisocyanate of a tackifying resin or mixture of tackifying resins having a ball and ring softening point of 60° to 140°C that has been predissolved in the diene polymer, and coreacting the diene polymer and the organic polyisocyanate in situ on the substrate web to give a pressure sensitive adhesive product.

The diene polymer and the organic polyisocyanate, plus any other ingredients coreactive with the diene polymer and the organic polyisocyanate that are used, are chosen so that the pressure sensitive adhesive polymer formed on the substrate web has the character of an elastomer.

Examples 1 through 10: The final pressure sensitive adhesive compositions were prepared by homogeneously mixing Part A, consisting in the main of the diene telechelic polymer and the tackifying resin, with Part B, consisting essentially of the polyisocyanate. In all of the examples, Part A was made by predissolving the tackifying resin in the diene telechelic polymer and then adding and homogeneously dispersing the balance of the ingredients. In Examples 3 and 10 the diene elastomer, Kraton 1107 employed was predissolved in the heptane and toluene solvents prior to being mixed with Part A.

In Examples 1 and 4 through 10, Part A was allowed to age for a period of time after the monoisocyanate, Monduro had been added to the diene telechelic polymer to allow essentially complete reaction therewith before mixing with Part B. In Examples 2, 3, and 10, the organic polyisocyanate of Part B was prereacted with part of the telechelic polymer to form an isocyanate-terminated prepolymer prior to the mixing of Parts A and B.

After the homogeneous mixing of Parts A and B, the final liquid pressure sensitive adhesive composition was cast onto a silicon resin-treated release paper in sufficient quantity to give the final in situ formed thickness indicated and conditioned in a circulating hot air oven for 10 minutes at 175°F to remove the solvent and initiate the in situ formation of the pressure sensitive adhesives.

The pressure sensitive adhesive-coated side of the release paper, after cooling to room temperature, was then laminated to a plastic substrate web, which in Examples 1, 2 and 4 through 10 was a 2 mil film of biaxially oriented polyethylene terephthalate, and which in Example 3, was a 2 mil film polypropylene used as received, and the resulting laminate conditioned for at least an additional 48 hours at ambient temperature (22° to 25°C) to complete the in situ formation of the pressure sensitive adhesive prior to removing the release paper and testing the pressure sensitive adhesive coated plastic web.

The tack, peel adhesion and creep resistance values for each example were obtained from test specimens that were prepared and tested as specified in test methods PSTC-6 for tack, ASTM D-903-49 for peel adhesion, and PSTC-7 for creep resistance. In the peel adhesion and creep tests, No. 304 stainless steel test panels were used and the laminated test specimens were conditioned 48 hours at 25°C prior to testing. Peel adhesion was determined with an Instron Model 1130 tester and the reported values represent the average force required for separation.

Creep resistance was determined with a 2 pound load and in a 90° ± 0.5°F environment. Prior to testing, the creep test specimens were preconditioned for 24 hours at the 90° ± 0.5°F test temperature.

Example 1 illustrates that a general purpose pressure sensitive adhesive of this process having excellent peel adhesion can be made with a blend of a hydroxyl-terminated polybutadiene and a reactive polypropylene ether triol. Noteworthy is the use in Example 1 of 15 parts of Kodaflex AD-2 and 5 parts of Santicizer 160 for a total of approximately 20 parts of vinyl resin-type plasticizers per 100 parts of the in situ formed elastomer component of the pressure sensitive adhesive to enhance tack, an expedient that is not possible with conventional pressure sensitive adhesives made with hydrocarbon polymers. This pressure sensitive adhesive composition, when utilized on a plasticized polyvinyl chloride web or

	Example, Parts by Weight									
	1	2	3	4	5	6	7	8	9	10
Part A										
Poly BD R-45M[1]	70	58.8	58.8	70					70	
Kraton 1107[2]			75							60
Poly BD CN-15[3]					70					
Hycar CTBN[4]						70				
Hycar MTBN[5]							70			
Poly BD CS-15[6]								70		
Poly BD R-45HT[7]										38
Super Beckacite 2000[8]	80			80	80	80	80	80	80	40
Foral 105[9]	32			40	40	40	40	40	22	104
Piccolyte Alpha 115[10]		100	35							
Piccolyte Alpha 85[11]			140							
Cellolyn 21[12]				10						
Mondur O[13]	0.9			1.9	0.5	0.5	0.5	0.5	0.9	2.7
Thanol SF 1500[14]	20			20	20	20	20	20	20	20
Water				2.5						
EBDC[15]	8			10					10	10
Kodaflex AD-2[16]	15	15	15						6.4	15
Santicizer 160[17]			15							
Shelflex 371N[18]									40	
Toluene			162	50					14	50
Heptane			52							
Methyl ethyl ketone			10							
Dibutyl tin dilaurate	0.5	0.5	1.0	0.5	0.5	0.5	0.5	0.5	0.5	0.5
Ethyl 330[19]			2							
Uvinul N-539[20]			4							
Isonol C-100[21]	2.8	0.5	0.5						5.6	0.5
Part B										
Arochlor 1254[22]		30	10							
Santicizer 160	5				15	15	15	15	12	
Hylene W[23]		9.56	9.56							
Isonate 390 P[24]	16.2			25.8	12.4	12.4	12.4	12.4	13.65	12
Piccolyte Alpha 115[10]										52
Foral 105[9]	18								22.4	
Poly BD R-45 M[1]		31.20	31.20							
Toluene	10		24							56
Poly BD R-45HT[7]										32
Properties										
Tack (inches)	8	4.2	1.5	15	2	15	15	8	6	1.5
Creep resistance (hr)	95	171	300	300					15	150
Gauge (mils)	1.5	1.5	1.5	1.5	4	5	4.5	5	1.5	1.5
Peel adhesion (pli)	9.5	3.5	3.2	6.5	12.5	2	1.5	8.5	1.5	6.8

[1]Poly BD R-45M – hydroxyl-terminated polybutadiene having equivalent weight of 1,250 and an average of 2.1 to 2.3 hydroxyl groups per molecule.
[2]Kraton 1107 – styrene-isoprene-styrene block copolymer having a Brookfield viscosity of 3,000 cp at a 25% total solids in toluene.
[3]Poly BD CN-15 – hydroxyl-terminated 85/15 butadiene acrylonitrile copolymer having an equivalent weight of 1,667 and approximately two hydroxyl groups per molecule.
[4]Hycar CTBN – carboxyl-terminated butadiene-acrylonitrile copolymer having 19.4% bound acrylonitrile, an average molecular weight of 3,270 and a functionality of 1.74.
[5]Hycar MTBN – thiol-terminated butadiene acrylonitrile copolymer having 24% bound acrylonitrile, an average molecular weight of 1,700 and a functionality of 1.6.
[6]Poly BD CS-15 – hydroxyl-terminated 75/25 butadiene-styrene copolymer having an equivalent weight of 1,538 and approximately two hydroxyl groups per molecule.
[7]Poly BD R-45HT – hydroxyl-terminated polybutadiene having an equivalent weight of 1,358 and an average of about two hydroxyl groups per molecule.
[8]Super Beckacite 2000 – polyterpene phenolic resin having a softening point of about 105°C.
[9]Foral 105 – pentaerythritol ester of saturated abietic acid having a softening point of about 105°C.
[10]Piccolyte Alpha 115 – alpha pinene resin having a softening point of about 115°C.
[11]Piccolyte Alpha 85 – alpha pinene resin having a softening point of about 85°C.
[12]Cellolyn 21 – hydroabietyl phthalate having a softening point of about 63°C.
[13]Monduro – octadecyl isocyanate.
[14]Thanol SF 1500 – propylene oxide adduct of glycerin having an equivalent weight of about 490.
[15]EBDC – liquid polyvinyl ether (tacky plasticizer).
[16]Kodaflex AD-2 – high molecular weight polymeric plasticizer having an acid number of 13, a refractive index, n25°C/D of 1.4824, and a specific gravity 20/20 C of 1.041.
[17]Santicizer 160 – butyl benzyl phthalate.
[18]Shelflex 371N – paraffinic mineral oil.
[19]Ethyl 330 – 1,3,5-trimethyl-2,4,6-tris(3,5-di-tert-butyl-4-hydroxybenzyl)benzene.
[20]Uvinul N-539 – octyl-2,2-diphenyl-1-cyanoacrylate.
[21]Isonol C-100 – N,N-bis[2-hydroxylpropyl] aniline.
[22]Arochlor 1254 – chlorinated biphenyl having 54% chlorine.
[23]Hylene W – 4,4′-methylene bis(cyclohexylisocyanate).
[24]Isonate 390 P – mixture of p,p′-methylene diphenyl diisocyanate and polymethylene polyphenyl isocyanate having NCO content of about 32%.

adherent surface, has demonstrated excellent resistance to plasticizer migration, showing less than a 20% decrease in peel adhesion on accelerated heat aging. By way of comparison, conventional pressure sensitive adhesives based on a hydrocarbon polymer will generally lose at least 50% of its original peel adhesion and, as a consequence, normally would not be suitable for such applications.

Example 2 illustrated the preparation of a pressure sensitive adhesive having an excellent balance of physical properties employing the prepol-mixing technique and sans any solvent. Because it is made with an aliphatic diisocyanate it exhibits better resistance to sunlight exposure than do the general purpose pressure sensitive adhesives of the prior art that are based on the hydrocarbon polymers.

The adhesive composition of Example 3 was applied to a polypropylene web used as received and having no special primer to obtain adhesion to the adhesive layer. Even so, it exhibited excellent tack and creep resistance and adequate peel adhesion with failure occurring principally to the metal rather than to the polypropylene web. It was observed that the pressure sensitive adhesive of Example 3 exhibits excellent tack at low temperatures, for example 40°F or lower, as compared to conventional polyacrylic ester-based pressure sensitive adhesives.

Further, the composition of Example 3 possesses excellent resistance to battery acid. The polypropylene web coated with the pressure sensitive adhesive composition of Example 3 when adhered to a rigid PVC substrate and preconditioned for 72 hours at 25°C and 10 minutes at 175°F, exhibited only $\frac{1}{32}$ to $\frac{1}{16}$ inch of edge separation (undercutting) after being in contact with battery acid (H_2SO_4) for 7 days at 175°F, indicating its eminent suitability for battery labels. The best commercially available pressure sensitive adhesive products recommended for battery labels (based on crosslinked acrylic ester copolymers) when similarly tested, had more than twice this amount of edge undercutting, viz, $\frac{4}{16}$ to $\frac{5}{16}$ inch.

Because of their ability to adhere to nonpolar surfaces and their resistance to acid environments, pressure sensitive adhesive compositions of the type exemplified in Example 3, in which nonpolar resins provide a major portion of the tackifying resins, constitutes one of the preferred commercial systems.

Example 3 also illustrates that these advantages can be obtained even when a considerable portion, viz, 43% of the pressure sensitive adhesive elastomer component is supplied by a diene elastomer. In this connection, it was observed that when a diene elastomer is employed, in the practice of this process, more consistent pressure sensitive adhesive properties are obtained when it is predissolved in the diene telechelic polymer and/or solvents, if used, rather than in the organic polyisocyanate.

Example 4 illustrates that water can be used, if desired, in the formulation of the pressure sensitive adhesives when special properties are required. The pressure sensitive adhesive of Example 4 because of its low tack permits the production of pressure sensitive products that can be slidably positioned on an adhered surface and yet give a high order of ultimate peel adhesion and creep resistance. While more than one equivalent weight of water per equivalent weight of the diene telechelic polymer and reactive polyether was present in Example 4, only about one equivalent weight was reacted because of the limited quantity of polyisocyanate, viz, 2 equivalents, employed which, because of the catalyst used, preferentially reacted first with the hydroxyl groups present. Examples 5 and 8 demonstrate that satisfactory pressure sensitive adhesives can be made from hydroxyl terminated, random diene copolymers, when desired, in place of the diene homopolymers, Example 5 being made with a dihydroxyl terminated copolymer of butadiene containing approximately 15% of randomly copolymerized acrylonitrile, while Example 8 was derived from a dihydroxy terminated butadiene copolymer containing about 25% of randomly copolymerized styrene.

Examples 6 and 7 further demonstrate the use of random copolymers in the pressure sensitive adhesives and, in addition, illustrate the suitability of carboxyl groups and mercaptan (thiol) groups for coreaction with the polyisocyanate in the in situ formation of the pressure sensitive adhesive after being cast on the substrate web.

Example 9 illustrates the versatility and formulating latitude possessed by the pressure sensitive adhesives. Because of this the adhesive technologist can formulate the pressure sensitive adhesive of this process to provide a wide spectrum of properties to meet the specification requirements of a number of different end-use applications. In particular, the pressure sensitive adhesive of Example 9 was formulated with a considerable quantity of a paraffin mineral oil to provide an adhesive having reduced tack and peel adhesion with reasonable retention of creep resistance, a balance of properties especially desirable for removal of pressure sensitive adhesive products. In contradistinction, conventional hydrocarbon pressure sensitive adhesives when so formulated exhibit less than one hour of creep resistance.

Example 10 illustrates a second formulation employing a diene elastomer in the pressure sensitive adhesives of this process and particularly, the increased factory processing speeds that can be achieved with its use. With the composition of Example 10, it is possible to apply an adhesive layer at a coating weight of 2 g/100 in^2 of substrate at speeds of up to 150 feet per minute employing as the solvent-evaporation means, high intensity infrared heaters having a length of only 6 feet, and a further run of 9 feet for cooling prior to wind-up. Conventional pressure sensitive adhesives based on solid elastomers, on the other hand, typically cannot be run at this coating weight at a speed much in excess of about 90 feet per minute when dried in the circulating hot air ovens of 100 to 150 feet length commonly used in the pressure sensitive adhesive industry.

Because of this reduced length of the heating means required by the process when solvents are employed, substrate web scrap losses are reduced typically to 3% or less as compared to the 7 to 10% that are common in the pressure sensitive adhesive production. Further, it is manifest that the reduced length of the heating means and/or increased speed that is typical in the production of the pressure sensitive adhesives significantly reduces the capital investment required to practice the process and hence lower fixed costs assessed to amortization.

Adhesive Strip for Automotive Trim

It has been proposed to provide ornamental trim strips made of plastic material for exterior decoration of automobile bodies and to attach such trim strips to an automobile body by means of an adhesive composition. However, the adhesives proposed for attaching such strips to the automobile body have certain limitations. Among the conditions which have to be considered in attaching trim strips to the automobile body are: the effect of water exposure on the adhesive in warm and in cold weather; exposure to high temperatures in summer and below freezing temperatures in winter; the toleration of surface contamination (such as dust, etc.) in attaching the trim strip to the body surface; resistance to peeling by acts of vandalism, and resistance to dislodgment by impact, as when a door carrying such trim strip is slammed closed, especially in below freezing temperatures. It is also desirable that if a trim strip is accidentally dislodged, it should be capable of reattachment by the original adhesive.

It has been proposed to attach such trim strips to automobile bodies by the interposition of a base strand or ribbon of cured elastic polyprene sponge or other cured elastic sponge material coated on both surfaces with a thin layer of pressure sensitive adhesive. Such strip superficially appears to adhere well to the body, but once removal is started, by pulling away its end, it can be removed relatively easily with separation at the adhesive interface. After being pulled away in this manner, or otherwise accidentally removed or pulled away, generally it cannot be satisfactorily reattached. Also, after exposure to water followed by freezing, such strips are deficient in low temperature adherence under impact.

K.H. Seto and J.T. Thompson; U.S. Patent 3,896,245; July 22, 1975; assigned to Protective Treatments, Inc. describe a composite structure comprising a substantially nonporous base strand or sheet of compounded plasticized elastomeric material having little or no resilience, and capable of being permanently deformed by pressure above about fifteen pounds per square inch but exhibiting little or no flow at 0.6 pound per square inch, or less, the base strand or sheet having a pressure sensitive material of unlike character diffused into or coated on one or more of its contacting surfaces, to enhance the adhesiveness of the composite structure to various surfaces.

Tensile strength of the base strand measured as a ribbon about 0.03 to 0.08 inch thick is at least 20 pounds per square inch at break. Such a composite coated strand or ribbon may be employed for attaching a trim strip to an automobile body. Such a composite strand can be applied to a waxed surface, and conforms to irregular body surfaces. It shows outstanding resistance to peeling and to the impact cleavage test.

The properties of the base strand are conveniently measured in the form of a rectangular strand about 0.4 inch on a side, and the properties of the base strand described herein refer to a strand of such form and thickness unless otherwise indicated. In such form the base strand, for example, is permanently deformable to half thickness by a force in the range 45 to 250 pounds per square inch, and after such deformation exhibits little or no spring back or resiliency. Thus, rebound under load one hour after such compression is in the range 0 to 7½% of original thickness, and thereafter, loaded by a weight of 1½ pounds over a 6 inch length of strand for 24 hours at 190°F flow of the strand is 0 to 0.02 inch measured by decrease in height.

Preferably the base strand has a rebound 1 hour after compression of 0 to 0.15 inch, more preferably 0 to 0.001 inch, and thereafter loaded by 1½ pounds over a 6 inch length for 24 hours at 190°F has 0 to 0.001 inch flow, measured by decrease in height. A preferred base strand composition comprises an intimate tacky mixture of elastomeric polymer or copolymer material, which may be in an uncured state or in less than a fully cured state, or may contain a mixture of uncured and partially cured elastomeric polymers, a low-volatile liquid plasticizer, and finely divided solids which are fibrous or which form thixotropic mixtures with the liquid plasticizer. In addition, the composition may contain other types of finely divided solids, for improving tensile strength and extrudability of the composition or as an extender, tack increasing ingredients, mixing aids, antioxidants, and other desirable ingredients.

Resistance to impact cleavage may be measured in an apparatus in which a strip of metal is attached by the adhesive strand to be tested to a plate of suitable material so that a predetermined length of the strip extends beyond the edge of the plate. This extending portion is struck by a weighted pendulum on its down stroke moving in the direction to separate the strip from the plate.

Referring to Figure 1.1a, the mounting frame **1** of the apparatus supports a rod **2** of low friction material, such as Teflon, on which a sleeve **3** of metal is freely journalled, and a rod **4** extends from the sleeve. This rod carries a weight **5** at its end. A plate **6** has attached thereto one or more samples **7** to be tested. As shown in Figure 1.1b, sample **7** consists of the metal strip **7'** mounted thereon by an adhesive strip **8** to be tested. The plate is rigidly held on the frame and the pendulum rod **4** is raised to the horizontal position and then released to move in the direction of the arrows so as to strike the projecting portion A of the test strip.

FIGURE 1.1: IMPACT CLEAVAGE TEST

a.

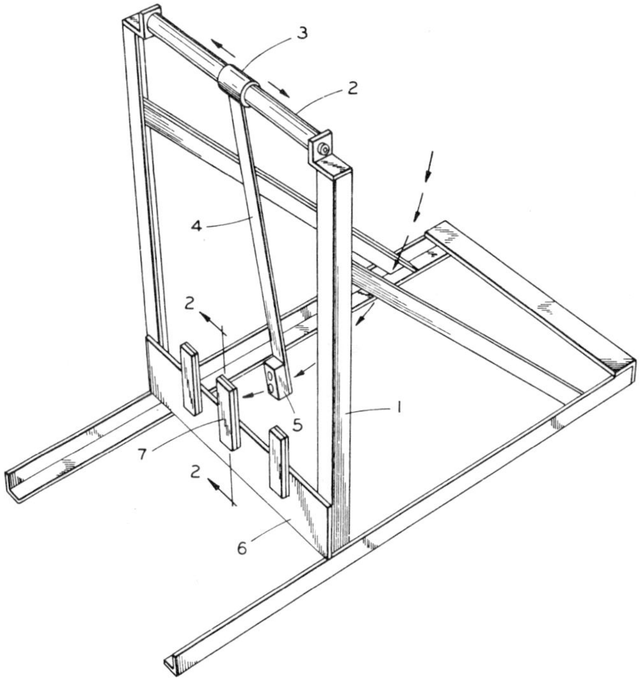

Impact Cleavage Testing Apparatus

(continued)

FIGURE 1.1: (continued)

b.

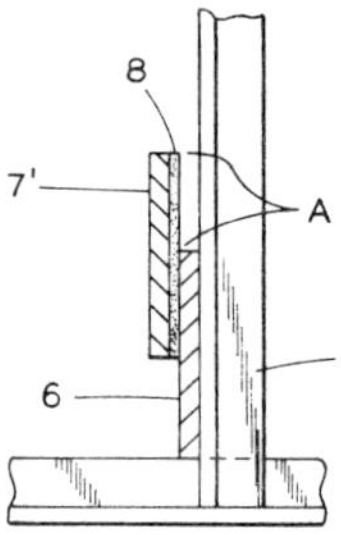

The Mounting of Sample Strips for Testing

c.

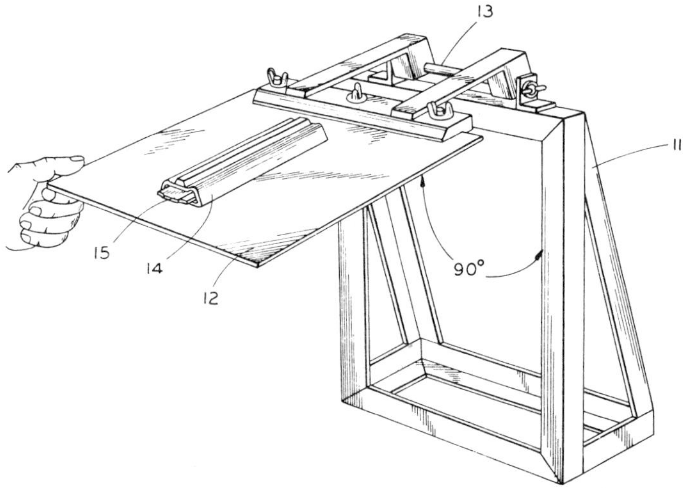

Partial View of a Slam Test Apparatus

Source: K.H. Seto and J.T. Thompson; U.S. Patent 3,896,245; July 22, 1975

Failure is indicated if the test sample shows any permanent displacement from the mounting plate **6**. In a low temperature impact cleavage test failure generally occurs by the sample being forcibly projected off of the plate **6**. The test specifications are as follows:

Pendulum impact	15 inch pounds
Strip dimensions	1 x 3 x ¼ inches
Exposed end of strip	1 inch
Pressure for attaching strip to plate	15 psi

Successful test requires withstanding five or more impacts at -20°F. This test is a measure of resistance to vandalism.

The slam test apparatus shown in Figure 1.1c is intended to simulate the force developed when an automobile door is slammed closed. The vertical frame **11** carries a plate **12** hinged at **13**, and the plastic trim strip **14** to be tested is adhered by a strand **15** to the plate **12** with its end three inches from the bottom edge of the plate **12**. The plate is raised to horizontal position and allowed to fall freely and impact the frame **11** at the bottom of its fall. In a successful test the trim strip must remain intact after five slams at -20°F. Failure generally occurs by the trim strip being projected rearwardly off the plate, apparently due to the developed rebound force. The process is illustrated by the following examples of base strands:

Example 1: An adhesive (A) is made from the following formulation:

Ingredients	Parts by Weight
Chlorinated butyl rubber	200
Isobutylene-divinyl aromatic polymer	200
Hydrogenated wood rosin	44
Tall oil fatty acid	8
Carbon black (140 parts)	82.3
Oil in carbon black	57.7
Asbestos powder	320
Processing oil	475
Platy talc	320
Lime (calcium oxide)	46

The hydrogenated wood rosin increases tackiness. The tall oil fatty acid acts as a wetting agent for the particulate materials. The lime binds any water present and is useful where the composition must withstand temperatures above 212°F, as for example, in testing. Total finely divided solids is about 1.9 times the elastomeric material and the asbestos is about 0.8 time the elastomeric material.

In the mixing procedure, the elastomers are charged into a jacketed rotary arm mixer and then the hydrogenated wood rosin is added and mixed in to a homogeneous mass, the mixing being carried out at about 160°F. Then the carbon black is added and mixed in to a smooth consistency. Then the asbestos and about one-fourth of the processing oil is added and mixed in. Now the hydrated silica, another fourth of the processing oil, and the talc and lime are mixed in. The balance of the processing oil now is mixed in and the mixture is cooled and discharged from the mixer.

The composition is extruded through a die to form a tape which is mounted on a pliable release backing, for example, a silicon coated kraft release paper. The mounted tape is then heat treated in an oven at about 175°F for 24 hours to relieve internal strains. Alternatively, if desired, the heat treatment may be at 210°F for 80 minutes. The heat treatment reduces cold flow and increases rebound.

In formulating a second adhesive (B), 400 parts of butyl rubber are used in place of the elastomers of (A), the amount of asbestos powder is increased to 400 parts, and the processing oil to 515 parts. When extruded into a rectangular base strand 0.4 inch on a side, it required a force of 110 pounds to compress a 6 inch length to half thickness. This calculates to 45.8 pounds per square inch. Rebound height after 1 hour is 0.230 inch, and rebound is 7.5 percent of original thickness. Yield strength is 8 pounds per square inch and Shore A hardness is 12.

In formulation (C) half the isobutylene-divinyl aromatic polymer is replaced by butyl rubber, the asbestos powder is increased to 400 parts and the processing oil to 515 parts.

Two further mixtures are prepared. In (D) the asbestos of (C) is replaced by an equal weight of a chain type silica, for example Cab-O-Sil and in (E) the asbestos of (C) is replaced by an equal weight of Bentone 34.

Properties	(C)	(D)	(E)
Initial thickness, inch	0.4	0.4	0.4
Rebound thickness at 1 hour, inch	0.227	0.214	0.210
Flow at 24 hours at 190°F, inch	0.003	0.000	0.001
Residual height at 24 hours, inch	0.224	0.214	0.209
Compressive load, pounds	110	110	100
Yield strength, psi	8	11	15
Shore A hardness	13	16	18

Example 2:

Ingredients	Parts by Weight
Butyl rubber PB 402	50
Butyl rubber PVB 3301	38
Thermoplastic rubber	12
Oronite 20 polybutene	25
Sun Oil 2280	25
Processing oil (in carbon black)	19.37
Carbon black (chain type with oil)	27.68
Asbestos 7RF 10 grade	80
Platy talc	62
Hydrated silica	11
Mixing aid	1.5
Amberol ST 137X	8

In this example the total liquid plasticizer content is 69.37 parts and the total finely divided solids is 180.7 parts. The carbon black is a chain type structure, commercially available in admixture with the processing oil, and with the asbestos makes up 107.7 parts. The Amberol ST 137X is a tackifier. The polybutene also serves as tackifier and improves the resistance of the elastomers to deteriorating weathering influences. Sun Oil 2280 is a processing oil and in admixture with the polybutene produces a plasticizer liquid of lower viscosity. If higher viscosity polybutene is used a greater amount of processing oil may be present to lower the viscosity.

The above mixture is extruded in the form of a ribbon about 0.05 inch thick and ⅝ inch wide onto a backing strip. Heat treatment of the extruded strip is omitted. The extruded strip is then coated on both sides with an acrylic polymer pressure sensitive material in a suitable solvent, for example, a 76% solution of Hycar 2100 X 20 in toluene and methyl ethyl ketone. Such coated ribbon is applied to a trim strip of polyvinyl chloride, cellulose acetate butyrate, or other suitable plastic material which may encapsulate a metallic luster film of metal or plastic. Absence of creep is an important property of the strand. Creep may be measured in an accelerated test by attaching a strand to a polyvinyl chloride trim strip 3⅝ inch long, adhering the strip to an acrylic finished metal panel by a pressure of about 5 pounds per square inch, and hanging a 200 gram weight vertically therefrom, the test being conducted in an air oven at 158°F for 7 days. Movement of the top edge is the creep. Zero creep in 7 days is desirable, and creep in the range up to 0.03 inch in 7 days is acceptable, a creep under 0.015 inch in 7 days being preferred.

In a comparative impact cleavage test involving the composition of Example 1 with a solution of Hycar 2100 X 20 acrylic polymer, and a commercially available neoprene sponge ribbon coated with a pressure sensitive adhesive, the neoprene ribbon failed on the first or second impact cleavage test, whereas the coated Example 1 ribbon withstood more than five impacts, and when testing was continued to failure, failure occurred in cohesion. In the low temperature slam test at -20°F, a strip of the commercial sponge neoprene adhesive which had been exposed to water failed on the first or second impact whereas the coated Example 1 ribbon similarly exposed to water and tested successfully resisted more than five impacts.

VINYL ACETATE COPOLYMERS

Ethylene-Vinyl Acetate-Trialkyl Cyanurate Copolymer Dispersions

E. Bergmeister, P. Ludwig, H. Wiest and K. Hafener; U.S. Patent 3,835,086; September 10, 1974; assigned to Wacker-Chemie GmbH, Germany describe a polymeric dispersion of a pressure sensitive adhesive or a pressure sensitive adhesive cement, containing: (1) 40 to 65% by weight of aliphatic and/or cycloaliphatic hydrocarbons, and (2) 60 to 35% by weight of a polymer, consisting of 40 to 70% by weight of vinyl acetate, 10 to 50% by weight of ethylene, 0.1 to 5% by weight of a monomer having at least two olefinic double bonds, and 0 to 40% by weight of an acrylic ester with an aliphatic alcohol having 3 to 18 carbons and/or a vinyl ester of a carboxylic acid with 3 to 18 carbons.

The polymeric dispersions of pressure sensitive adhesives are prepared by the polymerization of the olefinically unsaturated monomers in aliphatic and/or cycloaliphatic hydrocarbons in the presence of oil-soluble free radical initiators and stabilizers. The process is characterized in that at temperatures of 20° to 80°C and a pressure of 10 to 80 atm, the monomers are mixed-polymerized to give the desired mixed polymer.

The charging recipe to obtain the above range of monomer content in the polymer requires excess ethylene, as is well known. The amount of the excess depends on the polymerization conditions and can be readily determined. Ordinarily, 1 to 3 times the amount of ethylene monomer desired in the total weight of the polymer is charged. The excess ethylene is removed from the polymerization autoclave and recycled. The dispersions obtained are particularly suitable for the preparation of permanently sticky coatings. The good stickiness and good cohesive strength of the thermoplastic, stability to aging, which is contained in the dispersion is thereby of great advantage. In addition, dispersions with a high solids content can be prepared, so that no great amounts of dispersion media have to be removed in the processing.

In the examples, data are given on adhesive power, shearing resistance and hot-peeling resistance. The values were determined on adhesive films which were coated with 25 g/m^2 (solid content of dispersion), by the following methods.

Adhesive Power: Adhesive film strips, 25 cm long and 2.5 cm wide are laid at a length of about 12 cm on a glass plate and pressed on by rolling down ten times with a roller weighing 2.4 kg, without applying additional pressure. After a waiting time of one minute, the initial adhesive power is measured in kp/cm by peeling off at a peeling velocity of 300 mm/min. In the same manner the adhesive power after a storage time of 24 hours is measured.

Shearing Resistance: A 2.5 cm wide film strip is laid on a 2 cm long strip of stainless steel, V2A, and pressed on ten times with a roller weighing 2.4 kg. After a storage time of 30 minutes the film strip, hanging vertically downwards, is loaded with a 2 kg weight and the time to the total shearing off to the strip is determined in minutes.

Hot-Peeling Resistance: A film strip, 5 cm wide and 20 cm long is pressed on a glass plate with a pressing-on roller (2.4 kg, 10 times) and stored for 60 minutes at 50°C. Then the glass plate is fixed horizontally, with the film on the underside, in a drier at 50°C and the film end was loaded with a 25 gram weight. The peeling velocity is given in min/cm.

Example 1: Into an agitator-autoclave of stainless steel, rinsed before with inert gas and evacuated, the following substances were drawn in: 400 parts by weight of vinyl acetate; 100 parts by weight of butyl acrylate; 660 parts by weight of a benzene fraction, BP 60° to 95°C, with 1% by volume aromatic content; 280 parts by weight of a stabilizer solution containing 10% by weight of polyisobutyl vinyl ether having a K-value 120, in benzene (as above); 4 parts by weight of triallyl cyanurate, dissolved in the vinyl acetate; 4 parts by weight of tert-butyl perpivalate; and, 260 parts by weight of ethylene.

The mixture was agitated for 30 hours at 60°C. After cooling and depressurizing to normal, a dispersion was obtained with a polymer content of 40% by weight, which was stable for several months. The polymer dispersion had an ethylene content of 25%, based on the dry weight of the polymer. The polymer particles had an average size of 3 to 5 μ. The dispersion had a viscosity of 12,000 cp, measured with a Brookfield RVT at 10 rpm and dries to a transparent, strong, permanently sticky film. An adhesive film, produced from it by drying, has an initial adhesive power of 0.4 kp per 2.5 cm, an adhesive power of 1.0 kp per 2.5 cm, a shearing resistance of 60 minutes and a hot-peeling resistance of 60 minutes per centimeter.

For comparative purposes the identical formulation used above was repeated without the inclusion of triallyl cyanurate. This mixture was agitated for 30 hours at 60°C. After cooling and depressurizing to normal, a dispersion was obtained, which was stable for several months, and which dried to a permanently sticky, smudgy film. An adhesive film, produced from it, had an initial adhesive power of 0.35 kp per 2.5 cm and an adhesive power of 0.5 kp per 2.5 cm. The shearing resistance amounts only to 0.05 minute, the hot-peeling resistance to only 1 minute per centimeter.

Example 2: The following components were charged, according to Example 1, and agitated for 2 hours at 60°C: 2,000 parts by weight of vinyl acetate; 500 parts by weight of butyl acrylate; 1,260 parts by weight of benzene fraction, BP 60° to 95°C; 28 parts by weight of triallyl cyanurate; 5 parts by weight of tert-butyl perpivalate; 1,200 parts by weight of ethylene; and, 140 parts by weight of polyisobutyl vinyl ether with a K-value of 120.

Within a further 11 hours, 3,300 parts by weight of benzene, BP 60° to 95°C, and 30 parts by weight of tert-butyl perpivalate were added continuously at the same temperature. The dispersion obtained was stable, had a solid content of 40% by weight and dried to a transparent permanently strong sticky film. The polymer dispersion had an ethylene content of 25% based on the dry weight of the polymer. An adhesive film, prepared from it, had an initial adhesive power of 0.5 kp per 2.5 cm, an adhesive power of 1.0 kp per 2.5 cm, a shearing resistance of over 2 hours and a hot-peeling resistance of 240 min/cm.

Example 3: According to Example 1, the following components were drawn in and heated to 60°C: 3,045 parts by weight of a benzene fraction, BP 100° to 140°C, with 8% by volume of aromatic content; 1,680 parts by weight of vinyl acetate; 720 parts by weight of 2-ethylhexyl acrylate; 800 parts by weight of ethylene; 45 parts by weight of polyisobutylvinyl ether with a K-value of 120; 15 parts by weight of triallyl cyanurate; and, 30 parts by weight of tert-butyl perpivalate.

After 25 hours of heating under agitation, a stable dispersion with a solid content of 45% by weight was obtained. The polymer dispersion had an ethylene content of 25%, based on the dry weight of the polymer. The film, obtained by air-drying, was strongly, permanently sticky. The initial adhesive power of a PVC-film, finished with it, is 0.6 kp per 2.5 cm, the adhesive power 1.3 kp per 2.5 cm, the shearing resistance 50 minutes and the hot-peeling resistance 75 min/cm.

Maleic Anhydride-Alkyl Acrylate Copolymers in EVA Systems

According to a process described by *Y. Kosaka, M. Uemura, T. Fujiki and M. Saito; U.S. Patent 3,838,079; September 24, 1974; assigned to Toyo Soda Manufacturing Co., Ltd., Japan* a pressure sensitive adhesive composition is formed from 30 to 70% by weight of a carboxylated polymer product obtained by copolymerizing maleic anhydride and an alkyl acrylate in contact with an ethylene vinyl acetate copolymer, and from 70 to 30% by weight based on the total composition, of a tackifier and a viscosity reducing agent.

In preparing the pressure sensitive adhesive composition without a solvent, the first component of the modifier, such as the rosin, the second component of the modifier, such as coumarone-indene resin, and any other modifiers needed, are melted at 150°C and then the carboxylated polymer product is added. The mixture is homogeneously melt mixed, with stirring for 1 to 2 hours.

The process is based on the finding that when the above-described carboxylated polymer product is homogeneously mixed in its molten state, with at least the modifier, the mixture can be coated onto substrates, such as paper, cellophane, plastic films and sheets, cloth, etc., at a desirable temperature, to provide a pressure sensitive adhesive of superior bonding strength, as compared with currently available rubber base adhesives. In particular, this composition provides superior adhesion stability against creep at temperatures above room temperature. Very important in this process, is the realization that this composition requires no ordinarily volatile material during any part of the procedure. The adhesive composition has excellent bonding strength under a rather wide range of temperatures above room temperature. It can be used for bonding a wide variety of substrates including glass, ceramics, metals, woods, cellophanes and a wide range of plastics.

Adhesive strength and adhesive stability against shearing creep at temperatures higher than room temperature, were tested in accordance with the following methods. The test sheets were prepared by uniformly coating kraft paper with a composition which was melt blended at 150°C, using a 0.003 inch doctor blade while heating, or which was prepared as a 50% toluene solution at 100°C, using a 0.006 inch blade. The tests of the adhesive strength were conducted according to JIS Z1523. A 0.2 mm thick aluminum plate JIS Code No. A1P1-H½, was used as a substrate. In testing adhesive strength, the test was conducted at a cross head speed of 33 mm/min, instead of 300 mm/min as provided for in the JIS, because at 300 mm/min it is impossible to break the kraft paper.

The adhesive strength of the pressure sensitive adhesive kraft paper tapes and cellophane tapes, which are representative rubber based pressure sensitive adhesive tapes currently commercially available was 1,125 g/mm and 760 g/25 mm, respectively, at a cross head speed of 300 mm/min.

The test of the adhesive stability was conducted as follows. A 12.7 x 12.7 mm test sheet was bonded by finger pressure onto a substrate 0.2 mm thick of aluminum plate of JIS Code No. A1P1-H½, treated with chromic acid mixture. This test sheet was kept 5 minutes at the testing temperature. Then, the specimen was hung with a 500 g weight on the lower end of the test sheet, at the testing temperature, and the time until the test sheet separated from the substrate was measured as the value of the adhesive stability.

The adhesive stability against shearing creep at higher temperatures for the pressure sensitive adhesive kraft paper tape and cellophane tape, which are representative rubber based pressure adhesive tape, commercially available, was 8 minutes and 80 minutes respectively, at 40°C.

Example 1: A polymer, having a melt flow index of 95, was formed by copolymerizing 7% maleic anhydride and 40% butyl acrylate in contact with 53% of an ethylene-vinyl acetate copolymer having a vinyl acetate content of 19%, 53% of this product, 30% of hydrogenated rosin glycerin ester, 3% of coumarone-indene resin, with a ring and ball softening point of 120°C, 10% of liquid coumarone-indene resin, with a viscosity, as determined in accordance with the JIS Z8803, of 30 cs, 10% of dioctyl phthalate, and 500 ppm of 4,4'-thio-bis(3-methyl-6-tertiary-butyl phenol), were blended in a molten state at 150°C. The adhesive strength of this composition was 1,600 g/25 mm in accordance with the abovementioned method. The adhesive stability against shearing creep at higher temperatures as determined according to the abovementioned method, was 6 hours at 40°C.

Example 2: A polymer, having a melt flow index of 145, was prepared by copolymerizing 10% maleic anhydride and 55% octyl acrylate in contact with 35% ethylene-vinyl acetate copolymer having a vinyl acetate content of 42%. 33% of this product, 40% of hydrogenated rosin glycerin ester, 13% of coumarone-indene resin, with a ring and ball softening point of 120°C, 10% of liquid coumarone-indene resin with a viscosity, determined in accordance with the JIS Z8803, of 30 cs, 4% of dibutyl phthalate and 500 ppm of 4,4'-thio-bis(3-methyl-6-tertiary-butyl phenol) were melt blended at 150°C. The adhesive strength of this composition could not be determined under the condition mentioned above because of failure of the substrate of the test sheet. The adhesive stability of this composition was 1 hour and 20 minutes at 40°C.

Example 3: A product having a melt flow index of 92, was formed by copolymerizing 5% of maleic anhydride and 25% of 2-ethyl hexyl acrylate, in contact with 71% of ethylene-vinyl acetate copolymer having a vinyl acetate content of 19%, 53% of this product, 30% of p-t-butyl phenol formaldehyde resin, with a ring and ball softening point of 80° to 90°C, 2% of coumarone-indene resin with a viscosity of 30 cs, 7% of diisodecyl phthalate and 500 ppm of 4,4'-thio-bis(3-methyl-6-tertiary-butyl phenol) were melt blended at 150°C. The adhesive strength of this composition was 1,500 g/25 mm and the adhesive stability was 16 hours at 40°C.

Example 4: A product, having a melt flow index of 122, was formed by copolymerizing 6% of maleic anhydride and 44% of 2-ethyl hexyl acrylate, 50% of ethylene-vinyl acetate copolymer having a vinyl acetate content of 19%, 60% of that product, 27% of hydrogenated rosin glycerin ester, 9% of liquid xylene-formaldehyde resin with a viscosity, determined in 80% toluene solution at 20°C, of 60 cp, 4% of dioctyl phthalate and 500 ppm of 4,4'-thio-bis(3-methyl-6-tertiary butyl phenol) were melt blended at 150°C. The adhesive strength of this composition could not be determined under the conditions mentioned above because of failure of the substrate of the test sheet. The adhesive stability was 2 hours and 30 minutes at 40°C.

In related work *Y. Kosaka, M. Uemura, T. Fujiki, M. Kimura and M. Saito; U.S. Patent 3,896,067; July 22, 1975; assigned to Toyo Soda Manufacturing Co., Ltd., Japan* describe a polymer composition having substantial adhesiveness which is prepared by copolymerizing an alkyl acrylate and maleic anhydride in the presence of an ethylene-vinyl acetate copolymer and a hydrocarbon base selected from the group consisting of a hydrocarbon wax, a low molecular weight nonaromatic hydrocarbon resin, a chlorinated hydrocarbon wax and a chlorinated low molecular weight nonaromatic hydrocarbon resin having a molecular weight of 200 to 20,000. In the following examples parts and percentages are by weight unless otherwise specified.

Example 1: A graft copolymerization reaction was conducted with the following components in a 3 liter autoclave: 200 grams ethylene-vinyl acetate copolymer (vinyl acetate content 28%, melt index 400 g/10 min); 100 grams atactic polypropylene (MW 8,000); 30 grams maleic anhydride; 200 grams 2-ethyl hexyl acrylate; 100 grams ethyl acrylate; 2 ml t-butyl peroxy octoate; and, 1,000 grams benzene.

Ethylene-vinyl acetate copolymer and atactic polypropylene were charged into the autoclave which was purged with nitrogen gas. To the purged autoclave was added maleic anhydride, 2-ethyl hexyl acrylate, ethyl acrylate and benzene. After stirring the components for 1 hour at 90°C, a polymerization initiator was added and the mixture was copolymerized at 90°C for 6 hours. After the copolymerization reaction, the solvent and the monomers were recovered by steam distillation.

The resulting polymer composition was washed with water and dried whereby 586 grams of the polymer composition were obtained. 70 parts of the polymer composition were added to 30 parts of a hydrogenated rosin glycerin ester and the mixture was melted and mixed and was coated at 150°C on a kraft paper substrate to a thickness of 50 μ to prepare an adhesive tape. The viscosity of the molten composition was 100,000 cp. The properties of the adhesive tape are shown in the table below.

	- - - - -Adhesive Strength, JIS Z 1523- - - - -	
Substrate	Adhesive Agent Example 1	Rubber Type Adhesive Agent
Aluminum	2.3 kg/25 mm	0.8 kg/25 mm
Lauan	1.5 kg/25 mm	0.7 kg/25 mm
Water repellent corrugated board	2.0 kg/25 mm	0.7 kg/25 mm
Polyethylene	2.0 kg/25 mm	0.2 kg/25 mm
Glass	2.8 kg/25 mm	0.4 kg/25 mm

It is clear that the adhesive tape of Example 1 is superior to the rubber type adhesive tape. The adhesive strength of the polymer composition of Example 1 on an aluminum plate was 3.9 kg/25 mm.

Example 2: The graft copolymerization reaction of Example 1 was followed except that the chlorinated paraffin (40% chlorine content) was used instead of atactic polypropylene. The yield of the polymer composition was 600 grams. The resulting polymer composition and a hydrogenated rosin glycerin ester were melted and mixed to prepare the adhesive composition in accordance with the procedure of Example 1. The pressure sensitive adhesive properties of the adhesive composition are shown in the table below. The viscosity at 150°C was 47,000 cp.

Substrate	Adhesive Strength
Aluminum	1.5 kg/25 mm
Lauan	1.2 kg/25 mm
Water repellent corrugated board	1.3 kg/25 mm
Polyethylene	1.1 kg/25 mm
Glass	1.7 kg/25 mm

The adhesive strength of the polymer composition of Example 2 on an aluminum plate was 2.8 kg/25 mm.

Fumaric Diester-Vinyl Acetate Polymers

G. Guse and H.G. Pietsch; U.S. Patent 3,880,813; April 29, 1975; assigned to Beiersdorf AG, Germany describe pressure sensitive adhesive material for an adhesive composition adhering on contact which comprises a copolymer formed from the polymerization of a monomer mixture of:

(A) 50 to 80% by weight, based on the total weight of the monomers, of at least one diester of fumaric acid and a saturated monovalent aliphatic alcohol containing from 3 to 12 carbons;

(B) 18 to 40% by weight, based on the total weight of the monomers, of at least one vinyl ester of a saturated aliphatic monocarboxylic acid copolymerizable with (A);

(C) from 0.5 to 15% by weight, based on the total weight of the monomers, of at least one monoolefinic polar compound copolymerizable with (A) and (B) selected from the group consisting of monoesters of fumaric, maleic or itaconic acid with monovalent saturated aliphatic alcohols of from 1 to 8 carbons, maleic acid anhydride, crotonic acid, itaconic acid, and methylene glutaric acid; and

(D) from 0.1 to 10% by weight, based on the total weight of the monomers, of at least one monoolefinic compound copolymerizable with (A), (B), and (C), the compound having a self-crosslinkable epoxy group selected from the group consisting of glycidyl esters of acrylic and methacrylic acids, the epoxy group of (D) being at least partially crosslinked.

Example 1: In making the copolymer the following ingredients were used:

	Parts by Weight
Di-2-ethylhexyl fumarate (A)	64.0
Vinyl acetate (B)	30.0
Monoisopropyl fumarate (C)	5.0
Glycidyl methacrylate (D)	1.0
Acetone	13.3
Petrol (boiling range 30° to 85°C)	53.4

The mixture was placed in a vessel of glass or stainless steel which was provided with a stirrer, reflux condenser, thermometer and nitrogen feed (a gas inlet tube), and which could be heated in a water bath, oil bath or steam bath. The mixture of the monomers was dissolved in a mixture of acetone and benzene, and the mixture obtained saturated with nitrogen while stirring and warmed to the reflux temperature (48°C). 0.2 part by weight of dibenzoyl peroxide (75% in water) dissolved in a little acetone, was then added as initiator.

The viscosity of the solution gradually increases during the continued refluxing. After 11, 13 and 14 hours reflux, calculated from the point of introduction of the initiator, the solution was diluted, each time with 35 parts by weight petrol of boiling range 60° to 95°C. After a total reaction time of 15 hours, the solution had a copolymer content (solids content) of 34.6% by weight. The K-value of the copolymer was 67.4, measured on a 1% solution of the copolymer in toluene at 20°C.

For carrying out the K-value determination, the solution of the copolymer obtained as described was painted in a thin layer onto a paper strip provided with an antiadhesive coating (release paper), dried at room temperature and finally dissolved in the required quantity of toluene to give a 1% solution.

The viscous solution of the copolymer was coated, using a conventional coating apparatus, onto a polyethylene terephthalate foil 0.25 mm thick. Thereafter, the coated foil was passed through a heated drying channel and dried for 5 minutes at high temperature to drive off the solvent. Thereby, a simultaneous self-crosslinking of the copolymer took place, as the values in the following table show:

	Drying Temperature		
	70°C	100°C	130°C
Coating weight, g/m^2	28.4	28.1	28.1
Adhesive power, gram weight/cm	499	464	431
Holding power at 50°C, minutes	>257	913	>1100
Holding power at 100°C, minutes	12	33	1128

The values given in the table were determined after at least 24 hours storage of the tapes or foils. The coating weight was determined by weighing identical sized pieces of uncoated and coated substrate material, and expressing the results in grams per square meter.

For measuring the adhesive power, 10 to 20 mm wide strips of the adhesive foil were adhered under light pressure to sheets of polished and degreased stainless steel and rolled to and fro with a weighted roller (2 on band width, speed 10 meters per minute) five times. The test strips were pulled away from the steel sheet in a tensile test machine and the values given by that given in grams weight per centimeter (pull off angle 180°, pull off speed 30 cm/min).

For testing the heat resistance, or holding power, a 5 x 3 x 1 mm thick steel plate was polished in the long direction with emery paper and cleaned with hot petrol and toluene. A wide adhesive tape about 15 cm long and 2 cm wide was so adhered to the polished and degreased surface of the steel plate, that taken from the edge of one smaller side, a length of 2.54 cm lay on the steel plate.

The stuck-on portion of the tape was then pressed on by one to and fro rolling with a 2 kg weight roller. The free end of the strip was now hung up vertically and loaded with an 800 gram weight thereon (400 g/cm). The test was carried out at various temperatures (50° and 100°C) and the time for falling off of the strip (and weight) measured in minutes. As is evident from the table, the holding power rises strongly with the drying temperature.

Example 2: In making the copolymer the following ingredients were used:

	Parts by Weight
Di-2-ethylhexyl fumarate (A)	68.0
Vinyl acetate (B)	30.0
4-hydroxybutyl acrylate (D)	2.0
Acetone	13.3
Petrol (boiling range 30° to 85°C)	53.4

The mixture was dissolved in the solvent as described in Example 1 and the resultant solution saturated with nitrogen and warmed to reflux temperature (53°C). 0.3 part by weight of dibenzoylperoxide (75% in water), diluted in a little acetone, was added as initiator. On continued warming at reflux temperature, the viscosity of the solution gradually increased. After 3½ and 6½ hours, starting from the addition of the initiator, the solution was diluted each time with 35 parts by weight of petrol (boiling range 60° to 95°C). After a total reaction time of 11 hours, the solution so made of the copolymer had a solids content (polymer content) of 50.3% by weight. The K-value of the copolymer amounted to

39.9 (measured in toluene at 20°C). This solution of the copolymer was coated onto a 0.025 mm thick foil of polyethylene terephthalate and dried for 5 minutes at 70°C. This pressure sensitive adhesive foil showed in the dried uncrosslinked state an unusually high adhesive power, but, for certain areas of use, insufficient holding power. By the addition of a small quantity of a triisocyanate as a polyfunctional compound to the solution of copolymer, a crosslinking during drying at 70°C can take place after coating onto the substrate.

For this, there was used as triisocyanate, the addition product of 3 mols toluene diisocyanate with 1 mol of 1,1,1-trimethylolpropane in the form of a 37.5% solution in ethyl acetate. The quantities given in the following table are based on this solution, calculated on the solids content.

	Without Crosslinking Agent	With 1% Triisocyanate	With 2% Triisocyanate
Coating weight (g/m^2)	22.1	2 25.8	22.1
Adhesive power (g wt/cm)	739	497	367
Holding power at 50°C (min)	3	245	336
Holding power at 100°C (min)	2	17	26

If the same solution of the copolymer, mixed with 1% of the triisocyanate solution, was coated at greater thickness onto a polyvinyl chloride foil of 0.2 mm thickness and dried for 5 minutes at 70°C, the following values were measured: 54.8 g/m^2 coating weight; 2,101 g wt/cm adhesive power; 133 holding power at 50°C; and, 35 minutes holding power at 100°C.

Example 3: In making the copolymer the following ingredients were used:

	Parts by Weight
Di-2-ethylhexyl fumarate (A)	62.0
Vinyl acetate (B)	30.0
4-hydroxybutyl acrylate (D)	8.0
Acetone	13.3
Petrol (boiling range 30° to 85°C)	53.4

The monomer mixture was dissolved in a mixture of petrol and acetone, the solution saturated with nitrogen and heated to reflux temperature (50°C). 0.3 part by weight dibenzoyl peroxide (75% in water) dissolved in a little acetone, was then added as initiator.

During the continued heating the viscosity rose gradually. After 6, 8, 11 and 14½ hours, taken from the introduction of the initiator, the solution was diluted each time with 35 parts by weight petrol (boiling range 60° to 95°C). After a total reaction time of 15 hours, the solution had a solids content (polymer content) of 27.2%. The K-value of the copolymer amounted to 51.8 (measured in ethyl acetate at 20°C).

The viscous solution of the copolymer was coated onto a foil of polyethylene terephthalate and dried for 5 minutes at 70°C. The so manufactured pressure sensitive adhesive foil had in the uncrosslinked condition, an unusually high adhesive power. Its holding power at elevated temperature could be further improved by crosslinking with a triisocyanate (as in Example 2), as is evident from the table shown on the following page.

	Without Crosslinking Agent	With 4% Triisocyanate
Coating weight (g/m^2)	23.8	24.4
Adhesive power (g wt/cm)	773	348
Holding power at 50°C (min)	8	46
Holding power at 100°C (min)	3	17

HOT MELT ADHESIVES

A class of adhesive compositions known as thermoplastic or hot melt adhesives have enjoyed continually increasing usage in many industrial applications. These adhesives, or hot melts, as they are commonly designated, are solid or semisolid combinations of film forming resins, tackifying resins, rubbery polymers, plasticizers, waxes and similar materials which are added to the adhesive composition in order to impart various properties.

Adhesive bonds derived from hot melts are particularly useful because they are 100% solid materials and they quickly form tough adhesive bonds simply by cooling, i.e., no need for solvent evaporation, etc.

Adhesives derived from hot melts are particularly useful because of their very rapid set up or bonding time. Strong bonds are formed when the adhesive cools below its melt point and crystallizes. There is no long wait for solvent or water evaporation or for chemical curing as with other type adhesives. Depending on composition, hot melt adhesives can be prepared having good tack, high bond strength, good flexibility, low temperature properties, good environmental resistance, etc.

A typical class of hot melt adhesive composition utilizes polyolefin polymers as the base or carrier material. The polyolefin base is usually blended with other polymers and copolymers, resin tackifiers and with modifiers and additives. Generally, the polyolefin carrier is polypropylene (either atactic or crystalline or both) or polyethylene but that is usually done for cost considerations and it is not necessary that it be so restricted from a technical standpoint.

In spite of the tremendous growth in the use of hot melts and the advantages to be realized by their use, the extent of their utilization has nevertheless been limited by several serious shortcomings. Polyolefin based hot melts, for example, usually suffer from the lack of strong adhesion to nonporous metallic surfaces such as aluminum, steel and the like. They also tend to be quite brittle unless suitably modified. All hot melts generally suffer from low bond strengths at elevated temperatures.

When heated, even well below their melt points, they tend to soften and lose strength.

A number of recent studies utilizing a variety of polyolefin polymers and formulating techniques to overcome many of these problems are described in this chapter.

ETHYLENE-VINYL ACETATE

Oxidized Low Molecular Weight Polypropylene

According to a process described by *Y. Kosaka, H. Kuroki, N. Echida, K. Otomo and M. Uemura; U.S. Patent 3,896,069; July 22, 1975; assigned to Toyo Soda Manufacturing Co., Ltd., Japan* a hot melt composition is prepared by employing an ethylene-vinyl acetate copolymer, having a vinyl acetate content of 10 to 45% by weight, and melt index of 1 to 1,000 grams per 10 minutes and mixing at least one tackifier, wax or asphalt and an oxidized low molecular weight polypropylene.

In the preparation of the hot melt composition, the heat resistance (softening temperature) of the composition can be improved by adding an oxidized low molecular weight polypropylene having an acid value of 1 to 20 per molecule, in a ratio of 1:100 to 60:10 times by weight to the ethylene-vinyl acetate copolymer. The oxidized low molecular weight polypropylene is prepared by oxidizing low molecular weight polypropylene (500 to 10,000) with oxygen or an oxygen-containing gas.

The low molecular weight polypropylene itself may be prepared by telomerization of the olefin or thermal cracking of high molecular weight polypropylene, which is effective technologically and economically. In the thermal cracking of high molecular weight polypropylene, the polypropylene is heated to a temperature higher than 300°C in an inert gas or in a steam saturated inert gas, such as nitrogen, carbon dioxide, argon or the like. The low molecular weight ethylene-propylene copolymer which is obtained as a by-product may be used as the low molecular weight polypropylene.

The low molecular weight polypropylene having an average molecular weight of 500 to 10,000 (measured by the intrinsic viscosity method), which is obtained by the above procedure, is oxidized with oxygen or an oxygen-containing gas. The oxidized low molecular weight polypropylene having an excellent color may also be molecular by supplying steam to the oxidation step. The oxidized low molecular weight polypropylene has an average molecular weight (measured by intrinsic viscosity) of 500 to 10,000 and has an acid value of 1 to 20 per molecule.

The following components (weight percent) were heat melted and mixed at 120° to 200°C for 1 to 2 hours, and the physical properties of the resulting hot melt composition were tested. In the examples, the following abbreviations are used: EVA, ethylene-vinyl acetate copolymer; VA, vinyl acetate; MI, melt index; MW, molecular weight (measured by intrinsic viscosity).

Example 1:

EVA (VC content 28%, MI 400)	40%
Glycerine ester of hydrogenated rosin (R and B softening temperature 72°C)	40%
Oxidized low molecular weight polypropylene (MW 3500 Acid value 5)	20%

Properties of the composition

R and B softening temperature	126°C
Melt viscosity (160°C)	3300 cps
Peel strength (Al/Al)	9.5 kg./25 mm
Tensile strength	16.0 kg/cm
Elongation	1500%
Hardness (Shore-A)	52
Cold resistance	Good
Heat seal strength (OPP/OPP)	2.5 kg/15 mm

Note: Test methods

(1) R and B softening temperature
Japanese Industrial Standard K 2531

(2) Melt viscosity
Brookfield viscometer

(3) Peel strength 90° peeling
Substrate: 25 mm width 0.1 mm thickness aluminum foil.
Structure: Al foil/hot melt/Al foil comp. (0.2 mm thickness)
Peeling speed 150 mm/min.

(4) Tensile strength and Elongation
ASTM D 638-68

(5) Hardness: ASTM D 2240-68

(6) Cold resistance:
A hot melt composition which was melted, was coated on a corrugated board and another corrugated board was plied onto the coated surface under pressure. The product was kept at −30°C for 3 hours and the product was peeled after warming to room temperature. When the corrugated board was broken, the cold resistance was good. When the coated surface was broken, the cold resistance was bad.

(7) Heat seal strength
The hot melt composition was coated on OPP in a thickness of 20μ and another OPP was plied onto the coated surface. The plies were heat-sealed at 80°C under a pressure of 1 kg/cm^2 for 1 second. The product was peeled at a peeling speed of 200 mm/min.

Example 2:

EVA (VC content 19% MI 350)	40%
Hydrogenated dicyclopentadiene resin (R and B softening temperature 123°C)	20%
Oxidized low molecular weight polypropylene (MW 4900 acid value 15)	40%

Properties of the Composition

R and B softening temperature	138°C
Melt viscosity (160°C)	3800 cps
Peeling strength (Al/Al)	3.5 kg/25 mm
Tensile strength	36 kg/cm^2
Elongation	650%
Hardness (Shore - A)	75
Cold resistance	Good

Example 3:

EVA (VA content 28% MI 400)	40%
Hydrated rosin (R and B softening temperature)	40%
Oxidized low molecular weight polypropylene (MW 4000 acid value 3.5)	20%

Properties of the Composition	
R and B softening temperature	125°C
Melt viscosity (160°C)	2900 cps
Peeling strength (Al/Al)	9.5 kg/25mm
Tensile strength	14 kg/cm²
Elongation	1700%
Hardness (Shore - A)	51
Cold resistance	Good
Heat seal strength	3.0 kg/15 mm

As a reference, the low molecular weight polyethylene (MW 1,500, R and B 102°C) was substituted for the oxidized low molecular weight polypropylene. The properties of the hot melt compositions were as follows: the heat resistance, adhesiveness, softness and cold resistance of the compositions were quite inferior to those of the former composition.

R and B softening temperature	86°C
Melt viscosity	2800 cps
Peel strength	1.5 kg/25 mm
Tensile strength	20 kg/cm²
Elongation	350%
Hardness (Shore - A)	79
Cold resistance	Bad
Heat seal strength (OPP/OPP)	0.2 kg/15 mm

Amorphous Polypropylene and Petroleum Resin

A process described by *K. Masuda and K. Tasaka; U.S. Patent 3,860,543; Jan. 14, 1975; assigned to Mitsui Petrochemical Industries, Ltd., Japan* relates to hot melt compositions which are produced by mixing a petroleum resin (C) and an ethylene-vinyl acetate copolymer having a 15 to 40% by weight vinyl acetate content (melt index, 2.5 to 200) and/or an amorphous polypropylene (D) in a ratio by weight of C:D = 50:100 to 200:50, adding, if desired, a wax of a 40° to 80°C SP (softening point) and fusing the mixture by heating.

The petroleum resin (C) is produced by mixing fractions selected from C_4 to C_5 fractions (A) having a −15° to 45°C boiling point obtained by the reforming or cracking of petroleum and an aromatic hydrocarbon fraction (B) having a high degree of unsaturation and a boiling point in the range of 145° to 280°C, that is, an unsaturated aromatic hydrocarbon fraction prepared by separating lower olefins, i.e., those having from 2 to 7 carbon atoms, benzene, toluene and xylene from petroleum cracking oils in a ratio by weight of A:B = 5:100 to 80:100, and polymerizing the mixture using a Friedel-Crafts catalyst.

The low boiling fraction (A) used in the process is a C_4 to C_5 fraction which is obtained as a by-product at the cracking or reforming of petroleum, and has a boiling point in a range of from −15° to 45°C. The hydrocarbon fraction (A) having such a boiling point comprises a major proportion (more than 50% by weight) by weight of C_4 and C_5 olefins and diolefins such as 1-butene, 2-butene, isobutene, butadiene, 1-pentene, 2-pentene, cyclopentene, 1,3-piperylene, isoprene, cyclopentadiene, 2-methyl-1-butene, 2-methyl-2-butene and 3-methyl-1-butene, and, for example, has the following composition.

	Percent by weight
C_4–C_5 Olefins	40 - 70
C_4–C_5 Diolefins	20 - 60
C_4–C_5 Paraffins (for example, butane, pentane, cyclopentane, etc.)	10 - 30
Others	0 - 5

Unless otherwise indicated, in the following examples all hue or color values are Gardner Color values (ASTM D-154-58).

Example 1: One hundred grams of a residual oil (shown in Table 1) which had been produced by extracting 1,3-butadiene from a C_4 fraction having a -15° to 15°C boiling point range and containing unsaturated hydrocarbons was admixed with 300 grams of a hydrocarbon oil (bromine value 78) having a 145° to 240°C boiling point range containing unsaturated hydrocarbons which had been produced at the cracking of petroleum.

After adding 3.6 grams of boron trifluoride phenolate as the catalyst, the mixture was allowed to polymerize at 35°C for 3 hours at atmospheric pressure. After washing with alkali and then with water, unreacted components were removed by vacuum distillation (the distillation was carried out at 5 mm Hg at 180°C to remove unreacted paraffins and olefins, and low molecular weight polymers containing dimers and trimers) to produce a light yellow petroleum resin. The resin softening point is 80°C; the hue is 11; the bromine value is 18; and the average molecular weight is 653.

TABLE 1

Fraction	Raw fraction		De-diene fraction*	
Below 3 carbons	1%	by weight	2%	by weight
Iso-butane	1	do.	2	do.
n-Butane	2	do.	3	do.
1-Butene	18	do.	28	do.
Iso-butene	28	do.	43	do.
Trans-2-butene	8	do.	12	do.
Cis-2-butene	6	do.	9	do.
1,3-Butadiene	36	do.	1	do.
	100%		100%	

*Fraction where butadiene had been removed from the original C_4 fraction.

The hydrocarbon oil of a bromine value of 78 in this example had a boiling point of 145° to 240°C and the following composition: styrene, indene, other vinyl aromatic hydrocarbons of 8 to 12 carbon atoms, 55%; other aromatics, 37%; olefins, 2%; diolefins, less than 1%; and paraffins, 5%.

One hundred and sixty grams of the resin prepared by the abovementioned process and 160 grams of an ethylene-vinyl acetate copolymer (Evaflex 220; vinyl acetate content, 28.5%; melt index, 164) were fused together with 80 grams of a paraffin wax (softening point 145°F, average molecular weight of 630) as the additive in an oil bath at 160°C, and the fused mixture was stirred for 2 hours. The melt viscosity of the resulting hot melt composition was measured by a B-type viscosimeter. The adhesive property was measured as follows. The composition was applied to a series of sheets of kraft paper by an automatic applicator so as to form a coating layer 20μ thick. Then, the coated surfaces of two sheets of coated paper were adhered by a heat-sealing device. Conditions of the adhesion were as follows: temperature, 100°, 120° and 140°C; pressure, 1.33 kg/cm^2; and time, 1 second.

The miscibility of the composition was determined by observing the compatibility of the pressed sheet of the composition having a 1 mm thickness at normal temperature and that of a fused solution at 160°C. The smell was determined by observing the odor of the above described pressed sheet having a 1 mm thickness (at room temperature) and that of a fused solution at 130°C. The results are shown in Table 3.

Comparison Example 1: Three grams of boron trifluoride phenolate was added to 300 grams of a hydrocarbon oil (bromine value 78, same as was used in Example 1) containing unsaturated hydrocarbons and boiling in the range of 145° to 240°C which had been produced by the cracking of petroleum. The mixture was allowed to polymerize at 35°C for 3 hours at atmospheric pressure.

After washing the polymerized oil with alkali and then with water, unreacted components and low molecular weight reaction products were removed by vacuum distillation as in Example 1 to produce a light yellow petroleum resin. The petroleum resin had a softening point of 100°C, a bromine value of 17, an average molecular weight of 701 and a hue of 12. Using this resin, a hot melt composition was produced in the same manner as in Example 1. The results are shown in Table 3.

Example 2: Three hundred grams of a fraction (bromine value, 78, same as hydrocarbon oil of Example 1) boiling in the range of 145° to 240°C and 100 grams of a fraction boiling in a range of −15° to 45°C (see Table 2) which were obtained by the cracking of petroleum were mixed. After adding 4 grams of boron trifluoride phenolate, the mixture was polymerized at 30°C for 3 hours at atmospheric pressure. The resulting polymerized oil (raw oil) was treated in the same manner as in Example 1 to produce a light yellow petroleum resin.

TABLE 2

C_4 - C_5 fraction composition	
1-butene	10.6%
2-butene	9.6%
isobutene	15.3%
1,3-butadiene	18.3%
1-pentene	2.6%
2-pentene	3.2%
cyclopentene	2.1%
1,3-piperylene	3.0%
isoprene	6.3%
cyclopentadiene	4.7%
2-methyl-1-butene	3.7%
2-methyl-2-butene	2.1%
C_4-C_5 paraffins and other olefins (only trace amount of other olefins)	18.5%

The petroleum resin had a softening point of 96°C, a bromine value of 34, an average molecular weight of 678 and a hue of 10. A hot melt composition was produced in the same manner as in Example 1. The results are shown in Table 3.

Example 3: Three hundred grams of a fraction (bromine value 78, same as hydrocarbon oil in Example 1) boiling in the range of 145° to 240°C and 100 grams of a diene containing fraction boiling in the range of −15° to 15°C (raw fraction described in Table 1) which were obtained from a hydrocarbon oil containing unsaturated hydrocarbons produced at the cracking of petroleum were mixed. After adding 4 grams of boron trifluoride phenolate, the mixture was polymerized at 30°C for 3 hours at atmospheric pressure.

The resulting polymerized oil was treated in the same manner as in Example 1 to produce a light yellow petroleum resin. The petroleum resin had a softening point of 73°C, a bromine value of 32, an average molecular weight of 685 and a hue of 10. A hot melt composition was produced in the same manner as in Example 1. The results are shown in Table 3.

Comparison Examples 2 and 3: One hundred and sixty grams of Rosin WW (ASTM D-509; Gardner Color 19.0) or 160 grams of Ester gum H (hydrogenated product of glycerin ester of rosin, yellow resin which has an acid value of less than 15 and a softening point greater than 68°C) was used instead of the petroleum resin in Example 1. The results are shown in Table 3.

TABLE 3

	Example 1	Comparison 1	Example 2	Example 3	Comparison 2	Comparison 3
Resin	Petroleum resin in Example 1	Petroleum resin in Comparison 1	Petroleum resin in Example 2	Petroleum resin in Example 3	Rosin WW	Ester gum H
Melt viscosity (CPS)						
Temperature 120°C	13000	14000	13100	11100	7900	7800
140°C	6300	7200	6400	5600	5000	5100
160°C	3000	4100	3200	3100	3200	3300
Adhesive property g/25mm kraft paper						
Heat-sealing						
Temperature 100°C	1359	501	1260	1340	1704	1570
120°C	1357	611	1310	1372	1624	1614
140°C	1457	561	1395	1453	1858	1813
Smell						
(1) normal temperature, press-sheet	Very faint	Intensive	Faint	Very faint	Faint	Faint
(2) 130°C, fused solution	Very faint	Intensive	Faint	Very faint	Intensive	Faint
Compatibility						
(1) normal temperature, press-sheet	Transparent	Opaque	Transparent	Transparent	Transparent	Transparent
(2) 160°C, fused solution	Transparent	Opaque	Transparent	Transparent	Transparent	Transparent

Isopropenyltoluene Copolymers as Tackifiers

K. Yamada, H. Ueki, K. Masuda and K. Tasaka; U.S. Patent 3,875,095; April 1, 1975; assigned to Mitsui Petrochemical Industries, Ltd., Japan describe a hot melt composition comprising a mixture of a resin (C) prepared by copolymerizing 100 parts by weight of vinyl aromatic hydrocarbons (A) containing more than about 25% by weight isopropenyltoluene and about 5 to 250 parts by weight of a by-product fraction obtained in refining or cracking petroleum and having 4 to 5 carbon atoms, the proportion of the isopropenyltoluene in the total amount of the hydrocarbons (A) and the fraction (B) being more than 20% by weight, in the presence of a Friedel-Crafts catalyst and an ethylene-vinyl acetate copolymer (D) containing about 15 to 40% by weight vinyl acetate. Isopropenyltoluene is easily available from the dehydrogenation reaction of cymene, in which cymene

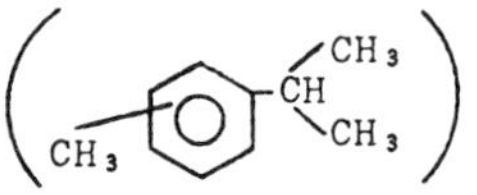

is dehydrated in the presence of a catalyst such as a metal oxide, for example, Cr_2O_3, Fe_2O_3, ZnO, etc., or alumina. The method and the conditions employed are described in Japanese Patent 2,373 (1957). Isopropenyltoluene is also readily available as by-product in the production of cresol by a cymene process described in U.S. Patent 2,628,983 or by dehydration of tolyldimethylcarbinol. The ortho isomer, meta isomer and para isomer of isopropenyltoluene can be used but a mixture of the isomers containing more than 40% of the meta isomer is preferable. Unless otherwise indicated, all parts, percents, ratios and the like are by weight.

Examples 1 through 13 and Comparison Examples 1 through 9: Polymerization reactions were conducted in a 1 liter autoclave using the raw materials, the solvents, and the catalysts shown in Table 1 under the reaction temperatures and reaction times also shown in Table 1. After inactivating the catalysts in the reaction products with methanol or an aqueous sodium hydroxide solution and then removing any insoluble matter therefrom by filtration, the reaction products were washed with water until the product became neutral.

By further concentrating the reaction products under reduced pressure, the resins were obtained. The yields for and the properties of the resins thus obtained are shown in Table 1 and Table 2.

TABLE 1

		Raw Material (g)		Catalyst (g)	Solvent (g)	Reaction Conditions	Resin Yield (%)
Example	1	IPT	70	$AlCl_3$ 1.07	Hexane 200	−25°C,3hr	72
		C_4–C_5 Fraction	30				
"	2	IPT	50	"	"	−30°C,3hr	55
		C_4–C_5 Fraction	50				
"	3	IPT	30	"	"	"	36
		C_4–C_5 Fraction	70				
"	4	IPT	30	BF_3-Phenol Complex 0.78	Toluene 100	0°C,3hr	87
		α-MST	70				
		C_4–C_5 Fraction	30				
"	5	IPT	70	BF_3-Phenol Complex 0.78	"	−7°C,3hr	89
		α-MST	30				
		C_4–C_5 Fraction	30				
"	6	IPT	30	BF_3-Phenol Complex 0.90	Toluene 150	−10°C,3hr	87
		α-MST	70				
		C_4–C_5 Fraction	50				
"	7	IPT	70	BF_3-Phenol Complex 0.90	"	−5°C,3hr	82
		α-MST	30				
		C_4–C_5 Fraction	50				

(continued)

TABLE 1: (continued)

		Raw Material (g)		Catalyst (g)	Solvent (g)	Reaction Conditions	Resin Yield (%)
Example	8	IPT α-MST C_4-C_5 Fraction	90 10 50	BF_3Phenol Complex 0.90	"	−20°C,3hr	86
"	9	IPT α-MST C_4-C_5 Fraction	70 30 70	BF_3-Phenol Complex 0.90	"	−10°C,3hr	80
"	10	IPT VT C_4-C_5 Fraction	70 30 30	BF_3-Phenol Complex 0.78	Toluene 100	−10°C,3hr	92
"	11	IPT VT C_4-C_5 Fraction	30 70 30	BF_3-Phenol Complex 0.78	"	"	91
"	12	IPT α-MST VT C_4-C_5 Fraction	45 45 10 30	BF_3-Phenol Complex 0.78	Toluene 100	−2°C,3hr	89
"	13	IPT C_5-Fraction	70 30	$AlCl_3$ 1.07	Hexane 100	20°C,3hr	80
Comparative Example	1	α-MST C_4-C_5 Fraction	70 30	$AlCl_3$ 1.47	Methyl-cyclo-hexane 70	−10°C,3hr	87
"	2	α-MST C_4-C_5 Fraction	50 50	"	Methyl-cyclo-hexane 100	−20°C,3hr	75
"	3	α-MST C_4-C_5 Fraction	30 70	"	"	−30°C,3hr	60
"	4	IPT α-MST C_4-C_5 Fraction	50 50 15	BF_3-Phenol Complex 0.69	Toluene 100	−5°C,3hr	90
"	5	IPT VT ST α-MST C_4-C_5 Fraction	20 55 18 7 30	BF_3-Phenol Complex 0.78	Mesity-lene 100	30°C,3hr	88
"	6	VT C_4-C_5 Fraction	70 30	$AlCl_3$ 1.07	Hexane 200	−20°C,3hr	84
"	7	St C_4-C_5 Fraction	70 30	BF_3-Phenol Complex 0.60	Toluene 100	−10°C,3hr	97
"	8	Rosin Ester, EG-H* (trade name, produced by Tokushima Seiyu K.K.)					
"	9	Petroleum Resin, Plas Tac** (trade name, produced by Mitsui Petrochemical Industries, Ltd.)					

*Ester gum, the glycerin ester of rosin.
**Polymerized petroleum fraction resin
NOTE:
IPT: Isopropenyltoluene comprising 5% o-IPT, 60% m-IPT and 35% p-IPT
α-MST: α-Methylstyrene
VT: Vinyltoluene
ST: Styrene

The composition of the C_4 to C_5 fraction was as follows.

1-Butene	10.6%
2-Butene	9.6%
Isobutene	15.3%
1,3-Butadiene	18.3%
1-Pentene	2.6%
2-Pentene	3.2%
Cyclopentene	2.1%
1,3-Pentadiene	3.0%
Isoprene	6.3%
Cyclopentadiene	4.7%
2-Methyl-1-Butene	3.7%
2-Methyl-2-Butene	2.1%
Saturated Hydrocarbons	18.5%

The composition of the C_5 fraction was as follows.

1-Pentene	4.8%
1,3-Pentadiene	9.1%
2-Methyl-1-Butene	7.2%
Cyclopentadiene	9.3%
2-Pentene	4.9%
3-Methyl-1-Butene and Isopentane	12.0%
2-Methyl-2-Butene	3.5%
Isoprene	14.7%
Others	34.5%

TABLE 2

		Resin Property			
		Softening Point (°C)	Hue (Gardner)	Molecular Weight	Bromine Value
Example	1	98	less than 1	760	10
"	2	101	"	870	14
"	3	91	2	870	22
"	4	95	3	770	13
"	5	97	3	750	11
"	6	91	2-3	710	18
"	7	92	3-4	750	13
"	8	103	3	850	12
"	9	108	1	870	15
"	10	90	2-3	770	16
"	11	98	3	780	14
"	12	95	2-3	800	12
"	13	98	4	750	8
Comparative Example	1	103	1-2	930	11
"	2	107	1	1290	18
"	3	93	3	1490	21
"	4	90	3	720	12
"	5	101	2-3	770	13
"	6	93	2	1200	13
"	7	87	2-3	1150	13
"	8	80	9	730	52
"	9	99	10	820	34

A mixture of 160 grams of each of the resins prepared in the above described examples and comparative examples, 160 grams of Evaflex 220 (ethylene-vinyl acetate copolymer containing 28% by weight vinyl acetate and having a melt index of 150), and 80 grams of a paraffin wax (140°F) was melted by heating and stirred for 2 hours in a stainless steel beaker in an oil bath of 160°C to provide a hot melt composition. The properties of the hot melt composition thus prepared were measured in the following manner and the results obtained are shown in Table 3.

(1) Adhesive Strength – The hot melt composition was coated in a thickness of 20 microns on an aluminum foil having a width of 25 mm and a thickness of 50 microns and also having a mirror-like surface using an automatic applicator. After the coated layers on the aluminum foils thus prepared were sealed together using a seal width of 15 mm, a sealing temperature of 140°C, a sealing pressure of 1 kg/cm^2, and a sealing time of 2 seconds employing a heat sealer, the peeling strength was measured at a tension speed of 300 mm/minute by means of an Instron tension tester.

(2) Hue – The hue of the hot melt composition melted by heating (160°C) was shown by a Gardner value.

TABLE 3

Example	Compatibility	Cloud Point, °C	Odor	Viscosity, cp at 140°C	Flexibility, 25°C	Thermal Stability	Adhesive Strength (g/25 mm)	Hue (Gardner)
1	A	60	A	4,700	A	A	1,040	1-2
2	A	63	A	4,900	A	A	870	3
3	A-B	62	A-B	6,000	A	A	1,200	3
4	A	124	A-B	5,100	A	A	600	2
5	A	68	A	4,900	A	A	930	2
6	A	122	A	5,000	B	A	870	2-3
7	A	63	A	4,800	A	A	1,170	3
8	A	101	A	5,400	A	A	840	2-3
9	B	65	A	4,200	A	A	1,070	2
10	A	65	A	4,700	A	A	950	3-4
11	A	137	A	4,900	A	A-B	490	3
12	A	68	A-B	4,500	A	A-B	990	3
13	A-B	70	A	4,800	A	A	850	4
Comp. Example								
1	D	>190	A-B	6,200	C	C*	1,090	2-4
2	C-D	>190	A	7,300	C	C*	780	3-4
3	C-D	>190	A-B	6,000	C	C*	1,130	4-6
4	B	160	A	6,000	A	B	320	4-6
5	B	168	A-B	6,500	B	B	1,070	4-6
6	B	168	B	5,500	C	C*	1,010	2-4
7	D-E	168	A-B	8,800	C	C*	800	4-5
8	A	64	B	4,000	A	B	1,300	10
9	A	64	C	4,700	A	B	1,150	11

*Phase separation.

(3) Compatibility – The transparency of the hot melt composition melted by heating (160°C) was evaluated by the naked eye and samples of good transparency to completely opaque due to the formation of a white turbidity were classified into five grades of A, B, C, D and E with A being the best grade and E being the poorest.

(4) Cloud Point – According to JIS K 2266

(5) Odor – Samples of weak odor to strong stimulative odor at heating to 160°C were evaluated using the grades of A, B, C and D with D being the most offensive in odor.

(6) Melt Viscosity – The viscosity of the sample at 140°C was measured using a Brookfield viscometer.

(7) Flexibility – A sheet of the hot melt composition having a thickness of 2 millimeters, a width of 15 mm and a length of 100 mm was bent with a bending tester (at a temperature of 25°C), a tensile force of 1.8 kg/cm^2, and a bending of 80° and the number of times the sheet could be bent until the sheet was broken was measured. A sample which could be bent more than 5,000 times was classified as A, the sample which could be bent 1,000 to 4,999 times as B, and a sample which could be bent less than 1,000 times as C.

(8) Thermal Stability – After the sample was heated to 200°C for 100 hours, the change of hue, the presence of carbonization (gellation), the presence of phase separation, the change of viscosity, and the change of adhesive strength were measured. These results were classified into the three grades of A, B and C as the entire evaluation. That is a grade of A showed that the changes were low and a grade of C showed that the changes were severe and the sample was unsuitable for practical use.

Chlorinated Polyphenyl, Chlorinated Polyisoprene and Nitroso Compound

M.H. Dollhausen and G. Hohmann; U.S. Patent 3,864,301; February 4, 1975; assigned to Bayer AG, Germany have found that a mixture consisting of (a) an ethylene-vinyl ester random copolymer; (b) a chlorinated olefin or diene polymer; (c) a chlorinated polyphenyl resin that is solid at room temperature; (d) an aromatic polynitroso compound; and (e) an organic solvent or solvent mixture can be processed to form adhesives which give bonds of much greater strength, particularly at elevated temperatures, coupled with a considerably greater resistance to organic solvents, oils, fats and other outside influences.

Particularly preferred combinations of components (a) through (e) are shown in Table 1. In order to obtain special effects, other substances, for example, natural or even synthetic resins, for example, phenol-formaldehyde resins, fillers such as silicates and/or carbon black or dyes and/or pigments, can be added to the adhesives according to the process.

In one special variant, in which the bonds obtained show optimum strength and resistance values, the adhesives according to the process can be used in conjunction with primer solutions which ensure maximum adhesion to metal. One preferred primer solution is the solution of a low molecular weight phenol-formaldehyde resin with a chlorinated olefin or diene polymer as mentioned under (b) in a solvent or solvent mixture of the kind mentioned under (e). The low molecular weight phenol-formaldehyde resin is a resin with a ratio of phenol to formaldehyde of 1:0.5 to 1.5 and a viscosity (60% in butanol) of 1,500 to 10,000 cp, as measured at 20°C with a Brookfield LVF viscosimeter.

The primer solution consists of from 5 to 40 parts by weight and preferably of from 5 to 15 parts by weight of phenol-formaldehyde resin, of 5 to 40 parts by weight, preferably 5 to 15 parts by weight of the chlorinated olefin or diene polymer mentioned under (b) and of 30 to 90 parts by weight, preferably 70 to 90 parts by weight, of a solvent of the kind mentioned under (e).

Examples: The adhesives specified in Table 1 were used for the following examples.

TABLE 1

Adhesive	A	B	C	D	E	F	
ethylene-vinylacetate copolymer*	8	8	8	4	8	8	parts by weight
chlorinated polyisoprene**	8	8	8	8	4	8	do.
chlorinated polyphenyl***	8	8	8	8	8	4	do.
carbon black†	2	2	2	2	2	2	do.
p-dinitrosobenzene	4	2	6	4	4	4	do.
xylene	55	55	55	55	55	55	do.
perchlorethylene	15	15	15	15	15	15	do.

*With a vinyl acetate content of 45% and a melt index in the range of 2 to 5, as measured in accordance with ASTM D 1238, condition E.

**With a chlorine content of 67% and a viscosity of 160 cp (20% in toluene), as measured at 20°C with a Brookfield LVF viscosimeter (spindle 1, 30 rpm).

***With a chlorine content of 52% and a softening point according to DIN 53180 in the range of 60° to 70°C.

†MT-carbon black (thermal carbon black).

Bonds of grade 9 S 20 K free-cutting steel with different rubber materials were produced with these adhesives. The structure and vulcanization conditions of the different rubbers used are shown in Table 2.

TABLE 2

Vulcanisate I

100.0 parts by weight of natural rubber;
20.0 parts by weight of zinc oxide;
2.0 parts by weight of stearic acid;
1.2 parts by weight of phenyl-β-naphthylamine;
30.0 parts by weight of ISAF-carbon black (oven soot);
1.0 part by weight of 2-mercaptobenzothiazole; and
3.5 parts by weight of sulphur
vulcanisation: 35 minutes at 140°C

Vulcanisate II

100.0 parts by weight of nitrile rubber (33% of acrylonitrile);
7.5 parts by weight of zinc oxide;
0.5 parts by weight of stearic acid;
40.0 parts by weight of SRF-carbon black (oven soot);
1.0 part by weight of collophony;
2.0 parts by weight of methylene-bis-thioglycolic acid butyl ester;
5.0 parts by weight of coumarone resin;
0.8 parts by weight of benzothiazyl-2-diethyl sulphenamide; and
1.7 parts by weight of sulphur
vulcanisation: 35 minutes at 140°C

(continued)

TABLE 2: (continued)

Vulcanisation III

100.0 parts by weight of butyl rubber;
5.0 parts by weight of zinc oxide;
60.0 parts by weight of APF-carbon black (oven soot)
3.0 parts by weight of pinewood tar;
1.5 parts by weight of zinc-N-diethyl dithiocarbamate;
2.0 parts by weight of mercaptobenzothiazole; and
2.0 parts by weight of sulphur
vulcanisation: 45 minutes at 140°C

Vulcanisate IV

100.0 parts by weight of styrene butadiene rubber;
5.0 parts by weight of zinc oxide;
1.0 part by weight of stearic acid;
60.0 parts by weight of HAF-carbon black (oven soot);
15.0 parts by weight of aromatic mineral oil plasticiser;
1.5 parts by weight of phenyl-β-naphthylamine;
1.4 parts by weight of benzothiazyl-2-cyclohexyl sulphenamide;
0.1 part by weight of tetramethyl thiuram monosulphide; and
1.8 parts by weight of sulphur
vulcanisation: 20 minutes at 150°C

The test specimens were produced in accordance with ASTM D 429-68 method B. Metal strips that had been sandblasted and degreased with trichloroethylene and strips of the particular vulcanizates to be bonded, cleaned with trichloroethylene, were coated with thin layers of the adhesives specified in Table 1. After drying for 3 hours at room temperature, the coated metal and vulcanizate strips were united with the adhesive layers in between and heated under light pressure for 40 minutes at 140°C.

After the test specimens had been stored for 24 hours at room temperature, the strength of the rubber-metal bonds was determined by the delamination test according to ASTM Test Method D 429-68 method B both at room temperature and at elevated temperatures. The results obtained are set out in Table 3. The strengths of bonds obtained by using a primer X of the following composition are also set out in Table 3.

	Parts by Weight
Chlorinated polyisoprene*	10
Phenol-formaldehyde resin**	8
Carbon black***	2
Methyl isobutyl ketone	80

*With chlorine content of 67% and a viscosity of 160 cp (20% in toluene), as measured at 20°C with a Brookfield LVF viscosimeter (spindle 1, 30 rpm)

**Nonmodified hardenable phenol-formaldehyde condensation product (60% in butanol) with a viscosity of 3,200 cp, as measured at 20°C with a Brookfield LVF viscosimeter (spindle 3, 30 rpm)

***MT-carbon black (thermal carbon black)

The primer X was applied in a thin layer to the sandblasted and degreased metal strips. After drying for 30 minutes, the adhesives specified in Table 1 were applied to the primer coated metal strips and to the rubber vulcanizates.

After another 3 hours, the test specimens were prepared as already described in accordance with ASTM D 429-68 method B. Adhesives of the following composition prepared in accordance with German Patent 1,295,123 were used for comparison in Examples 19 and 20.

Adhesive K	Parts by Weight
Ethylene-vinyl acetate copolymer*	11
Maleic resin**	2
Terpene-phenol resin***	2
Methylene chloride	35
Toluene	35
Petrol (BP 65° to 95°C)	15

*With a vinyl acetate content of 45% and a melt index of 2 to 5, as measured in accordance with ASTM D 1238, condition E

**With an acid number according to DIN 53183 of 16 and a melting point according to DIN 53181 of 130°C

***With an acid number according to DIN 53183 of 65 and a melting point according to DIN 53181 of 140°C

TABLE 3

Example	Vulcanisate	Adhesive	Primer	Bond strength (kp/cm²) 20°C	80°C	100°C	120°C
1	I	A	—	18.6*)	6.3	3.5	1.9
2	I	A	x	23.4*)	12.7	6.3	3.7
3	II	A	—	12.4			
4	II	A	x	19.3*)			
5	III	A	—	9.4			
6	III	A	x	14.2*)			
7	IV	A	—	8.2			
8	IV	A	x	14.0*)			
9	I	B	—	16.2*)	5.7	2.8	1.3
10	I	B	x	21.2*)	11.8	5.4	2.5
11	I	C	—	15.4*)	4.6	2.4	1.3
12	I	C	x	20.4*)	9.7	4.5	2.1
13	I	D	—	17.8*)			
14	I	D	x	22.6*)			
15	I	E	—	14.7*)			
16	I	E	x	18.2*)			
17	I	F	—	14.2*)			
18	I	F	x	18.8*)			
19	I	K	—	1.5	—	—	—
20	I	K	x	2.5	—	—	—

*Separation of the vulcanizate

Carpet Backing Formulation

D.D. Taft and T.H. Shepler; U.S. Patent 3,849,353; November 19, 1974; assigned to Ashland Oil, Inc. describe a hot melt composition which consists essentially of: from about 5 to 50% by weight of a copolymer of ethylene and vinyl acetate, and/or alkyl acrylate wherein the alkyl group contains 1 to 18 carbon atoms and containing from 40 to 85% by weight of ethylene and from 15 to 60% by weight of vinyl acetate and/or the C_1 to C_{18} alkyl acrylate; about 10 to 85% by weight of atactic polypropylene; and about 10 to 50% by weight of vulcanized rubber. The above quantities of copolymer, polypropylene and rubber are based upon the combined weight of the copolymer, polypropylene, and rubber and not necessarily upon the total composition which may contain other ingredients.

The hot melt compositions provide excellent adhesion between the pile loops and the primary backing and between the secondary backing and the carpet. This excellent adhesion can be observed from the force in pounds required to pull one of the pile loops loose from the primary backing and from the amount of force in pounds required to separate the secondary backing from the carpet.

For example, the hot melt compositions of the process require up to about 20 pounds of force to pull one of the pile loops loose from the primary backing whereas the conventional latex adhesives require only about 8 to 12 pounds of pull at equal application levels. The following examples illustrate the use of the compositions as carpet backing adhesives.

Example 1: The raw tufted carpet is a shag carpet composed of a jute base fabric weighing about 10 ounces per square yard and tufted with about 18 ounces per square yard of bulked continuous filament nylon carpet yarn which is needled into the base fabric at 5 needles per inch across the width of the base fabric and about 5 stitches per running inch of base fabric. The tuft loops extend 1¼" above the base fabric and the tufted bases extend 1/16" below the base fabric. The secondary backing material is woven polypropylene weighing about 3.5 ounces per square yard.

The raw tufted carpet is passed under and adjacent to an extruder where the bottom (nonpile) surface is coated with 20 to 26 ounces per square yard of a molten coating composition at a temperature of 350°F and having the following formula.

	Parts
Copolymer of 67% ethylene and 33% vinyl acetate, having an inherent viscosity of 0.78 (0.25% in toluene at 86°F), a melt index of 22 to 28 (ASTM D 1238-577), and a softening point of 240°F (ring and ball) (EVA 150)	3.55
Copolymer of 72% ethylene and 28% vinyl acetate, having an inherent viscosity of 0.94 (0.25% in toluene at 86°F), a melt index of 5 to 7 (ASTM D 1238-577), and a softening point of 310°F (ring and ball) (EVA 260)	3.55
Tall oil rosin	21.4
Atatic polypropylene	21.4
Vulcanized copolymer of 97% isobutylene and 3% isoprene (Rubber Research H602)	21.4
2,2'-methylenebis(4-methyl-6-tert-butylphenol)	0.4
$CaCO_3$ (No. 1 white)	28.3

The molten composition which is kept at 350°F is forced through the extruder head onto the back of the raw tufted carpet as it passes under the extruder head. The carpet then passes underneath a heated doctor blade which evenly distributes any excess adhesive that may be present.

The carpet then passes between two rollers at which point the secondary backing material from the upper roll is applied to the back side of the carpet. The carpet is then passed into a forced air cooling section 10 feet in length to set the coating composition. The primary and secondary backing materials are firmly bonded to each other, and the base of each loop in the pile is firmly bonded to the primary backing.

A pull of 15 to 18 pounds is required to pull one of the pile loops loose from the primary backing. The carpet has a T peel of more than 25 pounds. In addition the carpet demonstrated excellent resistance to delamination when subjected to repeated bending and straightening. In addition the percent elongation of a 40 mil film of the coating composition is greater than 300% as measured on an Instron Tensile Tester.

Example 2: Example 1 is repeated except that the coating composition has the following formula.

	Parts
Copolymer of 72% ethylene and 28% vinyl acetate, having an inherent viscosity of 0.59 (0.25% in toluene at 86°F), a melt index of 335 to 465 (ASTM-D-1238-577) and a softening point of 180°F (ring and ball)	9.0
Tall oil rosin	26.0
Atactic polypropylene	15.0
Vulcanized copolymer of 97% isobutylene and 3% isoprene	8.0
2,2'-methylenebis(4-methyl-6-tert-butylphenol)	0.2
$CaCO_3$ (No. 1 white)	42.0

The primary and secondary backing materials are firmly bonded to each other, and the base of each loop in the pile is firmly bonded to the primary backing. A pull of 18 to 22 pounds is required to pull one of the pile loops loose from the primary backing. The carpet has a T peel of about 18 to 20 pounds. In addition the carpet demonstrates excellent resistance to delamination when subjected to repeated bending and straightening. In addition the percent elongation of a 40 mil film of the coating composition is greater than 350% as measured on an Instron Tensile Tester.

Example 3: Example 1 is repeated except that the secondary backing material is jute weighing about 10 ounces per square yard. The primary and secondary backing materials are firmly bonded to each other, and the base of each loop in the pile is firmly bonded to the primary backing. To remove a pile loop from the primary backing requires a 15 to 18 pound pull. T peel is 20 pounds. The carpet has excellent resistance to delamination on repeated bending and straightening.

Example 4: Example 2 is repeated except that the secondary backing material is jute weighing about 10 ounces per square yard. The primary and secondary backing materials are firmly bonded to each other, and the base of each loop in the pile is firmly bonded to the primary backing. A pull of 18 to 22 pounds is required to pull one of the pile loops loose from the primary backing. The carpet has a T peel of about 25 pounds. In addition the carpet demonstrates excellent resistance to delamination when subjected to repeated bending and straightening.

Example 5: Example 1 is repeated except that the secondary backing material is nonwoven polyester polypropylene weighing about 4.2 ounces per square yard. The primary and secondary backing materials are firmly bonded to each other, and the base of each loop in the pile is firmly bonded to the primary backing. A pull of 15 to 18 pounds is required to pull one of the pile loops loose from the primary backing. The carpet has a T peel of about 30 pounds. In addition the carpet demonstrates excellent resistance to delamination when subjected to repeated bending and straightening.

OTHER POLYOLEFIN COMPOSITIONS

Amorphous Polypropylene

According to a process described by *L.D. Jurrens; U.S. Patent 3,900,694; August 19, 1975; assigned to Phillips Petroleum Company* hot melt adhesive compositions with high softening temperatures are prepared from a mixture of amorphous polypropylene having a molecular weight range of 500 to 15,000, an olefin copolymer of 95 to 98% by weight propylene and 5 to 2% by weight ethylene, and a polyterpene tackifier. The improved hot melt adhesive formulation can contain from about 25 to 35% by weight amorphous polypropylene, from about 35 to 45% by weight of the olefin copolymer and from about 20 to 40% by weight of a polyterpene tackifier and is particularly useful for sealing folding cartons and corrugated paperboard cartons.

Example 1: The formulations prepared consisted of 40% crystalline propylene-ethylene copolymer, melt flow 308 (ASTM D 1238-65T, Condition L), 30% by weight amorphous polypropylene and 30% by weight polyterpene tackifier. The propylene-ethylene copolymer was made by visbreaking a propylene-ethylene copolymer having a melt flow of 4 to a melt flow of 308 (as determined by ASTM D-1238-65T, Condition L). The crystalline copolymer of propylene and ethylene contained about 97% by weight propylene and about 3% by weight ethylene.

The polyterpene tackifier used was Nirez 1135. When added, the amount of antioxidant was 0.1% based on the crystalline propylene-ethylene copolymer plus the amorphous polypropylene. The antioxidant consisted of one part tetrakis[methylene(3,5-di-tert-butyl-4-hydroxyhydrocinnamate)] methane, 2 parts of a commercially obtained butylated hydroxytoluene, and 3 parts dilaurylthiodipropionate.

The tackifier, amorphous polypropylene, and the antioxidant were melted in a stainless steel beaker heated in an oil bath at 204°C for 30 minutes. The crystalline polypropylene-ethylene copolymer was added with agitation during the following 45 minutes and final blending to homogeneity required an additional 45 minutes with agitation. Formulations were prepared in which the amorphous propylene polymer had an average molecular weight of 7,500 (Formulation A) and in which the amorphous propylene polymer had an average molecular weight of 800 (Formulation B).

Example 2: A Nordson hot melt applicator, Model II, was used to test the performance of the hot melt adhesive compositions of this process and the control compositions in machinability and bond strength. The applicator is equipped with a large drum 12 inches wide and 39 inches in circumference which is then covered with kraft paper. The molten adhesive is extruded from a nozzle to the drum applying either an intermittent or continuous bead of the adhesive to the substrate. To make this useful for producing laminations, a modification was added so that two rolls of paper and a nip roller complemented the large drum. By varying the drum speed and raising or lowering the nozzle above the substrate, the length of time between application of adhesive to one substrate and application of the other substrate could be varied. Since it was desired that paper laminates be prepared at higher application temperatures than the 210°C maximum hose temperature of the commercial Nordson hot melt applicator, a modification of the applicator was adapted for some of the runs. This modification consisted of adding a 1 liter reactor equipped with an air driven stirring motor, electric heating mantle, thermocouples, and a stainless steel line with valve from the bottom of the reactor to the wheel laminating assembly, all of which was installed on a stand directly above the drum.

A comparison of the bond strength of the adhesive compositions of this process are compared with the bond strength of a commercially available amorphous polypropylene based hot melt adhesive in the data shown in Table 1. This also illustrates the improved bond strength at low pull speeds of laminates produced by bonding two strata with the hot melt adhesive of this process. Paper-adhesive-paper laminates are illustrated but any porous strata can be laminated in the same manner. A laminate comprising several strata can also be produced.

In the modified applicator with a line temperature of 232°C and at drum speeds of 45 and 9 rpm, paper tear or fiber tear was obtained when the laminates were pulled in the Instron testing machine at crosshead speeds of 10 inches and 0.2 inch per minute thus indicating excellent adhesion to paper. The formulations prepared in which amorphous propylene polymer having an average molecular weight of 800 was substituted for the amorphous polypropylene having an average molecular weight of 7,500 produced an adhesive of advantageously low viscosity which also showed good bond strength.

TABLE 1: BOND STRENGTHS OF PAPER/PAPER LAMINATES

Adhesive Composition	Applicator Type	Applicator Line, °C	Applicator Drum Speed rpm	Bond Strength[1] 10"/min	Bond Strength[1] 0.2"/min
Amorphous polypropylene avg. MW 7,500, crystalline polypropylene-ethylene copolymer, polyterpene (Formulation A)	modified Nordson[2]	232	45	P.T.[3] (0.64)[4]	P.T. (0.50)
	modified Nordson[2]	232	9	P.T. (1.25)	F.T. (0.88)
Amorphous polypropylene avg. MW 800, crystalline polypropylene-ethylene copolymer, polyterpene (Formulation B)	modified Nordson	193	45	F.T. (1.85)	F.T./A.F. (0.80)
	modified Nordson[2]	193	9	F.T./A.F. (1.55)	A.F. (0.09)
	Nordson	193	45	P.T. (1.64)	F.T./P.T. (1.08)
	Nordson	193	9	P.T. (1.19)	F.T./P.T. (1.08)
Amorphous polypropylene avg. MW 30,000, based commerical adhesive	Nordson	171	45	P.T./F.T. (1.31)	A.F. (0.14)
	Nordson	171	9	F.T./C.F. (1.42)	A.F. (0.12)

[1]ASTM D 1876-61T
[2]Containing means to facilitate use of adhesive mixtures at temperatures higher than 210°C
[3]P.T. = paper tear; F.T. = fiber tear; A.F. = adhesive failure; C.F. = cohesive failure 4 lb/in width

In contrast to the excellent adhesion of paper to paper observed with the adhesive of this process, the commercial hot melt adhesive based on an amorphous polypropylene with an average molecular weight of 30,000 showed good bond strength when pulled at 10 inches per minute but poor bonding at the lower crosshead speed of 0.2 inch per minute.

TABLE 2: SOFTENING POINTS OF HOT MELT ADHESIVE FORMULATIONS

	------Softening Point, °C------	
Adhesive Composition	**Ball and Ring**	**Static Load**
Amorphous polypropylene average MW 7,500, crystalline polypropylene-ethylene copolymer, polyterpene	146	149
Amorphous polypropylene average MW 800, crystalline polypropylene-ethylene copolymer, polyterpene	144	156
Amorphous polypropylene average MW 30,000 based commercial adhesive	92	80
Ethylene-vinyl acetate/wax/tackifier*	66	58

*Formulated with 40 parts Elvax 150, 30 parts Staybelite Ester 10, and 30 parts paraffin wax.

In the above table adhesive softening points of the compositions of this process are compared with the softening points of a commercial amorphous polypropylene based adhesive and a standard ethylene-vinyl acetate based hot melt adhesive. The temperature at which measurable softening and flow of the hot melt adhesive occurs defines the upper limit at which adhesive will be effective.

Since melting occurs over a range of temperature, it is difficult to measure exactly. The test employed to determine this value was ASTM D 36-66T (Ball and Ring Apparatus). The compositions of this process showed a softening point 42°C higher than the commercial amorphous polypropylene based adhesive and 78°C higher than the ethylene-vinyl acetate formulation.

Another test employed to determine softening temperature is the Static Load Test. In this test two one inch wide paper strips are bonded with a one inch overlap, resulting in a one square inch bond area. Duplicate samples are suspended in an oven and a 250 gram load added. The temperature of the oven is raised incrementally at 30 minute intervals. When the bond fails, the temperature is recorded. The results of this test are considered to more nearly represent the maximum performance temperature of a hot melt adhesive. It must be noted that the compositions of this process showed an increased performance advantage over the commercially available material using this test as compared with the Ball and Ring test.

Amorphous Polyolefin and Styrene-Butadiene Block Copolymers

V.K. Park; U.S. Patent 3,850,858; November 26, 1974; assigned to Eastman

Kodak Company has found that a blend comprising amorphous polyolefin, rubbery block copolymers of styrene and butadiene or styrene and isoprene, polybutylene, polyisobutylene, a tackifying resin, and crystalline polypropylene provides a hot melt pressure sensitive adhesive which can be applied without solvents and has a good combination of properties including good adhesive properties at both high and low temperature.

The hot melt adhesive compositions of this process, which are essentially 100% solids, have a Thermosel viscosity, Model RVT, in the range of 4,500 to 50,000 centipoises at 190°C and a softening point (ring and ball, ASTM D36-26) of at least about 245°F, preferably from about 280° to 310°F.

Example 1: A hot melt pressure sensitive adhesive is made in a 2 liter resin flask fitted with an anchor blade stirrer under nitrogen using the following components.

Amorphous polypropylene (viscosity is 3,000 cp at 190°C)	300 g (50% by weight)
Rosin ester (Foral 105)	90 g (15% by weight)
Polystyrene-polyisoprene-polystyrene block copolymer (crumb form of Kraton 1107)	88.5 g (14.75% by weight)
Polybutylene (Indopol H-1900)	60 g (10% by weight)
Polyisobutylene (Vistanex LM-MS)	54 g (10% by weight)
Crystalline polypropylene (inherent viscosity is 1.7)	6 g (1% by weight)
Antioxidant (Ethyl 702)	1.5 g (0.25% by weight)

The flask is heated with an electric heating mantle to a temperature of 190°C with stirring. The mixture is blended 4 hours at 190°C to obtain a homogeneous mass. The adhesive is then poured into a silicone lined box and allowed to cool. The viscosity is 4,850 centipoises at 190°C.

This adhesive is evaluated by adhering a rubber gasket weather strip to an acrylic enamel painted metal panel. The adhesive is placed in a Hot Shot No. 1 glue gun and heated to 325°F. Two beads of adhesive, one 6 inches long and one 2 inches long, and one round spot of adhesive approximately one inch in diameter are applied to the panel. After waiting two minutes the rubber strip of the same length is hand pressed into each bead of adhesive.

Three panels are prepared as above. The samples are then placed in an oven preheated to 165°F horizontally inverted with the rubber facing down. The effect of heat is observed on all panels during the 20 minute exposure. The movement of adhesive or falling of strips is recorded. If the results are satisfactory, the panels are retested at elevated temperature in increments of 25°F up to a temperature where the specimen fails. The adhesive passes 240°F.

The tensile strength of the adhesive is measured as follows. Panels are made with 6 one inch long pieces of weather strip pressed into the adhesive. The first weather strip piece is pressed into the adhesive after a lapse of one minute. The second, third, etc., pieces are pressed into the adhesive after 1.5, 2, 2.5, 3 and 5 minute periods. Tensile tests of these specimens, after being aged 24 hours at 72°F, are measured on an Instron tensile run at a rate of one inch per minute. These tests are run on duplicate samples and recorded. This product shows a tensile strength averaging 6.0 pounds per inch of pull.

Panels are prepared for low temperature properties the same as for the elevated temperature test. The panels containing the weather strip bonded to the panel with the adhesive are exposed at -20°F for 2 hours in a cold box. The panels are then taken out and shock spanked immediately after removal from the cold box. The bonds are examined for adhesion, adhesive fracture or brittleness and results are recorded. This adhesive shows good adhesion to the rubber and to the painted metal surface with no adhesive failure.

Example 2: A hot melt pressure sensitive adhesive is prepared the same as described in Example 1 except that only amorphous polypropylene, rosin ester (Foral 105), and polystyrene-polyisoprene-polystyrene rubber (Kraton 1107) and antioxidant (Ethyl 702) are used. The amounts used are 69.75 weight percent amorphous polypropylene, 25 weight percent Foral 105, 5 weight percent Kraton 1107, and 0.25 weight percent Ethyl 702 antioxidant. Testing of this adhesive as described in Example 1 shows the following results.

Tensile strength, pounds/inch	5.4
Elevated temperature resistance	Fail at 215°F
Cold shock test at -20°F	Good adhesion, no brittleness (pass)

This example shows good adhesive properties but poor elevated temperature resistance.

Example 3: A hot melt pressure sensitive adhesive is prepared in the same manner as described in Example 1 using 69.6 weight percent amorphous polypropylene, 30 weight percent hydrocarbon resin (Wingtack 95), 0.18 weight percent dilauryl 3,3'-thiodipropionate, and 0.18% Polygard HR antioxidants. This adhesive shows fair tensile strength but poor elevated temperature and low temperature resistance.

Example 4: A hot melt pressure sensitive adhesive is prepared in the same manner as described in Example 1 using 69.75 weight percent amorphous polypropylene, 25.0 weight percent Wingtack 95, 5 weight percent ethylene-vinyl acetate copolymer (Elvax 460), and 0.25 weight percent Ethyl 702 antioxidant. This adhesive shows poor tensile strength and poor elevated temperature resistance (fails at 215°F) and poor cold shock resistance.

Hot Melt Pressure Sensitive Adhesives

	Blend number							
	1 (control)	2	3	4	5	6	7	8
M-5, wt. percent	50	49	49	69.75	65	51	50	50
Wingtack 95, wt. percent		20	20	15.0				
Foral 150, wt. percent	15				15	15	15	15
Elvax 460, wt. percent								
Indopol H-1900, wt. percent	10		30	15.0	10	10	19	
Vistanex LM-MS, wt. percent	9.0	30			9.0	9.0		19
Crystalline polypropylene, wt. percent	1.0	0.75	0.75		1.0		1.0	1.0
Kraton 1107, wt. percent	14.75					14.75		
Bond test, lb./in.:								
1.0 min	6.5	4.7	3.9	5.0	5.2	6.4	4.9	5.2
3.0 min	6.2	4.7	4.0	4.8	4.8	6.2	4.8	5.1
5.0 min	6.0	4.6	4.0	4.8	4.1	5.9	4.6	4.6
Cold shock test at -20° F.:								
Adhesion	Good	Good	Good	Fair	Fair	Good	Good	Fair
Brittleness	Pass	Pass	Fail	Fail	Fail	Pass	Fail	Fail
Elevated temp. (thermo-resistance):								
165° F	do	do	Pass	Pass	Pass	do	Pass	Pass
190° F	do	do	do	do	do	do	do	do
215° F	do	Fail	Fail	Fail	do	Fail	do	do
240° F	do				Fail		Fail	Fail

Example 5: Hot melt pressure sensitive adhesives are prepared and evaluated in the same manner as in Example 1. The composition and the results of evaluation as adhesives for adhering a rubber gasket to painted metal are given in the above table. All of these blends are deficient in either elevated temperature or low temperature properties.

α-Methylstyrene-tert-Butylstyrene-Olefin Terpolymers

M.G. Henk and G.P. Sartler; U.S. Patent 3,903,060; September 2, 1975; assigned to The Standard Oil Company describe resin compositions which are made by copolymerizing α-methylstyrene, tert-butylstyrene and a monoolefin. The resins can be produced in a variety of physical forms, ranging from hard resins of up to 400°F ring and ball softening point to low molecular weight products which are viscous liquids at room temperature. The resins having a softening point up to about 350°F have excellent solubility in a variety of aliphatic and aromatic organic solvents, those in the range of 200° to 350°F being particularly preferred. These resins are particularly suitable for use in hot melt adhesive compositions and paper coating formulations.

The resins can be readily produced by catalytic polymerization of the aforementioned monomers with BF_3, BF_3 complexes, aluminum chloride, and similar Friedel-Crafts type acidic catalysts. This polymerization is conveniently carried out in an inert organic solvent at a temperature within the range of -60° to 100°C. The softening points and molecular weight of the resins produced can be readily controlled by varying the temperature at which the monomer mixture is contacted with the catalyst, by varying the ratio of catalyst to monomers and by varying the ratio of monomer to solvent according to techniques known to the art.

Generally lower temperatures produce resins of higher softening points and higher molecular weights. Higher temperatures produce resins of relatively lower softening points and lower molecular weights, while, at a given temperature, the higher the concentration of the reactants the lower the softening point and the higher the concentration of catalyst, the lower the softening point.

The utilization of α-methylstyrene-tert-butylstyrene-olefin terpolymers produced as set forth above have been found to impart excellent properties of heat stability, color, odor, and performance to the ethylene-vinyl acetate/petroleum wax compositions used as hot melt adhesives.

Examples 1 through 11: A series of six α-methylstyrene-tert-butylstyrene-olefin terpolymers and five α-methylstyrene-tert-butylstyrene copolymers were prepared by either batch or continuous polymerization. The data relating to the polymerization and characterization of the product are presented in the table below.

For the batch polymerization method an approximately 1,000 gram sample of a 80 weight percent solvent was percolated through a bed of anhydrous $CaCl_2$ and then transferred to a three-necked flask having a stirrer and a thermocouple. The stirred solution was then cooled to below the desired initiation temperature. The cooling bath was then removed and the solution was allowed to slowly warm to the desired initiation temperature. Upon reaching the initiation temperature, BF_3 catalyst was added and the reaction was allowed to go to completion.

	- Relative Parts by Weight-			Initiation Temp. (°C)	Change Temp (°C)	- - - - - Physical Properties of Resin- - - - -		Molecular Weight††††		
Ex.	AMS*	TBS**	Olefin***			R & B†	Cloud Point††	Mn	Mw	Mz
1	3	1	0.005i	-9	41	253	350	1,079	2,710	5,700
2	3	1	0.05i	-9	37	242	364	968	2,740	6,100
3	3	1	0.05b	-10	45	235	362	951	2,590	5,900
4	3	1	0.05b	-8	32	243	322	951	2,430	5,000
5	3	1	0.05b	-10	25	249	256	1,058	2,390	4,600
6	3	1	0.04i	-10	50	234	332	913	1,830	–
7	2	1	None	-10	46	242	332	1,089	2,830	–
8	2	1	None	-8	43	239	334	929	2,820	8,200
9	3	1	None	-8	44	218	450	869	2,930	8,600
10	3	1	None	-7	32	240	370	960	2,730	9,400
11	3	1	None	-7	33	243	440	924	2,620	7,400

Note: Examples 1 through 5 use toluene solvent and Examples 6 through 11 use cumene solvent.
*Alpha-methylstyrene
**Tert-butylstyrene
***Either isoprene (i) or 2-methyl-2-butene (b)
†Ring and Ball Softening Point, ASTM E28-51T
††Cloud Point of a blend of AMS/TBS copolymer/ethylene-vinyl acetate copolymer/paraffin wax in weight ratio of 100/100/300. (Elvax 260 containing 28% vinyl acetate)
†††Gel Permeation Chromatograph

During polymerization the time and temperature were monitored. Upon completion the BF_3 and solvent were removed by distillation with the use of nitrogen sparge at atmospheric pressure until the solution temperature reached approximately 200°C. A house vacuum (≈25 torr) was used between 200° and 245°C until devolatilization visually stopped. The remaining volatiles in the resins as measured by gas chromatography were generally less than 0.1%.

For the continuous polymerization method a sample in excess of 10,000 grams of monomer/solvent feed was prepared in a manner similar to the batch runs. The feed was pumped at a rate of 60 cc per minute through a cooling coil submerged in a constant temperature bath and then through a tube reactor where it was contacted with the BF_3 catalyst. The solvent and catalyst removal after polymerization were similar to the previous batch experiments.

Examples 1 through 6 are typical of the softening point and cloud point values achieved at a 3/1/0.05 ratio of α-methylstyrene/tert-butylstyrene/olefin. These values are more closely analogous to the typical values for a 2:1 ratio of α-methylstyrene:tert-butylstyrene as seen in Examples 7 and 8 than the values typical of a 3:1 ratio as seen in Examples 9 through 11, thus exemplifying the ability of trace amounts of olefin to increase the wax compatibility of an α-methylstyrene-tert-butylstyrene polymer at far lower levels of tert-butylstyrene than previously known.

Although the mechanism by which this is accomplished is not fully understood, one possible interpretation is that the presence of the trace amounts of olefin enhance the tert-butylstyrene ability to copolymerize with increased randomness thus decreasing the frequency of long chain homopolymer blocks of wax-incompatible poly-α-methylstyrene.

Example 12: The terpolymer resins of the previous examples when blended in the following formulation compare favorably with commercial α-methylstyrene-toluene copolymer coating resins. Generally the (3/1/0.05) terpolymer having a ring and ball softening point, ASTM E28-51T, of 240°F and a cloud point of 330°F will exhibit superior 60° gloss, TAPPI T653 ts-63, having an initial value as well as a value after 7 days in excess of 20. Both coefficients of friction (static and kinetic), ASTM D 1894-63, pick temperature and block temperature, TAPPI T652 05-61, are comparable.

However, the seal strength, TAPPI T642 SU-64, will be superior for the terpolymer reaching values in excess of 300 lb/10 cm. The moisture vapor transmission rates are similar, while the terpolymer displays excellent heat stability in that no separation occurs after 100 hours at 350°F with only slight color change.

Component	Parts by Weight
Terpolymer	100
Ethylene-vinyl acetate copolymer*	100
Paraffin wax	300
Microcrystalline wax	167

*Elvax 260 containing 28% vinyl acetate

Alkoxystyrene-Acrylonitrile Copolymers

According to a process described by *H.W. Hill, Jr. and C.W. Moberly; U.S. Patent 3,846,511; November 5, 1974; assigned to Phillips Petroleum Company* resins useful in adhesive formulations are prepared by polymerization of an alkoxystyrene with an olefinically unsaturated nitrile in a solution or dispersion of a rubber. The polymeric compositions can be applied to surfaces as a hot melt adhesive or solvent cement.

Examples of suitable α-alkoxystyrenes include α-methoxystyrene, α-ethoxystyrene, α-isopropoxystyrene, α-butoxy-2-methylstyrene, α-propoxy-2-ethyl-5-methylstyrene, and α-hexoxy-3,4-diethylstyrene. The α,β-unsaturated nitriles that can be copolymerized along with the α-alkoxystyrenes defined above include acrylonitrile and methacrylonitrile.

In order that the resins of this process will have the most desirable balance of properties, 25 to 35 weight percent of the resin should be derived from the rubber; 40 to 45 weight percent of the resin should be derived from the α-alkoxystyrene; and 25 to 30 weight percent of the resin should be derived from acrylonitrile or methacrylonitrile. However, in view of the lower reactivity of the α-alkoxystyrene as contrasted with the acrylonitrile or methacrylonitrile, it is sometimes beneficial to charge to the reactor a somewhat higher ratio of α-alkoxystyrene to acrylonitrile or methacrylonitrile.

Example 1: Resins were prepared by copolymerizing α-methoxystyrene and acrylonitrile in the presence of a rubbery polymer. Solprene 1205 rubber, a butadiene/styrene block copolymer with 75/25, respectively, parts by weight composition and a Mooney viscosity of 47 ML-4 at 212°F (ASTM D 1646-63) as the rubbery component, was dissolved in cyclohexane in the amounts shown in the following table. In Runs 1, 2 and 3, respectively, 1.90, 3.75 and 6.25 grams of Solprene 1205 rubber were dissolved in 75 milliliters of cyclohexane in 10 ounce bottles. After addition of rubber and solvent the bottles were sealed with extracted rubber gaskets and perforated crown caps.

Run	1	2	3
Solprene* 1205 rubber, g	1.90	3.75	6.25
Cyclohexane, ml	75	75	75
alpha-Methoxystyrene, g	9.0	9.0	9.0
Acrylonitrile, g	3.5	3.5	3.5
Azobisisobutyronitrile, g[a]	0.5	0.5	0.5
Yield, g[b]	10.8	12.7	15.5
Conversion of Monomers, %	71	72	74
Rubber-derived content of resin, wt. %	18	29	40
alpha-Methoxystyrene-derived units, wt. %	50	43	37
Acrylonitrile-derived units, wt. %	32	28	23

Lap Shear Test

Product	Lap Shear Strength, psi[c]
Product of Run 1	Too brittle
Product of Run 2	592
Product of Run 3	262
Cyclolac TD-1001 ABS resin	318
Solprene* 411 rubber	481
Neoprene rubber	363

[a]0.25 gram added at beginning and 0.25 gram added after 5.5 hours.
[b]In 24 hours total polymerization time at 70°C.
[c]ASTM D 1002-64.

The bottles were alternately evacuated and pressured with nitrogen through a hypodermic needle extending through the gasket to provide a nitrogen atmosphere in the bottles. The bottles were tumbled in a polymerization bath containing water as the heat transfer medium at 70°C until the rubber was dissolved. The bottles were removed from the bath, uncapped, and 9.0 grams α-methoxystyrene, 3.5 grams acrylonitrile and 0.25 gram azobisisobutyronitrile were added to each bottle.

The bottles were flushed as before with nitrogen and replaced in the polymerization bath for 5.5 hours at 70°C. The bottles were removed from the bath and 0.25 gram azobisisobutyronitrile was added to each opened bottle, which was again purged of air with nitrogen. Polymerization was continued for 18.5 hours for a total of 24 hours by tumbling the bottles at 70°C. As the polymerization proceeded, the reaction medium took on the appearance of a latex, and the final products in the bottles were relatively fluid liquids with the appearance of latex.

The contents of each bottle were poured into methanol to precipitate the resin. The resin was filtered out on a fluted paper and chopped in a Waring Blendor in methanol, filtered to collect the product, and dried in a vacuum oven at about 150°F (66°C). The products weighed 10.8, 12.7 and 15.5 grams, respectively, which correspond to conversions of the monomers, assuming complete recovery of the rubber charged, of 71, 72 and 74 weight percent. The weight percentage of Solprene 1205 rubber incorporated in the products were calculated to be 18, 29 and 40.

Example 2: To test the resinous products as potential components of hot melt adhesives, films of the products were melted between aluminum coupons, and lap shear strength values were obtained, the values being compared with those of other resins. Films of approximately 10 mils thickness were pressed between plates at 350°F (177°C) for the experimental resins, and Cycloac TD-1001, a commercial ABS (acrylonitrile/butadiene/styrene) resin. A section of the film of each was placed between 1 inch wide aluminum strips held in a jig to produce a 1 inch overlap of the strips with the melted resin sandwiched between.

Solprene 411 rubber, a branched block 70/30 (by weight) butadiene/styrene copolymer, and neoprene rubber are shown in comparative lap shear strength tests. For these rubber polymers, the aluminum coupons with polymers sandwiched between were prepared by applying rubber cement, a solution of the rubber, to each aluminum strip, evaporating the solvent to leave the strip coated with rubber, and bonding the strips together held in a jig in a press at 350°F (177°C). The values for lap shear strength were determined by placing the aluminum strips in the jaws of an Instron tester and pulling them apart to obtain values for lap shear strength in pounds per square inch as shown in the above table.

As shown in the table, the resins produced in Runs 2 and 3 exhibited substantial lap shear strength. In fact, the resin produced in Run 2 exhibited higher lap shear strength than did the ABS resin, the Solprene 411 rubber, or the neoprene rubber, each of which is useful in hot melt adhesive compositions. The resin produced in Run 1, in which insufficient rubber was employed, was too brittle.

Boric Acid as Viscosity Stabilizer in Ethylene-Propylene Adhesives

According to a process described by *A.M. Snow, Jr. and R.E. Uschold; U.S. Patent 3,882,065; May 6, 1975; assigned to E.I. du Pont de Nemours & Co.* the viscosity stability of hot melt adhesives containing an elastomeric linear EPDM, branched ethylene/propylene copolymer, or propylene/hexene-1 dipolymer is markedly improved by inclusion in the adhesive composition of boric acid, boric anhydride, or a boric acid ester, in combination with a phenolic antioxidant.

In the examples that follow, all parts and percentages are by weight unless otherwise indicated. Viscosity measurements are as taken with an RVT Brookfield Thermosel viscometer using a No. 21 spindle at 20 rpm unless otherwise indicated. Ring and Ball softening points are determined following the procedures of ASTM E-28-58T.

Example 1: (A) Formulation of Hot Melt Compositions – The synthetic resin used is a synthetic polyterpene resin having a Ring and Ball softening point of 100° ± 5°C, known as Wing-Tack 95. The petroleum wax used is a fully refined paraffinic wax having a melt point of 61°C (ASTM Test Method D-87). The branched elastomeric tetrapolymer used has the following monomer unit composition: ethylene, 72.75%; propylene, 23%; 2,5-norbornadiene, 0.15%; and 1,4-hexadiene, 4.1%. This tetrapolymer has an inherent viscosity of 1.25 at 30°C (measured on a solution of 0.1 gram of tetrapolymer in 100 ml of tetrachloroethylene), and a Mooney viscosity (ML-1 + 4 at 121°C) of 19.

A 2 gallon agitated, oil-jacketed mixing kettle is heated to about 177°C and charged with 100 parts of Wing-Tack 95 and 100 parts of refined paraffinic wax. When these components have melted, 40 parts of tetrapolymer are slowly added. Sequentially, 2 parts of maleic anhydride grafted polyethylene, 2.4 parts of 2,6-di-tert-butyl-4-methylphenol, and 2.4 parts of boric acid powder are added.

When the resulting stirred mixture has become homogeneous, it is poured into a container and allowed to cool. For test purposes, a portion of the mixture is collected on a stainless steel tray, cooled, and cut into 1 inch squares when solid. Typical results of viscosity measurements are recorded in Table 1 on the following page.

TABLE 1

COMPOSITION		PROPERTIES		
Ingredient	Parts	Hours at 190° C.	Visc (cps)	% Change
Branched Tetrapolymer	40.0	0	1545	—
Polyterpene Resin	100.0	24	1512	−2.1
Wax	100.0	48	1480	−4.2
Maleic Anhydride Grafted Polyethylene	2.0	72	1482	−4.1
2,6-di-t-butyl-4-methyl phenol	2.4	96	1492	−3.4
Boric Acid Powder	2.4			
	246.8			

(B) Comparative Example – For comparison, a hot melt adhesive composition is prepared using the ingredients and following the procedures of (A) except that boric acid and the hindered phenol are not added to the adhesive composition. Results are shown in Table 2. This example shows melt instability when boric acid and a hindered phenolic antioxidant are not included in the hot melt composition.

TABLE 2

COMPOSITION		PROPERTIES		
Ingredient	Parts	Hours at 190° C.	Visc. (cps)	% Change
Branched Tetrapolymer	40.0	0	1285	0
Polyterpene Resin	100.0	0.75	1207	−6.1
Wax	100.0	1.5	1175	−8.6
	240.0	4.75	1095	−15
		6.75	925	−28
		24	512	−60

(C) Comparative Example – For comparison, compositions are prepared using the branched tetrapolymer, polyterpene resin, and wax of (A). One composition contains only boric acid, and the second composition contains only the hindered phenolic antioxidant. Results are shown in Table 3.

TABLE 3

Ingredient		Parts	Ingredient		Parts
Branched Tetrapolymer		40	Branched Tetrapolymer		40
Polyterpene Resin		100	Polyterpene Resin		100
Wax		100	Wax		100
2,6-di-t-butyl-4-methyl phenol		3	Boric Acid Powder		2.4
		243			242.4
Hours at 190° C.	Visc. (cps)	% Change	Hours at 190° C.	Visc. (cps)	% Change
18	1095	—	0	1345	—
44	953	−13	24	957	−29
68	758	−31	48	685	−49
92	545	−50			

This example shows that boric acid or hindered phenolic antioxidant, by itself, somewhat improved viscosity stability of the hot melt but not to the degree obtained when both were present as in (A).

Example 2: To demonstrate selection of other alpha-olefin copolymers, tackifying resins, boric acid esters, and hindered phenols, compositions were prepared following the procedures of Example 1 (A). Resulting hot melt compositions are given in Table 4. These compositions were tested for viscosity stability as described in Example 1 (A) with the results recorded in Table 5.

TABLE 4

Composition	1	2	3	4	5	6	7	8	9	10	11	12	13
Branched Tetrapolymer(1)	40	40	40	40	40	40	40	40	—	—	—	—	40
Propylene/1-Hexene Copolymer(2)	—	—	—	—	—	—	—	—	40	40	—	—	—
Ethylene/Propylene Terpolymer(3)	—	—	—	—	—	—	—	—	—	—	40	40	—
Polyterpene Resin(1)	100	100	—	100	100	100	100	100	100	100	100	100	—
Petroleum hydrocarbon resin(4)	—	—	100	—	—	—	—	—	—	—	—	—	100
Wax	100	100	100	100	100	100	100	100	100	100	100	100	100
Boric Acid Powder	—	2.4	2.4	2.4	1.2	0.6	2.4	0.6	2.4	—	2.4	—	2.4
Tributyl Borate	2.4	—	—	—	—	—	—	—	—	—	—	—	—
2,6-di-t-butyl-4-methyl phenol	2.4	—	2.4	2.4	1.2	0.6	0.6	2.4	2.4	—	2.4	—	2.4
Hydroquinone	—	2.4	—	—	—	—	—	—	—	—	—	—	—
Maleic anhydride grafted polyethylene	—	2.0	2.0	2.0	2.0	2.0	2.0	2.0	—	—	2.0	—	2.0

1 As identified in Example 1.
2 Block copolymer prepared by slurry polymerization of propylene and 1-hexene in presence of γ-$TiCl_3$ and an organoaluminum compound; 27% hexene; 73% propylene; MI = 70 (ASTM-D-1238).
3 72% ethylene, 25.2% propylene, 2.8% 1,4-hexadiene; ML-4 of 60 at 250°F.
4 Composition 3: Ring and Ball softening point 119°C; Piccotac A.
Composition 13: Ring and Ball softening point 97°C; Piccotac B.

The tables show that in the presence of tributyl borate or boric acid, together with a hindered phenolic antioxidant, the melt viscosities of hot melts containing propylene/1-hexene copolymer and ethylene/propylene/1,4-hexadiene copolymer as well as hot melts containing the branched tetrapolymer are improved.

Compositions 2, 4 and 13 are particularly suited for use in case sealing due to their high degree of viscosity stability.

TABLE 5

Composition	Hours at 190°C. 0	24	48	72	96	Total Change[1] (%) in 96 hrs.
1	1227	1215	1232	1202	1142	−7
2	1510	1505	1507	1525	1450	−4
3	1945	1970	1962	2088	2192	+12.7
4	1545	1512	1480	1482	1482	−3.4
5	1480	1397	1360	1320	1262	−14.7
6	1477	1342	1230	1120	847	−42.6
7	1350	1268	1260	1230	1140	−15.6
8	1355	1262	1168	1030	707	−47.8
9	101	96.5	92.0	86.0	—	−15.0[2]
10	93.5	41.0	26.0	20.0	—	−78.6[2]
11	38400	36350	32200	30750	30000	−21.9
12	39800	—	—	—	—	−89[3]
13	1760	1722	1730	1762	1792	+1.8

[1]Sample 11 at 1 rpm; Sample 12, No. 27 spindle at 5 rpm.
[2]Total Change (%) in 72 hours.
[3]After 44 hours read 4,350; run discontinued.

Thermoplastic Polymer and Chelate of Aminoacetic Acid

N.S. Marans; U.S. Patent 3,862,099; January 21, 1975; assigned to W.R. Grace & Co. describes a thermoplastic adhesive comprising an intimate mixture of 100 parts of a thermoplastic polymer and 0.01 to 10 parts of a chelate of an aminoacetic acid or an aminopropionic acid and a polyvalent metallic ion, the chelate being free of acid hydrogen ions.

The thermoplastic adhesive is especially useful for laminating metal foils to plastic films (to prepare laminates useful for packaging foods and other items), for sealing side seams of metallic cans, for sealing packages, and the like. The composition is formed by mixing the metallic chelate and the thermoplastic polymer in a Brabender mixing apparatus for a period of about 5 minutes. A mixing temperature of 200°C is preferred for polypropylene, high density polyethylene, and polyvinyl chloride. A temperature of about 150°C is sufficient for low density polyethylene and for ethylene-vinyl acetate copolymers.

In no instance did the addition of a chelate to any of the thermoplastic polymers tested give polymer decomposition or crosslinking. On the contrary, the torque, while operating the Brabender apparatus at a constant speed remained constant, thereby indicating an absence of decomposition or crosslinking which would have changed the torque. Where mixing polyvinyl chloride about 3%, based on the weight of the polyvinyl chloride, of a commercially available polyvinyl chloride stabilizer (Advastab T-17M) was included in the mix because decomposition occurred in a trial run, using polyvinyl chloride without chelate and without stabilizer.

The composition can be used as a hot melt adhesive. It can be used to bond wood to wood, wood to metal, wood to concrete, wood to plastic, metal to metal, metal to plastic, plastic to concrete, ceramic material to ceramic material, wood to ceramic material, wood to glass, glass to ceramic material, plastic to glass and plastic to paper.

Example 1: One hundred parts of high density polyethylene and one part of an ethylenediaminetetraacetic acid chelate of titanium having the formula (EDTA)Ti was blended in a Brabender mixing apparatus for 5 minutes at 200°C, a temperature above the polymer's melting point, to form an improved thermoplastic adhesive of this process. The improved thermoplastic adhesive was removed from the mixing apparatus, recovered and designated Adhesive 1.

A 1.08 gram portion of Adhesive 1 was pressed using a Carver press at 200°C, a temperature above the polymer's melting point, into a sheet having an area of 36 square inches. The sheet was cooled below its melting point and a portion of it was used to heat seal two overlapping tin-free ether-washed steel coupons to each other. Each coupon was 13 mils thick and the lapped (sealed) area was 1 square inch.

The lap shear (a measure of bonding strength) of the sealed steel coupons was measured on an Instron testing apparatus and found to be 722 psi. This contrasts with a value of 91 psi obtained with high density polyethylene from the same lot but without the chelating agent.

Example 2: A number of runs were made using the general procedure of Example 1. The procedure was modified as indicated in the following table which reports the results of the runs. Among such modifications are:

Using for each run the chelate and polymer listed for such run. (Control runs were made as indicated in the table without the addition of a chelate and some runs were made using the chelate or polymer of Example 1.)

Using in some runs (as indicated in the table) ether-washed aluminum coupons rather than tin-free ether-washed steel coupons.

Using a mixing (blending) temperature and a pressing temperature of 150°F where blending or pressing low density polyethylene.

Improved Adhesives

Run No.	Chelate	Polymer	Lap Shear (psi)
1	None	High Density PE	91
2	NaCaNTA[1]	High Density PE	172
3	(NH_4)ZnNTA[1]	High Density PE	280
4	NaPbNTA[1]	High Density PE	184
5	SnEDTA[2]	High Density PE	222
6	NaFeEDTA[2]	High Density PE	195
7*	None	PVC	97
8*	TiEDTA[2]	PVC	132
9*	SnEDTA[2]	PVC	167
10*	FeNTA[1]	PVC	287
11*	NaCuNTA[1]	PVC	339
12*	Na_2CuEDTA[2]	PVC	300
13*	NaCaNTA[1]	PVC	296
14*	(NH_4)ZnNTA[1]	PVC	158
15*	NaMnNTA[1]	PVC	372
16*	NaPbNTA[1]	PVC	249
17*	PbEDTA[2]	PVC	164

(continued)

Run No.	Chelate	Polymer	Lap Shear (psi)
18*	$Na_2MnEDTA$[2]	PVC	304
19*	NaFeEDTA[2]	PVC	340
20*	Zn_2EDTA	PVC	160
21*	TiEDTP[3]	PVC	310
22*	Ba_2EDTP[3]	PVC	505
23*	FeNTP[4]	PVC	484
24*	$Co_3(NTP)_2$	PVC	388
25*	$(R_2CHNHCH_2CH_2NHCHR_2)Ba$[5] COO COO	PVC	600
26*	$(R_2CHNHCH_2CH_2NHCHR_2)Ti$[6] COO COO	PVC	540
27	$(R_2CHNHCH_2CH_2NHCHR_2)_3Fe_4$[7] COO COO	High density PE	710
28**	None	Low density PE	587
29**	SnEDTA[2]	Low density PE	908
30**	NaPbNTA[1]	Low density PE	800
31	None	PVAc	263
32	SnEDTA[2]	PVAc	326
33	NaFeEDTA[2]	PVAc	345
34	Cd_2EDTA[2]	PVAc	402
35*	None	PVC	102
36*	HFeEDTA[2,8]	PVC	0
37*	$Na_2CdEDTA$[2]	PVC	420
38*	$H_2CdEDTA$[2,8]	PVC	1
39*	HBaNTA[1,8]	PVC	5
40*	NaBaNTA[1]	PVC	460
41*	NaCaNTA[1]	PVC	281
42*	$[OOC(CH_2)_y]_2NCH_2CHCH_2N[(CH_2)_yCOO]_2Ba_2$[9] OH	PVC	400
43*	$[OOC(CH_2)_y]_2NCH_2CHCH_2N[(CH_2)_yCOO]_2Ti$[9] OH	PVC	385
44*	$[OOC(CH_2)_y]_2NCH_2CHCH_2N[(CH_2)_yCOO]_2Sn$[9] OH	PVC	375

Note: In runs 1 through 34, and 42 through 44, tin-free steel coupons were used. In runs 35 through 41, aluminum coupons were used.

*3% of a stabilizer (Advastab T-17M) based on the weight of the polyvinyl chloride was included in the mix.

**Mixing (blending) temperature and pressing temperature were 150°C.

[1]NTA means the nitrilotriacetate anion.

[2]EDTA means the ethylenediaminetetraacetate anion.

[3]EDTP means the ethylenediaminetetrapropionate anion.

[4]NTP means the nitrilotripropionate anion.

[5]R_2 is 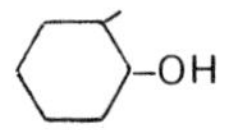

[6]R_2 is 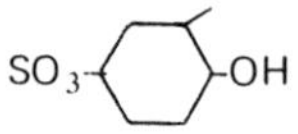

(continued)

[7]R_2 is OOC-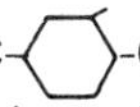-OH

[8]This run shows the adverse effect of free hydrogen ions.

[9]Y is 1.

Acrylic Acid Grafted Polyolefins

According to a process described by *K.W. Bartz, J.J. Higgins, A.J. Berejka and A.J. Di Cresce; U.S. Patent 3,868,433; February 25, 1975; assigned to Exxon Research and Engineering Company* thermoplastic adhesive compositions such as hot melts containing conventional components are improved considerably by the inclusion of acid or acid derivative grafted polyolefins particularly acrylic acid grafted polypropylene, polyethylene and/or ethylene-propylene polymer substances.

The grafting reaction is initiated by a free radical initiator which is preferably an organic peroxygen compound. Especially preferred peroxides are tert-butyl benzoate, dicumyl peroxide, 2,5-dimethyl-2,5-di-tert-butylperoxy-3-hexyne (Lupersol 130), α,α'-bis(tert-butylperoxy)diisopropylbenzene (VulCup R), or any free radical initiator having a 10 hour half-life temperature over 80°C.

Referring to Figure 2.1, an extruder **1**, having a feed zone **2**, a reaction zone or chamber **3**, and a final metering zone **4** is utilized to carry out a preferred form of the grafting process. In effect, polypropylene of a predominantly isotactic crystalline nature is introduced into a hopper **5** in the feed zone of the extruder. The extruder screw **6** in the feed zone can be of various conventional designs such as a feed portion **7**, a transition portion **8** and a first stage metering portion **9**. In the feed zone, the polypropylene is heated by heaters **10** to a barrel temperature in the range of 400° to 650°F, preferably 400° to 550°F.

It is one of the advantages of this process that fairly low temperatures can be used to accomplish outstanding modification. In processes utilizing O_2 as an initiator, much higher barrel temperatures, for example, about 600° to 800°F are required and control is awkward. In processes utilizing heat alone, even higher temperatures and reaction times are necessary.

The extruder screw has a root (sometimes called core) starting at the initial boundary of the reaction zone with a reduced cross-sectional area **11**. This provides additional volume for the reaction zone. When polymer under pressure reaches the reaction zone, the increased available volume results in a pressure drop, that is, decompression, so that particular mass of polymer is not subjected to the ordinarily high pressures in the extruder.

An injection line **12** connects the reaction zone to a source of initiator, preferably a peroxide. In some instances the peroxide will be combined with an active monomer. For the purposes of this specific example, the monomer is acrylic acid and the iniator is VulCup R.

Injection of initiator or initiator and monomer at this point, where low pressures in the reaction zone prevail, provides thorough dispersion of the initiator in polypropylene over an extremely short period of time and appreciable scission or degradation of the polypropylene results. Appropriate controls of the polypropylene feed rate and screw speeds are maintained.

FIGURE 2.1: EXTRUDER DESIGN FOR PREPARATION OF ACRYLIC ACID GRAFTED POLYOLEFINS

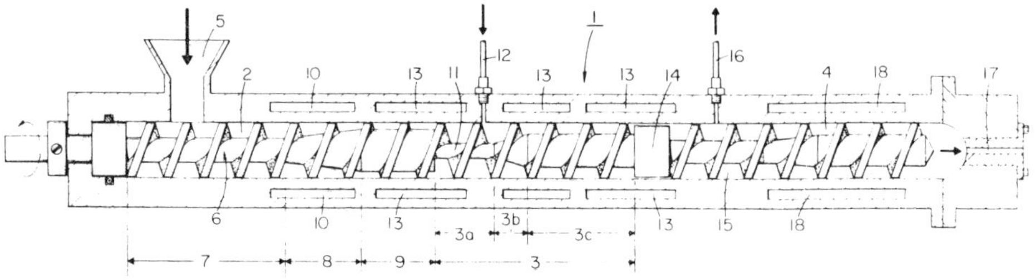

Source: K.W. Bartz, J.J. Higgins, A.J. Berejka and A.J. Di Cresce; U.S. Patent 3,868,433; February 25, 1975

The process can be conveniently operated to give high throughputs with good quality. In this case, the initiator and acrylic acid are added as a liquid blend to the reaction zone **3**. When only degradation is desired, initiator alone or initiator dissolved in a solvent is introduced into the reaction zone. It has been found that appreciable degradation of the polypropylene occurs when the back pressure against the liquid mixture of initiator and acrylic acid in injection line **12** is less than about 100 psig, preferably about 0 psig.

The pressure in the injection line, therefore, provides one indicia that the polypropylene feed rate and screw speed are being appropriately controlled for the particular products desired. The resulting graft copolymers have been appreciably degraded and changed in molecular weight distribution as compared to the base polymer. This is demonstrated by the fact that the graft copolymers have a lower die swell than the polypropylene base stock used in making the copolymer. Lower molecular weights are also indicated by changes in melt flow rates.

The portion of the extruder heated by heaters **13** will have a temperature of from about 160° to 450°F, preferably 250° to 350°F. The important thing is that the polymer be substantially in a melt phase during the reaction. The extruder screw **6** in the latter portion of the reaction zone can have any desired root cross-sectional area desirable to provide for pumping and ancillary mixing if desired and to allow residual reactants to complete their reaction.

It is to be noted that some homopolymerization of the acrylic acid (or any other monomer) to form polyacrylic acid also occurs. But this usually does not exceed 30% of the total acrylic polymer formed, particularly at the low monomer concentrations. Preferably, the decompression portion **3a** of the screw is immediately followed with transition zone **3b** of gradually increasing screw root cross-sectional area followed by a metering zone **3c** of constant cross-sectional screw root area. Thereafter, the extruder screw has a melt seal (also called cap or blister) **14** which prevents the free escape of initiator and acrylic acid from the reaction zone.

Screw **6** also has a second decompression portion **15** following blister **14**. Vent line **16** optionally provided with vacuum, is positioned above **15** to remove gases. When operating without vent **16** the blister may be omitted. The graft copolymer and homopolymer blend is then passed through metering zone **4** where it is extruded from a die **17** at the end of extruder **1**. The extruder barrel temperature is heated by heaters **18** in metering zone **4** to 350° to 450°F. The grafted polyolefins such as polypropylene made by the preferred technique described show exceptionally high adhesive strength at moderately elevated temperatures, e.g., about 250 psi Al to Al lap shear adhesions at 250°F. The hot bond strength is considerably higher than can be obtained in conventional hot melts.

Simply blending the modified grafted polymer in with the hot melt components to give a final hot melt composition will improve the adhesive properties considerably. For instance, a 60/40 blend of 6% acrylic acid modified polypropylene and 40% of Vistalon 3708 which is a commercially available ethylene/propylene elastomer had a 600 psi lap shear aluminum to aluminum bond strength at 72°F. The blend also had about 13 pounds per inch at 180° peel adhesion to aluminum foil. The following example illustrates the process.

Example: Several different materials were evaluated in lap shear adhesion tests and tension adhesion tests. These materials were an acrylic acid modified polypropylene containing 6 weight percent acrylic acid (≈50 MFR) prepared according to the method described above, a 60/40 blend of that polypropylene with Vistalon 3708, a commercially available ethylene/propylene diene monomer elastomer and an Elvax 150 ethylene-vinyl acetate copolymer. The results are summarized below in Table 1.

TABLE 1: LAP SHEAR AND TENSION ADHESION RESULTS WITH ACID MODIFIED POLYPROPYLENE

	Acrylic Acid Modified Poly-propylene	60/40 Blend Acid pp-VISTALON 3708	Ethylene-Vinyl Acetate Elvax 150
Lap Shear Adhesion			
(Al to Al pulled at 0.2 in./min.) at 72° F, psi:	1250 boundry layer failure	585 cohesive failure	500 mixed cohesive-adhesive failure
at 250° F, psi:	250+	105 cohesive failure	0 melted and flowed
Tension Adhesion			
(Al to Al pulled at 0.2 in./min.) at 72° F, psi:	1040 boundry layer failure	480 cohesive failure	550 adhesive failure

In the above table, Elvax 150 is an ethylene-vinyl acetate copolymer containing 33% vinyl acetate. This particular grade is recommended for high specific adhesion to aluminum and other nonporous surfaces. Two hundred and fifty pounds was the maximum force which could be detected by the load cell on the tensile tester encased in a convenient to use heated chamber. In some earlier work involving the use of carboxy terminated polyisobutylene to modify epoxies, lap shear adhesion results for a control of a commercial epoxy adhesive were 790 psi at 72°F and 100 psi at 250°F. Test specimens were formed by applying molten polymer to preheated pieces of aluminum. Only hand pressure was used to achieve adequate surface contact.

Coal Tar Pitch and Ethylene-Acrylic Acid Copolymer

A process described by *W.E. Smith, Jr.; U.S. Patent 3,882,066; May 6, 1975; assigned to United States Steel Corporation* provides compositions containing by weight between about 0.1 and 75% of a coal tar pitch and between about 99.9 and 25% of a copolymer of ethylene and an ethylenically unsaturated carboxylic acid containing from 3 to 8 carbon atoms. To this composition may be added up to about 50% by weight of the usual finely divided commercial filler materials.

The advantages of the compositions over other thermoplastic adhesives include: high tensile shear and peel strengths at temperatures between about -67° and 145°F; improvement in moisture resistance and to bacterial attack; and decreased sensitivity to dirty surfaces.

A pitch suitable for the purposes of this process is preferably that derived from coal tar. Useful pitches range in ring and ball softening points between about 70° and 200°C, a preferred range being between about 100° and 150°C.

The copolymers of ethylene with unsaturated carboxylic acids, suitable for purposes of this process are known. A method for preparing copolymers is described in U.S. Patent 3,132,120. The ethylenically unsaturated carboxylic acid component of the resin may comprise a monocarboxylic or dicarboxylic acid having from 3 to 8 carbon atoms such as acrylic acid, methacrylic acid, ethacrylic acid, fumaric acid, maleic acid, itaconic acid, crotonic acid and the like.

Useful copolymers may vary in melt index from about 0.1 to 400, the preferred range being between about 5 and 40. The acid content of useful copolymers may range between about 0.1 and 75% by weight of the copolymer. The acid range of commercially available copolymers with acrylic acid is between about 0.1 and 40% by weight of the copolymer. The preferred acid range of the latter is between about 2 and 20%, since above about 20% the products increase in crystallinity.

In the above classification, useful commercially available ethylene-ethylenically unsaturated carboxylic acid copolymers include ZETAFIN QX 3623.7; QX 3623.22 and QX 3623.24. These resins have about the same acrylic acid content, namely, about 6.8%. However, the melt indexes vary, the respective melt indexes being about 5, 20 and 40.

Other useful copolymers are Union Carbide EA 9000, EA 9300 and EA 9320 which vary in both acrylic acid content and melt index. The respective acid contents are about 14, 20 and 16%. The respective melt indexes are about 5, 50 and 5. Another useful copolymer is EMAZZ. This resin is an ethylene-maleic acid copolymer, having an acid content of about 70%. In the following examples the components are given in percentages by weight. The designated properties were determined by standard ASTM methods, more particularly, D1002-53T (tensile shear strength) and D903-49 (peel strength).

Examples 1 through 6: ZETAFIN QX 3623.7 is a commercially available proprietary copolymer of ethylene and acrylic acid. It has an acid content of about 6.8% by weight and a melt index of about 5. In Example 1, ZETAFIN QX 3623.7 alone was placed in the nip of a rubber mill, comprising two horizontal rolls turning toward each other, with a narrow slot therebetween. The rolls were preheated to a temperature of about 290°F and the resin was slowly added to the rotating rolls. After the resin pellets were melted the mill was operated for an additional 20 minutes. After removal, the product was pressed into sheets for testing.

In Examples 2 through 6, after pellets of the same resin were melted, coal tar pitch was added and the blend was mixed for 20 minutes and thereafter pressed into sheets for testing. In Examples 2 through 4, the proportions were varied for a coal tar pitch having a ring and ball softening point of about 150°C. In Examples 5 and 6, the proportions were varied for a coal tar pitch having a ring and ball softening point of about 110°C.

Comparing Examples 1 and 2, the tensile shear and peel strengths were much better for the resin pitch composition than for the resin alone. Comparing the results for Examples 2 through 6, a lower result at one temperature is usually accompanied by a higher result at another temperature. Comparing Examples 4 and 6, comprising 50:50 compositions with different pitches, the tensile shear strengths are comparable. The peel strengths with the 110°C melting point pitch are somewhat lower than with 150°C pitch.

The respective application temperatures are 250° and 350°F. Hence, with 110°C pitch, maximum joint strength is developed at a lower temperature. The results are tabulated in Table 1. Some of the test results at -67°F are shown with an asterisk in both Tables 1 and 2. Here, a true value could not be obtained, since the yield strength of the steel substrates was exceeded before the adhesive joints broke.

TABLE 1

Components Weight Per cent	Ex. 1	Ex. 2	Ex. 3	Ex. 4	Ex. 5	Ex. 6
Pitch 150°C	—	25	37	50	—	—
Pitch 110°C	—	—	—	—	25	50
ZETAFIN QX 3623.7	100	75	63	50	75	50
Properties						
Tensile Shear Strength						
psi at –67°F	2900*	3000*	2850*	2900*	3100*	2950
73°F	1675	1900	1950	1300	1900	1200
145°F	725	800	1050	450	725	500
Peel Strength (180°)						
ppi at –67°F	20	50	45	55	10	30
73°F	35	45	40	30	30	25
145°F	15	45	25	35	30	20

Examples 7 through 12: Example 2 was repeated as Examples 7 through 11, to illustrate the effect on the composition of a change in the ethylene-acrylic acid copolymer. In Examples 11 and 12 the compositions were the same. In Example 12, however, the blend was extruded. The components were mixed for 20 minutes in a twin-shell blender and then fed to an extruder preheated to between 240° and 300°F. The mixture was extruded as a 0.010 inch thick film and test samples prepared therefrom. The results are tabulated in Table 2.

More particularly, for ZETAFIN QX 3623.7 in Examples 2 through 6, there was substituted ZETAFIN QX 3623.22 in Examples 7 and 8 and ZETAFIN QX 3623.24 in Examples 9 through 12. ZETAFIN QX 3623.22 is an ethylene-acrylic acid copolymer having an acid content of about 6.8% and a melt index of about 20. ZETAFIN QX 3623.24 is an ethylene-acrylic acid copolymer having an acid content of about 6.8% and a melt index of about 40.

Tensile shear and peel strengths of the respective examples varied somewhat from each other but not to an appreciable extent. The results in Examples 7 through 12 were somewhat poorer than in Examples 2 through 6, more particularly, in the peel strength at -67°F. The type of blending, milling or extrusion produces some variation in the results.

TABLE 2

Components Weight Per cent	Ex. 7	Ex. 8	Ex. 9	Ex. 10	Ex. 11	Ex. 12
Pitch 150°C	25	—	25	—	50	50
Pitch 110°C	—	50	—	50	—	—
ZETAFIN QX 3623.22	75	50	—	—	—	—
ZETAFIN QX 3623.24	—	—	75	50	50	50
Properties						
Tensile Shear Strength						
psi at −67°F	3000*	3000*	2700	2900	2300	2250
78°F	1650	1050	1375	1025	1100	1700
145°F	575	400	375	275	250	450
Peel Strength (180°)						
ppi at −67°F	5	5	5	1	20	5
73°F	30	20	35	35	40	35
145°F	25	20	20	15	25	15

Water-Moistenable Vinyl Pyrrolidone-Vinyl Acetate Product

A water-moistenable hot melt applicable adhesive composition is a bonding agent which may be applied to one surface utilizing a hot melt technique and another surface using a water-moistening technique, so that the two surfaces are bonded together by means of the composition.

By way of example, such a composition may be applied by the hot melt technique as gumming for labels, sealing tapes and envelope flaps at a point of manufacture where the intricate equipment required for hot melt application is available, the labels, sealing tape and envelope flaps ultimately being applied by the consumer to other surfaces using the much simpler water-moistening technique, for example, simply by licking or otherwise moistening the adhesive coating prior to application to the other surface.

The hot melt application technique is utilized where the required equipment is available because it is a rapid and economical technique; the water-moistening technique is utilized precisely because it doesn't require such elaborate equipment.

J. Sirota, R.S. Forsyth and D. Fergesen; U.S. Patent 3,888,811; June 10, 1975; assigned to National Starch and Chemical Corporation describe a water-moistenable hot melt applicable adhesive composition characterized by the absence of blocking at 75% relative humidity under a 22 psi load at 22°C for 24 hours. The composition consists essentially of a blend of at least one water-sensitive vinyl pyrrolidone/vinyl acetate copolymer, at least one water-soluble solid polyethylene glycol based wax having an average molecular weight of at least 4,000, and at least one water-insoluble wax.

The water-soluble wax is generally selected from the group consisting of polyethylene glycol, the reaction product of polyethylene glycol and a diepoxide, and mixtures thereof. Preferably the water-soluble wax is a mixture of polyethylene glycol having an average molecular weight of at least 4,000 and a reaction product of polyethylene glycol and a diepoxide, the reaction product having an average molecular weight of at least 20,000. In such a preferred water-soluble wax mixture, the weight ratio of the polyethylene glycol and the reaction product is generally from about 1:1 to 2:1, and preferably about 1.5:1.

The following represents a preferred formulation for the adhesive composition and provides good heat stability and commercially acceptable blocking resistance in combination with a suitable compromise combination of adhesion, viscosity, machinability, moistening speed, moistening tack and tear speed.

PREFERRED FORMULATION

Polyvinyl Pyrrolidione/Vinyl Acetate Copolymer	1000 parts by wt.
CASTORWAX	40 parts by wt.
CARBOWAX 4000	25 parts by wt.
CARBOWAX 20M	16 parts by wt.
Antioxidant	14 parts by wt.
Saccharin (for flavor)	1.2 parts by wt.

A conventional antioxidant, such as 2,6-di-tert-butyl-para-cresol may be used in the formulation.

Example 1: A variety of polyvinyl pyrrolidone/vinyl acetate copolymers having various monomer ratios were prepared and mixed with varying amounts of water-soluble and water-insoluble waxes and additive ingredients as indicated below in Table 1. The properties of the samples thus formulated were tested (substantially according to conventional procedures described below) with the results indicated in Table 1.

In the humidity blocking test for block resistance, the coating composition is hot melt applied to a white wove envelope stock substrate (½" x 1"), and a glass plate is used to press the gummed surface of the substrate down upon a blank sheet of paper supported by a temperature and humidity conditioned urethane foam.

TABLE 1

SAMPLES	1	2	3	4	5	6[B]	7[C]	8[D]	9[E]
PVP/VA monomer ratio									
10/90									42
25/75								30	
30/70	1000	1000	1000	1000	1000				
40/60						90			
45/55							77	60	42
CW 600						10			
CW 1000								10	
CW 4000	25	25	25	25	41				
DOW E 6000				16					
DOW E 9000			16						
DOW E 20,000		16							
CW 20M	16								
Triacetin							23		
Glycerol									16
CASTORWAX	40	40	40	40	40				
Antioxidants	14	14	14	14	14		1	1	1
Saccharin	1.2	1.2	1.2	1.2	1.2				
Humidity Blocking									
75% RH	OK	OK	OK	OK	OK	100% tear	100% tear	100% tear	100% tear
80% RH	OK	OK	OK	OK	OK				
83% RH	OK	OK	OK	OK	sl. Pick				
Viscosity at 175°C,cps	4650	4440	4000	3440	2750				
Remoistening Speed, sec.	15	13	14	13	13				
Remoistening % Tear	100	100	100	100	100				
Heat Stability at 175°C									
Color	Light	Light	Sl. Darker	Light	Light				
1st Skin	No	No	No	No	No				
day Viscosity	4700	4500	4100	3500	2800				
2nd Color	Light	Light	Sl. Darker	Light	Light				
day Skin	V. Slight	V. Slight	No.	No.	No.				
Viscosity	4800	4600	4250	3600	2950				
Color	Light	Light	S.Darker	Light	Light				
3rd Skin	Thin	Thin	V. Slight	V. Slight	No				
3rd Viscosity	5,000	4750	4300	3750	3000				

LEGEND
[A]In Samples 1-5:
[B]Formulation 26 of U.S. 3,462,342
[C]Example 1 of U.S. 3,345,320
[D]Example 5 of U.S. 3,345,320
[E]Example 6 of U.S. 3,345,320

The test block is evaluated after 24 hours at the specified level of relative humidity (75, 80 or 83%) and 22°C temperature, and the amounts of tab, pick and percent tear are recorded. Nonblock or OK indicates completely free release of the gummed surface from the blank sheet. Tab indicates a superficial attraction usually accompanied by a zipping noise during release of the surfaces, and is not as serious as pick.

Pick indicates small, isolated, pin point areas of tear, usually totaling less than 5% of the total surface area. Percentage tear is used to indicate tears occurring over a larger percentage of the total surface area, usually over 5%. Viscosity is measured in centipoises at 175°C (the average hot application temperature) using a Brookfield Thermosel.

The heat stability test measures melt stability with time under hot application conditions. An eight ounce wide mouthed capped jar filled with 100 grams of hot melt is kept at 175°C in a noncirculatory oven for three days. Observations are made every 24 hours and any obvious change from initial condition of the melt is recorded, for example, color changes, skinning (a skin formation on the surface of the melt) or viscosity changes.

Remoistening speed is measured using coated substrates of the type described in connection with the humidity blocking test above the coatings being moistened in a single stroke with a sponge wet with water and pressed down on a large sheet of similar white wove envelope stock. A small one-quarter inch edge (bent back as a tab) is lifted slowly and the time in seconds recorded when tear first shows.

Remoistening percent tear is measured by a technique similar to that used in measuring remoistening speed, except that the moistened coating applied to the envelope stock is allowed to dry for 24 hours before lifting of the tab and a measurement is made of the percentage of the total coated surface torn.

The results indicate clearly that a commercially acceptable level of humidity blocking resistance is not achieved by PVP/VA copolymers in combination with the water-soluble waxes of low molecular weight or the triacetin or glycerol additives suggested by the prior art. On the other hand, the tripartite adhesive compositions consisting essentially of at least one PVP/VA copolymer, at least one water-soluble solid wax of high average molecular weight (at least 4,000), and at least one water-insoluble solid wax were block-free not only at the 75% RH level, but also at the 80% RH level and in most cases the 83% RH level.

Example 2: Generally speaking in this example the total wax content and the ratio of water-soluble to water-insoluble solid waxes in the preferred formulation are both varied over a wide range, and each of the resulting coating compositions are evaluated and found satisfactory. More specifically, based on 1,000 parts by weight of the copolymer, the wax content of the preferred formulation is varied to provide compositions with wax contents as indicated by the X-marked blocks in Table 2 below, each of the resultant coating compositions being tested for humidity block resistance, remoistening speed and remoistening percent tear and found satisfactory.

TABLE 2

WATER-INSOLUBLE WAX, parts	WATER-SOLUBLE WAX, parts								
	10	20	30	40	60	80	100	120	140
20	X	X			X		X		X
40		X		X		X		X	
60			X		X		X		
80				X		X			

J.W. Huebschmann and J.D. Peperone; U.S. Patent 3,852,231; December 3, 1974; assigned to Moore Business Forms, Inc. describe a hot melt, water-remoistenable adhesive composition containing a water-sensitive vinyl pyrrolidone-vinyl acetate copolymer, a plasticizer material which may include two different components, one of which operates as an adhesion promoter, and a wax component selected from the group consisting of Hoechst Wax S, peat wax, lignite wax and Castorwax.

The composition may also include, in its preferred form, an antioxidant, an antiblock ingredient and/or a taste improving material. Useful composition formulations are illustrated in the following examples.

FORMULATION NO. 1

Component	Parts by weight
Vinyl pyrrolidone-vinyl acetate copolymer having a weight ratio of vinyl pyrrolidone moiety to vinyl acetate moiety of 1:1	50
Hoechst Wax S	20
Glycerol tribenzoate Plasticizer	25
Polyethylene glycol (MW = 4,000) Adhesion Promoter	3.8
Malcolm Nicol P-800 Anti-Blocking Agent	1
Naugard 445 substituted diphenyl-amine Antioxidant	0.2

FORMULATION NO. 2

Component	Parts by Weight
Vinyl pyrrolidone-vinyl acetate copolymer having a weight ratio of vinyl pyrrolidone moiety to vinyl acetate moiety of 2:1	50
Hoechst Wax S	20
Glycerol tribenzoate Plasticizer	25
Polyethylene glycol (MW = 4,000) Adhesion Promoter	3.8
Malcolm Nicol P-800 Anti-Blocking Agent	1
Naugard 445 substituted diphenyl-amine Antioxidant	0.2

FORMULATION NO. 3

Component	Parts by Weight
Vinyl pyrrolidone-vinyl acetate copolymer having a weight ratio of vinyl pyrrolidone moiety to vinyl acetate moiety of 1:2	50
Hoechst Wax S	20
Glycerol tribenzoate Plasticizer	25
Polyethylene glycol (MW = 4,000) Adhesion Promoter	3.8
Malcolm Nicol P-800 Anti-Blocking Agent	1
Naugard 445 substituted diphenyl-amine Antioxidant	0.2

FORMULATION NO. 4

Component	Parts by Weight
Vinyl pyrrolidone-vinyl acetate copolymer having a weight ratio of vinyl pyrrolidone moiety to vinyl acetate moiety of 1:1	44.8
Hoechst Wax S	25
Glycerol tribenzoate Plasticizer	25
Polyethylene glycol (MW = 4,000) Adhesion Promoter	5
Naugard 445 substituted diphenyl-amine Antioxidant	0.2

FORMULATION NO. 5

Component	Parts by Weight
Vinyl pyrrolidone-vinyl acetate copolymer having a weight ratio of vinyl pyrrolidone moiety to vinyl acetate moiety of 1:1	40
Hoechst Wax S	18
Glycerol tribenzoate Plasticizer	40
Polyethylene glycol (MW = 4,000) Adhesion Promoter	1
Malcolm Nicol P-800 Anti-Blocking Agent	0.8
Naugard 445 substituted diphenyl-amine Antioxidant	0.2

FORMULATION NO. 6

Component	Parts by Weight
Vinyl pyrrolidone-vinyl acetate copolymer having a weight ratio of vinyl pyrrolidone moiety to vinyl acetate moiety of 1:1	60
Hoechst Wax S	15
Dipropylene glycol dibenzoate Plasticizer	24
Malcolm Nicol P-800 Anti-Blocking Agent	0.8
Naugard 445 substituted diphenyl-amine Antioxidant	0.2

FORMULATION NO. 7

Component	Parts by Weight
Vinyl pyrrolidone-vinyl acetate copolymer having a weight ratio of vinyl pyrrolidone moiety to vinyl acetate moiety of 1:1	47
Hoechst Wax S	27
Dipropylene glycol dibenzoate Plasticizer	13
Polyethylene glycol (MW = 4,000) Adhesion Promoter	12
Malcolm Nicol P-800 Anti-Blocking Agent	0.8
Naugard 445 substituted diphenyl-amine Antioxidant	0.2

The foregoing formulations are all examples of compositions which have been prepared in accordance with the process and accordingly include a wax component and a plasticizer. For comparative purposes, the following two formulations have been prepared without the inclusion of the wax component.

FORMULATION NO. 8

Component	Parts by Weight
Vinyl pyrrolidone-vinyl acetate copolymer having a weight ratio of vinyl pyrrolidone moiety to vinyl acetate moiety of 1:1	55
Glycerol tribenzoate Plasticizer	33.8
Polyethylene glycol (MW = 4,000) Adhesion Promoter	10
Malcolm Nicol P-800 Anti-Blocking Agent	1
Naugard 445 substituted diphenyl-amine Antioxidant	0.2

FORMULATION NO. 9

Component	Parts by Weight
Vinyl pyrrolidone-vinyl acetate copolymer having a weight ratio of vinyl pyrrolidone moiety to vinyl acetate moiety of 1:1	50
Glycerol tribenzoate Plasticizer	39.8
Polyethylene glycol (MW = 4,000) Adhesion Promoter	7
Malcolm Nicol P-800 Anti-Blocking Agent	3
Naugard 445 substituted diphenyl-amine Antioxidant	0.2

In Table 1, the above formulations have been evaluated on the basis of performance for purposes of hot melt, water remoistenable adhesive compositions. In interpreting the table, the various properties are defined as follows.

Viscosity	Measured in centipoises on a Brookfield Thermosel Unit.
Setting Time	Time in seconds for a 0.001" thick film to become tack-free after being applied to a sheet of paper.
Heat Stability	Composition is conditioned at 300°F in an open container, after which viscosity is remeasured and color and skin formation are observed.
Blocking	Composition is conditioned at 85% relative humidity and 73°F for 24 hours after which the amount of fiber tear resulting from the application of a 1 psi load is observed.

(continued)

Tack Time	Time in seconds for the film to become tacky after being wet with water.
Time to Bond	Time necessary to develop fiber tearing bond after being wet with water.
Adhesion	Amount of fiber tear after being bonded for 24 hours.

TABLE 1: REMOISTENABLE ADHESIVE COMPOSITIONS

Raw Materials	Formulation Numbers								
	1	2	3	4	5	6	7	8	9
PVP/VA 1:1 ratio	50	—	—	44.8	40	60	47	55	50
PVP/VA 2:1 ratio	—	50	—	—	—	—	—	—	—
PVP/VA 1:2 ratio	—	—	50	—	—	—	—	—	—
Wax Component	20	20	20	25	18	15	27	—	—
Plasticizer Component	25	25	25	25	40	24	13	33.8	39.8
Adhesion Promoter	3.8	3.8	3.8	5	1	—	12	10	7
Anti-Blocking Agent	1	1	1	—	0.8	0.8	0.8	1	3
Antioxidant	0.2	0.2	0.2	0.2	0.2	0.2	0.2	0.2	0.2
Properties									
Viscosity-300°F cps.	2000	4500	1000	1100	1500	3500	1800	2500	2200
Setting time-sec.	<1	<1	<1	<1	<1	<1	<1	<60	<40
Heat stability - 100 hrs. at 300°F.	No Change	No Change	No Change	No Change	No Change	No Change	No Change	No Change	No Change
Blocking 85% RH at 73°F.	None	None	None	None	None	None	None	Yes	None
Remoistenability									
Tack time - sec.	5	10	3	6	5	8	7	2	10
Time to set - min.	1	2	1	1.5	1.8	1.4	1.2	1	2
Adhesion - %	100	100	100	100	100	100	100	100	25

It can be seen from Table 1 that the absence of the wax component results in a tremendous increase in the time required for the composition to set (Setting time). Also, in Formulation No. 8, the absence of the wax component apparently increased the blocking tendencies of the adhesive composition. A particularly preferred hot melt, water remoistenable adhesive composition is set forth in the following example.

FORMULATION NO. 10

Component	Parts by Weight
Vinyl pyrrolidone-vinyl acetate copolymer having a weight ratio of vinyl pyrrolidone moiety to vinyl acetate moiety of 1:1	43.0
Hoechst Wax S	21.5
Glycerol tribenzoate Plasticizer	21.5
Polyethylene glycol (MW = 4,000) Adhesion Promoter	8.5
Malcolm Nicol P-800 Anti-Blocking Agent	4.0
Urea (as a Taste Improver)	1.0
Naugard 445 substituted diphenylamine Antioxidant	0.25
Advestab CH-201 Complex Phosphite Antioxidant	0.25

ADDITIONAL FORMULATIONS

Block Polyester-Polyamides and Polystyrene

W.J. Jackson, Jr. and W.R. Darnell; U.S. Patent 3,907,733; September 23, 1975;

assigned to Eastman Kodak Company describe hot melt adhesive compositions capable of maintaining high adhesive strength upon aging as well as the ability to maintain high bond strengths under relatively high temperature conditions.

This is accomplished by blending from about 5 to 40% (preferably 5 to 25%) by weight of a vinyl aromatic polymer with about 95 to 60% (preferably 95 to 75%) by weight of a polyester-polyamide. The polyester-polyamide may be prepared by reacting a crystalline polyester prepolymer and an amorphous polyamide prepolymer. The polyester-polyamide contains from about 45 to 95% (preferably about 50 to 80%) of segments derived from a crystalline polyester and from about 5 to 55 weight percent (preferably 20 to 50) of amorphous polyamide.

The polyester reactant is considered to be a prepolymer in view of the fact that it copolymerizes with the other reactants to form a block copolymer. The polyester must be capable of contributing a crystalline structure to the resulting polyester-polyamide as is evidenced by a crystalline melting point as determined by differential thermal analysis (DTA) and/or differential scanning calorimetry (DSC) methods.

Moreover, the polyester should have a melting point higher than 145°C and preferably in the range of 200° to 300°C and an inherent viscosity of at least 0.1 and preferably in the range of from 0.2 to 0.5 when measured at 25°C using 0.50 gram of polymer per 100 ml of a solvent consisting of 60% by weight phenol and 40% by weight tetrachloroethane.

Examples: Parts and percentages are by weight and temperatures are in degrees Centigrade unless otherwise specified. All inherent viscosities are determined at 25°C using 0.50 gram of polymer per 100 ml of a solvent consisting of 60% by weight phenol and 40% by weight tetrachloroethane. In the examples, all polyester-polyamides are prepared by the following procedure.

The amorphous polyamide segment of the block polyester-polyamide is first prepared by heating and stirring the diamine(s), the dimer acid, and in some cases, a second aliphatic dicarboxylic acid at 190° to 220°C for about 1 hour at 100 to 150 psi. The reaction vessel is then vented, and the calculated amount of crystalline polyester segment (I.V. 0.1 to 0.5) is added. The reaction vessel is then heated slowly to 260° to 280°C over a period of an hour under nitrogen at 40 to 60 psi and maintained under these conditions for about 30 minutes.

The reaction vessel is then vented and vacuum is applied. The reaction is continued under full vacuum (0.2 to 5 mm of Hg) for 2 to 4 hours. The molten polyester-polyamide is then discharged under pressure into water and quenched. The inherent viscosity of the amber polymers obtained is about 0.4 to 0.8.

The polyester-polyamide-vinyl aromatic polymer blends are prepared by mixing 2 to 3 mm particles of the two polymers and melt extruding them at 200° to 240°C into a small rod. The rod is then chopped into ⅛" pellets which are used to prepare the adhesive bonds. The vinyl aromatic polymer used in the adhesive blends of Tables 1, 2 and 3 is polystyrene having an I.V. of 0.72. The T-peel strengths are determined on 6 mil, tin-free steel (TF-steel) in accordance with ASTM D 1876 but using precut (1" wide) specimens. The metal specimens are first degreased by washing three times in a detergent solution of Alconox and rinsing in water each time. The metal specimens are then rinsed in acetone and allowed to air dry. To prepare an adhesive bond, a specimen is heated sufficiently on a hot plate for the blend to melt (applied as 20 mesh to ⅛" particles or by rubbing a large piece back and forth). A hot plate temperature of 280°C is used. After the molten polymer is evenly spread on the specimen over a 2" length with a wooden spatula, the top piece of metal is pressed against the sample. The sample is then turned over on the hot plate with the second metal strip against the hot plate surface.

Heating is continued while the wooden spatula is rubbed back and forth over the adhesive area until a smooth, even glue line is obtained (about 20 seconds). Then the specimen is placed on the laboratory bench and rubbing continued about 10 seconds longer while the adhesive cools. This method of forming the adhesive bonds gives reproducible results similar to those obtained by clamping the specimens together and heating in a forced air oven 20° to 50°C above the polymer melting point. Glue lines are about 3 to 5 mils in thickness.

All bond aging tests are carried out in a forced air, convection heated oven at 100° ± 1°C for 24 hours. The specimens are allowed to cool to room temperature (25°C) before testing. The polyamide block of the block polyester-polyamides used in the examples of Table 1 is derived from dimer acid and hexamethylenediamine. This table illustrates the effect of polystyrene content on the bond aging characteristics of polyester-polyamide-polystyrene blends.

TABLE 1

Example	Poly(ester-amide) Polyester Segment Composition	Wt.%	Final I.V.	Polystyrene Content, Wt. %	T-Peel, Lb./In. Width, on Steel at 23°C. After Aging Unaged	1 Day at 100°C.
1 (Control)	Poly(ethylene terephthalate)	50	0.60	0	28	16
2	Poly(ethylene terephthalate)	50	0.60	5	27	19
3	Poly(ethylene terephthalate)	50	0.60	10	26	21
4	Poly(ethylene terephthalate)	50	0.60	20	27	28
5	Poly(ethylene terephthalate)	50	0.60	40	21	17
6 (Control)	Poly(ethylene terephthalate)	60	0.65	0	24	4
7	Poly(ethylene terephthalate)	60	0.65	5	30	7
8	Poly(ethylene terephthalate)	60	0.65	10	37	11
9	Poly(ethylene terephthalate)	60	0.65	20	30	16
10	Poly(ethylene terephthalate)	60	0.65	40	23	15
11 (Control)	Poly(ethylene terephthalate)	70	0.67	0	4	2
12	Poly(ethylene terephthalate)	70	0.67	5	24	6
13	Poly(ethylene terephthalate)	70	0.67	10	27	6
14	Poly(ethylene terephthalate)	70	0.67	20	29	9
15	Poly(ethylene terephthalate)	70	0.67	40	18	9

(continued)

TABLE 1: (continued)

Example	Poly(ester-amide) Polyester Segment Composition	Wt.%	Final I.V.	Polystyrene Content, Wt. %	T-Peel, Lb./In. Width, on Steel at 23°C. After Aging Unaged	1 Day at 100°C.
16 (Control)	Poly(ethylene terephthalate)	80	0.65	0	5	3
17	Poly(ethylene terephthalate)	80	0.65	5	19	5
18	Poly(ethylene terephthalate)	80	0.65	10	25	6
19	Poly(ethylene terephthalate)	80	0.65	20	30	11
20	Poly(ethylene terephthalate)	80	0.65	40	12	5
21 (Control)	Poly(ethylene terephthalate)	95	0.66	0	2	1
22	Poly(ethylene terephthalate)	95	0.66	5	13	10
23	Poly(ethylene terephthalate)	95	0.66	20	18	12
24 (Control)	Poly(tetramethylene terephthalate)	50	0.63	0	1	1
25	Poly(tetramethylene terephthalate)	50	0.63	5	12	12
26	Poly(tetramethylene terephthalate)	50	0.63	10	15	10
27	Poly(tetramethylene terephthalate)	50	0.63	20	13	7
28	Poly(tetramethylene terephthalate)	50	0.63	40	3	2
29 (Control)	Poly(tetramethylene 80/20 terephthalate/isophthalate)	60	0.68	0	5	3
30	Poly(tetramethylene 80/20 terephthalate/isophthalate)	60	0.68	5	12	8
31	Poly(tetramethylene 80/20 terephthalate/isophthalate)	60	0.68	10	16	10
32	Poly(tetramethylene 80/20 terephthalate/isophthalate)	60	0.68	20	15	11
33	Poly(tetramethylene 80/20 terephthalate/isophthalate)	60	0.68	40	10	7
34 (Control)	Copoly(80/20 tetramethylene/ethylene terephthalate)	60	0.64	0	5	2
35	Copoly(80/20 tetramethylene/ethylene terephthalate)	60	0.64	5	15	6
36	Copoly(80/20 tetramethylene/ethylene terephthalate)	60	0.64	10	18	8
37	Copoly(80/20 tetramethylene/ethylene terephthalate)	60	0.64	20	21	12
38	Copoly(80/20 tetramethylene/ethylene terephthalate)	60	0.64	40	16	7
39 (Control)	Poly(hexamethylene terephthalate)	60	0.72	0	2	1
40	Poly(hexamethylene terephthalate)	60	0.72	5	15	10
41	Poly(hexamethylene terephthalate)	60	0.72	10	18	12
42	Poly(hexamethylene terephthalate)	60	0.72	20	17	10
43	Poly(hexamethylene terephthalate)	60	0.72	40	5	4

The examples of Table 1 illustrate the improved T-peel strengths which are obtained in all cases after accelerated bond aging of the polyester-polyamide-polystyrene blends of the process when compared to the T-peel strengths of the respective polyester-polyamides alone. These examples also illustrate that, in most cases, the T-peel strengths even before bond aging of the block polyester-polyamides on the steel substrate are significantly improved by the addition of only a very small amount of polystyrene.

Examples 44 through 48 of Table 2 illustrate the improved bond aging characteristics which are obtained when polystyrene is blended with a block polyester-polyamide which is comprised of 60 weight percent of polyester blocks of polyethylene terephthalate and 40 weight percent of polyamide blocks derived from dimer acid and ethylenediamine, instead of hexamethylenediamine. The I.V. of the block polyester-polyamide is 0.64.

TABLE 2

Example	Polystyrene Content, %	T-Peel Lb./In. Width, on Steel at 23°C. Unaged	After Bond Aging 1 Day at 100°C.
44 (Control)	0	16	3
45	5	19	7
46	10	23	12
47	20	20	13
48	40	15	8

Examples 49 through 53 of Table 3 which follows illustrate the improved bond aging characteristics which are obtained when polystyrene is blended with a block polyester-polyamide which is composed of 60 weight percent of polyester blocks of polyethylene terephthalate and 40 weight percent of copolyamide blocks derived from hexamethylenediamine and a 75:25 weight ratio of dimer/sebacic acids. The I.V. of the block polyester-polyamide is 0.69.

TABLE 3

Example	Polystyrene Content, %	T-Peel, Lb./In. Width, on Steel at 23°C. Unaged	After Bond Aging 1 Day at 100°C.
49 (Control)	0	20	4
50	5	24	7
51	10	27	14
52	20	25	13
53	40	26	10

Examples 54 through 63 of Table 4 illustrate the effects on the T-peel strength at 23°C before and after bond aging for one day at 100°C on tin-free steel (TF-steel) of (a) the I.V. of the polystyrene, including copolymers, and (b) the composition of the polystyrene copolymer.

The block polyester-polyamide used in the blends of these examples contains 60 weight percent of polyethylene terephthalate segments and 40 weight percent of segments of the amorphous polyamide of dimer acid and hexamethylenediamine. The I.V. of the block polyester-polyamide is 0.65.

Formaldehyde is usually shipped commercially as a 37% aqueous solution, but concentrations as high as 50% are common. Even at the highest concentration, considerable diluent, water, is being shipped at high freight rates. When the formaldehyde is intended for use in urea-formaldehyde adhesives, it is common practice to add urea to these solutions, not only to stabilize the formaldehyde against polymerization at reduced temperatures, but also to make it possible to reduce the water content and provide a higher payload. Mixtures containing 50 to 51% formaldehyde with 20 to 21% by weight urea and 29 to 30% water are common.

When the mixture is received at the place of use, sufficient additional urea is added to provide the optimum mol ratio, and the mixture is then heated, (viz, the "resin cook") to form the partially polymerized urea-formaldehyde resin which is then suitable for use as an adhesive.

Paper and other base material, such as wood, are often coated with pigment-adhesive coating compositions to obtain products having a good brightness, color, gloss, appearance and printing properties. The coating compositions are generally prepared by mixing an aqueous pigment slurry containing clay, titanium dioxide, calcium carbonate and/or other pigments with a conventional water-base adhesive or mixture of these adhesives, containing small amounts of additives as antifoam agents, bactericides, dispersing agents, etc., in order to obtain a smooth homogeneous mixture.

The composition is then spread on paper or paperboard by any of a variety of methods. These include the use of an air doctor coater, roll coater, trailing blade coaters, and involve size press coatings, calender stack coatings, and water box applications.

The adhesives used in preparing the known adhesive coating compositions are generally water-base adhesives containing casein and other proteins, starches including hydroxyethylated starches, oxidized starches and other modified starches, polyvinyl alcohol or latexes, e.g., a Dow 636 latex (a styrene-butadiene copolymer latex), a Rohm & Haas B-15 latex (an acrylic resin latex), as well as combinations of these materials. Casein including modified caseins and other proteins, e.g., soy protein, and starches comprise the most commonly used adhesives in paper coating. This chapter provides a number of formulation modifications to optimize the performance of urea-formaldehyde, phenol-formaldehyde and starch-based adhesives for various wood and paper bonding processes.

PHENOL-FORMALDEHYDE RESINS FOR PLYWOOD

Polymerized Lignosulfonate

Phenolic resins generally are not used alone as plywood adhesives. Various fillers and modifiers have to be added to the resin to control the properties and different formulations have to be used for different conditions. For example, the formulation for southern pine veneers differs from the adhesive used for Douglas Fir. The adhesives generally are formulated just prior to use because of poor stability of the adhesive when intermixed with fillers or other additives. The fillers generally have a tendency to settle or float out of the resin or the mixed adhesives thicken or thin out rapidly upon standing.

The formulation of the adhesive is not only time consuming but is performed by personnel lacking sufficient chemical training to be able to make the necessary adjustments in the formulation to obtain optimum results under the variation in operating conditions normally encountered. Thus, it would be very desirable to have a phenolic plywood adhesive which could be formulated at the resin plant and made available to the mills in a form which would be used without requiring further processing at the mill. It is further desirable to have a phenolic adhesive which would be operative for different veneers and under the different conditions normally encountered without having to continually adjust the adhesive by addition or deletion of different additives or modifiers.

V.F. Felicetta and D.E. Wenzel; U.S. Patent 3,886,101; May 27, 1975; assigned to Georgia-Pacific Corporation describe a ready-to-use phenolic adhesive for the manufacture of plywood containing a polymerized lignosulfonate. The lignosulfonate is polymerized with an aldehyde or an epoxide to an extent that from 30 to 80% of the product is insoluble in an aqueous 5% sodium carbonate solution.

Example 1: A fermented calcium base spent sulfite liquor was alkaline treated by addition of about 9% sodium hydroxide based on liquor solids and heating for 20 hours in a boiling water bath. After the treatment, the alkaline treated liquor had a pH of 10. The alkaline treated material was diluted until it contained about 40% solids and then condensed with 10% formaldehyde, based upon the alkaline treated solids. The condensation was carried out by heating the mixture in a vessel for 2½ hours placed in a 130°C oven.

After condensation, the resulting gel was tray dried at 130°C and ground to pass through 100 mesh screen. 37% of the total product was insoluble in 5% sodium carbonate solution which if based upon the lignosulfonate content would represent an insolubility of about 50%. The swelling of the insolubles after washing with water was 50 grams wet gel per gram of insolubles.

In the determination of the amount of insolubles and the swelling of the insolubles, 0.5 gram of the product ground to pass through a 100 mesh screen was added to 45 milliliters of 5% sodium carbonate solution and the mixture heated in a boiling water bath for 30 minutes with the mixture being mixed periodically. The resulting mixture was then centrifuged and the solution decanted from the residue or gel. The residue was then washed 4 times by adding distilled water to the gel, in an amount about equal to the solution decanted, mixing and centrifuging the mixture to separate the insolubles from the wash water for each washing. After the last washing, centrifugation, and decantation the wet gel was weighed. After weighing, the gel was dried for 16 hours at 105°C.

The dried residue was weighed and then used for determination of the insolubility of the product in the sodium carbonate solution and the swelling of the insolubles. The swelling of the insoluble constituents as expressed is the ratio of the weight of the wet insolubles or gel to the dry weight of the insolubles and was obtained by dividing the dry weight of the insoluble constituents into the wet weight of the gel or insoluble constituents obtained after the last washing and decantation.

The above product was used in the preparation of a phenol-formaldehyde adhesive used for plywood preparation. The product was added to a mixture of phenol and formaldehyde during the preparation of phenol-formaldehyde resin to be used as an adhesive. The polymerized lignosulfonate was added in an amount of about 12% of the weight of the phenol and formaldehyde used in preparation of the adhesive. The mixture was heated under alkaline conditions until a viscosity of about 1,500 centipoises (25°C) was obtained.

The resin thus obtained was used in the preparation of 5-ply ½-inch plywood test panels from Douglas Fir veneer containing about 2% moisture. The adhesive containing about 40% solids was applied by spray in the amount of about 78 pounds per 1,000 square feet of double glue line. After spraying and lay-up, the panels were allowed to stand 10 minutes before prepressing for 5 minutes at 150 psi. The prepressed panels were than allowed to stand for up to about 24 hours before hot pressing for 6¼ minutes at 175 psi and 285°F. Wood failure tests of the panels were carried out by the APA vacuum pressure test for exterior plywood. The assembly time or the time from lay-up to hot pressing and the wood failures are shown in the table below.

Assembly Time, hours	Percent Wood Failure
½	89
2	72
6	93
24	87

Example 2: A concentrated fermented calcium base liquor was converted to a sodium base by the simultaneous addition of sulfuric acid and sodium hydroxide in the proportion of 1:2 on a molar basis to precipitate the calcium as calcium sulfate which was removed by centrifuging. The sodium base spent sulfite liquor was alkaline treated by addition of about 4% by weight of sodium hydroxide based on liquor solids and heating the mixture for 12 hours at 95°C. After the heat treatment, the mixture had a pH of about 10.

The mixture was diluted with water to a concentration of about 40% solids after which formaldehyde was added in an amount of 7% of the solids in the alkaline treated spent sulfite liquor. The formaldehyde-containing liquor was heated at about 95°C for about 17½ hours in which time the product gelled. The material was then dried in a ribbon blender at a temperature of from 135° to 160°C.

After drying, the material was ball milled so than 99% of the material could pass through a 100 mesh screen. The ground material had an insolubility of 43% in a 5% sodium carbonate solution. The swelling of the insolubles after washing with water was 48 expressed as the ratio of the weight of the wet insolubles to the dry weight of the insolubles.

The above polymerized lignosulfonate was used in preparation of adhesives used for plywood manufacture. The polymerized product was added to a mixture of phenol and formaldehyde in an amount to represent about 10% of the weight of the phenol and formaldehyde used in the resin preparation. The phenol and formaldehyde were reacted in the formation of the resin using the normal procedures for reacting the phenol with the formaldehyde with periodic additions of caustic. After preparation, the resin was tested as an adhesive for exterior grade plywood. Douglas Fir veneer containing about 2% moisture was used in preparation of 5-ply ½-inch plywood. The adhesive was applied by spray in an amount of about 78 pounds per 1,000 square feet of double glue line and after the assembly the panels were allowed to stand for 15 to 30 minutes before prepressing for 1 minute at 175 pounds per square inch. The prepressed panels were then allowed to stand for different periods of time prior to hot pressing for 6¼ minutes at 175 pounds per square inch and 285°F. Wood failure tests of the panels were evaluated using the American Plywood Association Vacuum Pressure Test procedure for exterior plywood. The assembly times, from lay-up to hot pressing, and the wood failures obtained are shown in the table below.

Assembly Time, hours	Percent Wood Failure
½	100
2	100
6	94
24	87

Kraft or Soda Black Liquors

According to a process described by *T.U.E. Enkvist; U.S. Patent 3,864,291; February 4, 1975* alkaline, water-soluble adhesives based on kraft or soda black liquors from the pulping industry are combined with adhesives made from phenols and formaldehyde, formalin, paraformaldehyde or other formaldehyde-containing or developing substances. The adhesive is used for gluing plywood and particle board.

A proportion which can be used is, for instance, 1 part by weight of phenol and 4 parts of organic matter in a kraft or soda black liquor. Adjusting of the pH value of the adhesive, measured potentiometrically after dilution 1:10 with water, to about 8 to 10 can be important. At the heating of the kraft black liquor component only small amounts of formaldehyde are necessary. This addition of formaldehyde or formalin, however, suffices for elimination of the bad odor of the liquor.

Example 1: Black Liquor Adhesive – To 250.4 grams of black liquor from kraft digestion of a softwood mixture containing two-thirds of Scotch Pine (*Pinus sylvestris L.*) and one-third of spruce (*Picea abies L.*), of dry content 51.2% and a content of organic substance, determined according to the analysis method 58:64 of the Central Laboratory of Helsinki, Finland (CCA 23:48), 38.2% and density 1.33, 39.4 milliliters of 35% formalin were added. The mixture was heated 3 hours at 70°C using a reflux tube. Thereafter it was diluted with 101.5 milliliters of water and swiftly evaporated in vacuo to a dry content of 39.4%. The resulting glue, 1,068 Asv, showed a density of 1.21, and a viscosity (Ford cup No. 2) of 19 cp at 20°C. After standing 4 days at room temperature, the viscosity was still 19 cp.

Phenol Adhesive – 30 grams of phenol, 28.5 grams of 4 molar sodium hydroxide solution and 48.8 milliliters of 35% formalin were mixed and heated for 3 hours at 90°C using a reflux tube. 32.1 milliliters of water were then added. The calculated dry content was thereafter 36.0%, and the amount of phenol used 20.3% of the weight of the adhesive (1,068 B fen). Density was 1.10, viscosity of a fresh sample was 27.0 cp, and after 4 days at room temperature, viscosity was still 27.0 cp.

177.9 grams of the black liquor adhesive 1,068 Asv and 54.6 grams of the phenol adhesive 1,068 B fen were mixed, the amount of phenol used constituting 19.1% of the sum of the phenol and the organic matter from the black liquor. 5.3 milliliters of 35% formalin were added to this mixture. The thus obtained mixed adhesive (1,068 AB form) had a dry content of 38.5%; d = 1.18; viscosity for fresh adhesive, 24.9 cp; after 24 hours at room temperature, 27.0 cp.

To 5 grams of this mixed adhesive were added 6 drops (0.22 milliliter) of 12 N sulfuric acid and 1.2 grams of an additive mixture "K", containing 42% Vavanite Powder, 21% quebracho extractive, 10.5% of wheat flour and 26.5% of chalk (calcium carbonate), all of them in a finely divided state. The pH was then 9.4, measured potentiometrically in a sample diluted 1:10 with distilled water. The mixed adhesive does not give any reaction for hydrogen sulfide on lead acetate paper after acidification with hydrochloric acid 1:3 and slight heating. Sulfide ions, $S^=$ and SH^-, are thus absent. Gluing of plywood by heating a 3 sheet, 4-millimeter thick test sample for 6 minutes at 145°C resulted in a strength on air dried plywood of xx112xx124 kilograms per square inch (25 by 25 mm).

After standing overnight in a vessel with water on the bottom, the moisture increase of the plywood was 4.3 and 4.7%, respectively; strength xx176x152 kilograms per square inch. (The x's mean that one or both of the glued joints in the test sample were resistant against loosening by hand.) In the case of the samples which had been standing in moist air, the kg value as well as the loosening testings were made after the samples had been immersed in water 18 to 20 hours at room temperature. Number of x's: 7 of 8 possibles.

Example 2: The black liquor adhesive 1,070 Csv was prepared analogous to 1,068 Asv (Example 1) from 243.3 grams black liquor and 38.7 milliliters of 35% formalin by heating 3 hours at 70°C and adding 64.1 milliliters of water. Dry content thereafter was 39.9%; d = 1.23. No evaporation in vacuo was used. The phenol adhesive 1,070 B fen was prepared analogous to 1,068 B fen in Example 1.

212 grams of the black liquor adhesive 1,070 Csv were mixed with 69.6 grams of the phenol adhesive 1,070 B fen and 33.9 milliliters of 35% formalin. In the resulting mixed adhesive, 1,070 BC, the amount of phenol used was 20.1% of the sum of phenol and organic matter of the black liquor. The dry content was 38.5%; density was 1.18; viscosity at room temperature for a fresh sample (Ford cup No. 2), 37.7 cp; after 11 days at room temperature, 273 cp. Testing for hydrogen sulfide on lead acetate paper (See Example 1) was negative.

To 5 grams of this mixture, 10 drops (0.20 milliliter) of 9.5 N lactic acid and 1.2 grams of the additive mixture K were added. The pH after dilution 1:10 was 8.95. Gluing 6 minutes at 145°C as in Example 1 on air dried birch plywood xx112xx112 kilograms per square inch, moisture increase, 4.2 and 4.2%; strength xx140x171 kilograms per square inch; 7 x's of 8 possible.

Pulp Mill Paper Making Waste Sludge as Extender

A.L. Lambuth and W.M. Hearon; U.S. Patent 3,907,728; September 23, 1975; assigned to Boise Cascade Corporation describe thermosetting resinous adhesives, useful particularly as plywood or lumber laminating adhesives which comprise a phenol-, resorcinol-, urea-, or melamine-formaldehyde resin and an extender.

The extender comprises an aqueous cellulosic pump used in relative proportions of from 1 to 50 parts by weight extender for each 100 parts by weight resin, dry solids basis. Water is included in an amount sufficient to impart to the compositions a viscosity of 1,000 to 12,000 centipoises. The cellulosic pulp extender preferably comprises the pulp mill papermaking waste sludges containing the discarded cellulosic fines separated from the pulp products of the mill. Its inclusion has the effect of holding the water at the glue line and accordingly of improving significantly the assembly time characteristics of the adhesive. The adhesive products of the process are illustrated by the following examples where parts are expressed as parts by weight on a dry solids basis, and wood failure values are given in percent, 85% or better being an acceptable level in the tests employed.

Example 1: To test the adhesive capabilities of a typical phenol-formaldehyde resin-full chemical pulp mill cellulosic sludge of the process, there was prepared a mixture of 1,000 grams of aqueous alkaline phenol-formaldehyde resin, and 150 grams of papermill sulfite sludge. The resin (Monsanto PF 541) contained 41% resin solids and had a pH of 11.2. The sludge was a sulfite pulp sludge refined to a fiber length of at least 90% less than one-eighth inch by 15 passes through a Morden papermaking refiner. It contained 31% solids of which about 60% comprised cellulosic fibers.

The mixture was mixed until homogeneous. 100 grams of water was then added and the mixing continued until a homogeneous mixture again was obtained. The mixing was carried out in a Hamilton Beach high speed mixer.

The resulting glue contained 32% phenol-formaldehyde resin solids. Its pH was about 11.5. It was spread in an amount of 60 pounds MDGL on Douglas Fir plywood veneers which then were laid up into 3-ply assemblies using assembly times varying between 3 and 45 minutes. The assemblies were pressed for 2½ minutes at 285°F and 200 psi.

The resulting five-sixteenths inch plywood panels were then subjected to the standard American Plywood Association vacuum pressure test for glue bond strength as measured by wood failure on shear. As indicated in the test data given below, all samples passed the test with significantly high values of wood failure, even at the extreme limits of assembly time.

	Average of 10 Panels	
Assembly Time (Min.)	Load (PSI)	Wood Failure (%)
3	260	93
15	110	100
30	140	99
45	130	100
Average at 4 assembly times	173	98

Example 2: The procedure of Example 1 was followed, but using a somewhat less advanced phenol-formaldehyde resin, i.e., Monsanto PF 544 containing 41% resin solids, dry solids basis, and having a pH of 11. The same sulfite-full chemical sludge was employed as in Example 1, but at a solids content of 32%. The sludge was refined by 10 passes through a Morden papermaking refiner. The results are shown in the table on the following page.

Assembly Time (Min.)	Average of 10 Panels Load (PSI)	Wood Failure (%)
3	150	88
15	285	92
30	170	98
45	240	95
Average at 4 assembly times	211	93

Dried Animal Blood as Foaming Agent

C.N. Cone and J.M. Steinberg; U.S. Patent 3,905,921; September 16, 1975; assigned to Pacific Adhesives Company, Inc. describe a rapidly foamable, fast setting phenolic resin liquid plywood glue for use particularly in a plywood production line including an in-line continuous foaming unit. The glue comprises an aqueous alkaline solution of a fast setting, highly advanced, phenol aldehyde resin and a dried animal blood foaming agent. Glyoxal or other aldehyde is included to control the viscosity of the glue. Upon agitation in air or other gas, the resulting glue is uniformly foamable to a density of the order of 0.2 gram per cubic centimeter in a time of the order of 2 seconds. In the following examples, parts and percentages are by weight.

Example 1: This example compares the application of the rapidly foamable, fast settling phenolic resin glue containing a highly advanced phenolic resin, in both foamed and unfoamed condition, with a commercial phenolic resin glue containing phenolic resin having a conventional degree of advancement, in the manufacture of plywood.

The adhesive composition of the process was prepared as follows. 240 parts water and 50 parts blood having a water solubility at 72°F of 50% by weight were mixed in a mechanical mixer for 2 minutes. 1,000 parts of aqueous alkaline phenol-formaldehyde resin (Borden W-166S) having a viscosity of 5 poises, a pH of about 11 and a resin solids content of 40% by weight were added and the mixing continued for 2 minutes. The composition was then ready for application.

The conventional phenolic resin adhesive composition was prepared as follows. 220 parts water at 70°F, 100 parts alder bark filler, and 34 parts wheat flour filler were mixed for 5 minutes. 75 parts aqueous alkaline highly advanced phenol-formaldehyde liquid resin (Borden W-166S) having a viscosity of 5 poises, a pH of about 11 and a solids content of about 40% were added and the mixing continued for 2 minutes.

30 parts 50% aqueous caustic soda then was added and the mixing continued for another 2 minutes. 15 parts soda ash was added and the mixing continued for another 20 minutes. 540 parts of the same phenol-formaldehyde resin was added and the mixing continued for 5 minutes. This gave the finished conventional adhesive composition. The two adhesive compositions prepared in the foregoing manner were then applied to the manufacture of plywood.

In one test procedure a portion of the adhesive composition of the process which contained blood as a foaming agent was introduced into the in-line continuous foamer described above and thereby converted in a foaming time of about 3 seconds to a uniform foam having a density of 0.2 gram per cc. This was extruded onto the surfaces of wood veneers which then were laid up into plywood assemblies. For purposes of comparison, another portion of the blood-resin glue of the process was roller spread on wood veneers which then were laid up into plywood making assemblies in the usual manner. The conventional phenol-formaldehyde resin glue could not be foamed in the in-line foaming unit in the dwell time of the unit. It was roller spread on plywood veneers which then were laid up into plywood assemblies in the usual manner.

In all three cases the assemblies comprised 3-ply Douglas Fir panels of 7/16 inch thickness pressed 2 per press opening in a hot press with plates at 300°F for 5¼ minutes. Test pieces were cut from the resulting panels and subjected to the standard Plywood Association shear test. In each case groups of panels having assembly times of 4, 14 and 32 minutes were tested. This spans the range of assembly times ordinarily used in the plywood mill. The test results were averaged. The glue spread rate for the blood-resin glues of the process was 26 pounds per 1,000 square feet of single glue line; that for the conventional phenolic resin glue, 34 pounds. The pounds of resin applied per 1,000 square feet thus was approximately the same in both cases. The results of the test are given in the following table.

Plywood Shear Test Results, (average)

Glue	Procedure	Load, psi	% Wood Failure
Blood resin	Foamed and extruded	246	97
Blood resin	Not foamed, roller spread	202	65
Conventional phenolic resin glue	Not foamed, roller spread	34	02

The foregoing results clearly indicate several things. First, the blood-resin glue is rapidly foamable under conditions in which the conventional phenolic-resin glue is not foamable at all, although both glues contain the same phenolic-resin of high degree of advancement. Second, the blood-resin glues can be cured adequately in the press for only 5¼ minutes. The usual press time required is 6¼ minutes or longer. This represents a 20% saving in press time. Third, the blood-resin composition foamed has a noticeably faster cure than does the same blood-resin mixture unfoamed. In a similar manner there are prepared and applied the glues of the process using, in place of the phenol-formaldehyde resin, a cresol-formaldehyde resin, a resorcinol-formaldehyde resin, a phenol-acetaldehyde resin, or a phenol-furfural resin.

Example 2: This example illustrates the unique suitability of dried animal blood as a rapid foaming agent for foaming phenolic-resin solutions containing phenolic-resins of a high degree of advancement. In the preparation of the test compositions, "x" parts of the test agent was mixed with 300 parts of water for 2 minutes. The resulting mixture was then added to 1,000 parts of aqueous alkaline phenol-formaldehyde resin having a solids content of 40% by weight, a viscosity of about 5 poises and a pH of about 11 and mixed for another 2 minutes. The resulting composition then was passed continuously through an in-line foamer where it was mixed with a stream of air. In order to foam the mixture exhaustively, the streams were recirculated until the liquid would take up no more air. This required a time of up to 3 minutes.

The stream when it emerged from the foamer contained large bubbles of incorporated air. These bubbles were separated from the stream and a measured volume of the foamed material, free of large bubbles, was weighed to determine its density (grams per cubic centimeter). The results are given in the table below.

Test No.	FOAMING AGENT Kind	Amount (Parts by weight)	Foamed Density (g/cc.)
1	Blood-30% water soluble	50	.198
2	"	30	.320
3	Blood-40% water soluble	30	.221
4	Blood-50% water soluble	50	.121
5	"	40	.172
6	"	30	.251
7	"	20	.427
8	Blood-70% water soluble	30	.290
9	Blood-90% water soluble	30	.301
10	Blood-water insoluble	30	.484
11	Soybean flour	100	.834
12	Animal glue	50	.990
13	Alkyl aryl sulfonate type foaming agent ("Ultra Wet 60L")	30	.712
14	"	60	.594
15	Sodium salt of an alkyl aryl sulfonate type wetting agent ("Ultra Wet SK")	30	.780
16	Oleic acid ("Neofat 9404")	30	.846
17	Stearic acid ("Neofat 854")	30	1.03
18	Aliphatic amine type wetting agent ("Armeen DMHT")	30	1.08
19	Amine type wetting agent ("Armflow 65")	30	1.05
20	Sodium alkyl aryl polyether sulfate type wetting agent ("Triton 770")	30	.984
21	" (Triton X200")	30	.985
22	Palmolive soap	30	.876

The test results set forth in the table clearly indicate the unique suitability of dried animal blood as a foaming agent for highly advanced aqueous alkaline phenol-formaldehyde resins. Even when exhaustively foamed over a period of several minutes, the listed nonblood foaming agents, which all are of recognized potency, did not produce a foam having the density of 0.5 or less necessary for successful application in the in-line application for foamed glue to plywood production.

The test results further indicate that animal bloods of all grades of water solubility, i.e., percent soluble in water at 72°F in 5 minutes, share this unique property. Thus, even blood which has been heated until it is totally insoluble is a reasonably efficient foaming agent for the intended purpose. Bloods having solubility in the range of 30 to 90% by weight are preferred.

Disubstituted Bis(Aryl Amine) Curing Agents

G.F. Baxter, H.G. Freeman and G.T. Tiedeman; U.S. Patent 3,896,081; July 22,

1975; assigned to Weyerhaeuser Company describe resin products having particular utility as rapid curing adhesives for wood and other materials. The products are made by reacting aldehyde condensation polymer containing reactive alkylol groups, such as a phenol-formaldehyde condensation polymer, with disubstituted bis(aryl amine) having the formula:

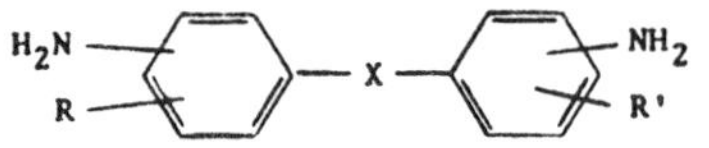

where R and R' are alkoxy, amino, hydrogen, hydroxyl, nitro, carboxyl or C_1 to C_4 alkyl and X is $-[C(Y_1)(Y_2)]_z-$, $-SO_2-$, $-NH-$, $-O-$, $-S-$, $>C{=}C<$, or $-S-S-$ where Y_1 and Y_2 are hydrogen or alkyl radicals having from 1 to 3 carbon atoms and Z is an integer ranging from 0 to 2, to obtain an amine modified polymer.

When the resulting reaction product is blended with an appropriate curing agent, such as an aldehyde, the compositions cure very rapidly at ambient temperature. When pieces of wood or other materials are spread with the preferred adhesives employing the resin compositions of this process and brought into contact with another wood surface the bond strength develops within minutes. The durability, strength and flexibility of the adhesives under adverse weathering conditions are excellent.

The following amines are exemplary of those that may be used in the preparation of these resins: 3,3'-dimethoxybenzidine, 4,4'-oxydianiline, o-tolidine, 4,4'-diaminodiphenylsulfone, 3,3'-diaminobenzidine, 4,4'-diaminodiphenylamine, 4,4'-diaminostilbene, 4,4'-methylenedianiline and 4,4'-diaminodiphenyldisulfide. The following examples illustrate this process; parts and percentages are by weight unless otherwise specified.

Example 1: This example illustrates the fast cure rates of the resin compositions of this process. Cure rate was determined by means of gel time. Gel time was determined by weighing out a 10-gram aliquot of the first component, adjusting the pH to the desired point, and mixing the second component with the first component. The time elapsed from mixing to gelling of the composition is termed gel time.

In each instance, the compositions were formulated by mixing an amine modified phenol-formaldehyde polymer with additional formaldehyde. The phenol-formaldehyde polymer was prepared by mixing 42.06 parts by weight phenol, 4.51 parts water, 11.35 parts flake paraformaldehyde (91%) and 0.46 part calcium acetate monohydrate. The mixture was brought to reflux (approximately 109°C) in about 60 minutes at a uniform rate and held at reflux for 120 minutes.

To separate 100-gram aliquots of this prepolymer was added 0.305 mol of the respective amines shown in the table on the following page. The mixture in each case was refluxed for 2.25 hours, cooled, and 26.3 grams of methanol stirred into the mixture. A 10-gram aliquot of the amine modified resin was then weighed out, the pH adjusted, and 2.5 milliliters of 55% formaldehyde in methanol-water solution added. The mixture was stirred rapidly until gelled, and the time recorded.

The solution pH has some effect on the gel time of the resins. This pH effect is dependent primarily on the particular amine used but also on the solvent used and the concentration of the amine. The following table lists the amines used, the gel time and the pH range over which the gel times were obtained. Though the cure rates of many of the resins are pH dependent, the determination of the optimum pH for a desired gel time can be easily determined by a skilled technician.

Compound	Gel Time	pH Range
3,3'-dimethoxybenzidine	Less than 100 seconds	More than 8
4,4'-oxydianiline	Less than 100 seconds	More than 5
o-tolidine	Less than 100 seconds	More than 11
4,4'-diaminodiphenylsulfone	About 100 seconds	2-8
4,4'-diaminodiphenylamine	Less than 100 seconds	4.5-11.0
3,3'-diaminobenzidine	Less than 100 seconds	7-10
4,4'-diaminodiphenylmethane	Less than 100 seconds	3-10

Example 2: To 100 grams of high solids phenol-formaldehyde prepolymer prepared as described in Example 1 was added 0.305 mol of 3,3'-dimethoxybenzidine. The mixture was refluxed for 2.25 hours, cooled, and 26.3 grams of methanol stirred into the mixture. A 10-gram aliquot of the amine modified resin was weighed out, and 5.0 grams of a hardener added, the hardener comprising a solution of 55% formaldehyde in methanol and water thickened with a small quantity of refined chrysotile asbestos (96.4% formaldehyde solution and 3.6% asbestos). The mixture was stirred rapidly and used in a cross-lap test.

The cross-lap test is conducted by spreading the adhesive composition on the central area of a piece of Douglas fir wood 1 inch wide by approximately ¾ inch thick and 2¾ inches long, a similar piece of Douglas fir wood is immediately laid over the first with the grain direction at right angles. A measured quantity of adhesive may be used or an excess may be applied with the surplus resin allowed to squeeze out of the joint, the latter method being used for the examples given here.

As soon as the cross-lap is laid on the first piece, a pressure of 40 pounds is applied for the desired length of time. The joint is then broken in tension and the bond strength recorded. At the end of 12 minutes press time the tensile strength necessary to separate the cross-lap was 65 pounds per square inch.

Normally wood failure begins to occur at a value of about 150 psi. For many purposes, however, a bond strength well below this value is wholly satisfactory. It should also be noted that strength usually continues to increase significantly for at least 24 hours after initial assembly.

Example 3: A resin was made in similar fashion to that of the preceding example except that the amine compound of the resin was 0.305 mol of 4,4'-diaminodiphenyldisulfide. The resin was used in a cross-lap test with 2 parts by weight of the resin being mixed with 1 part of the asbestos thickened aldehyde hardener. After 12 minutes of press time a tensile strength of 45 pounds per square inch had developed in the glue joint. The gel time of the adhesive composition was 21 seconds at pH 6.6.

Finely Divided Aluminum Powder as Acid Scavenger

It is known that many common wood adhesives, such as urea-formaldehyde resins and some phenol-formaldehyde resins, cure most rapidly and provide superior bonds when cured at a low pH, i.e., less than 3.5. It is also known, however, that the use of adhesives having such a low pH to bond wood members can result in gradual degradation of wood along the glue line when the cured adhesive is exposed to hydrolytic conditions, e.g., if the wood has a high moisture content at the time the adhesive is applied or if the bonded structure is subsequently exposed to a moist environment for a prolonged period.

Such degradation, which can be particularly acute when the adhesive is cured at a pH of less than 2.5, apparently results from gradual leaching of acid from the cured adhesive and consequent hydrolysis of the wood. When bonded structures are sheared along the glue line, such degradation usually is evidenced by very low shear strengths coupled with shallow wood failure over a large percentage of the area adjacent the glue line.

Experience indicates that a glue line pH of 3.5 is a common safe minimum for long term wood stability. (ASTM D2559-66T-6.1 specifies that wood adhesives shall have a pH of no less than 2.5, while *Military Specifications for Wood Adhesives,* Mil-A-397B-3.7, specifies a minimum pH of 3.5.) However, where bonded wood structures are likely to be subject to conditions of high temperature and high humidity, even higher pH's may be necessary to prevent degradation completely.

According to a process described by *G.T. Tiedeman and M.F. Gillern; U.S. Patent 3,872,051; March 18, 1975; assigned to Weyerhaeuser Company* incorporating a minor amount of finely divided aluminum or iron into acidic adhesive compositions of the type described above prevents or markedly reduces degradation of cellulosic substrates . The metal, which survives curing of the adhesive, apparently functions as a latent acid scavenger to react with and consume the hydrogen ions liberated when the cured adhesive is subjected to hydrolytic conditions, i.e., is exposed to moisture.

It is preferred that the amount of metal employed in the adhesives be such that when they are subjected to hydrolytic conditions after curing, the adhesives will exhibit a pH of at least 3.5, and most preferably above 4.0. The following example illustrates the process; all parts and percentages are by weight unless otherwise indicated.

In the following example, the results of Block Shear Tests on adhesives are reported in terms of shear strength and percent wood failure. The following procedures were employed in these tests. A liquid adhesive was spread on the face of one of a pair of matched, longitudinally grained Douglas fir strips (¾ inch by 2½ inches by 12 inches). The two strips were then superimposed face to face and clamped together at a selected pressure and for a selected time at 72°F.

The resulting laminate was sawed transversely to produce five specimens which were trimmed to dimensions of 1½ inches by 2 inches by 2 inches. Each of these specimens was then sawed across the grain to the glue line to remove one ¼ inch by ¾ inch by 2 inches strip from each lamina (the strips being removed from opposite ends of the specimen). The laminae in the resulting specimens were thus offset by ¼ inch, the remaining glue line having an area of 3.0 square inches (1.5 inch by 2.0 inches).

Two specimens were dry conditioned at about 70°F and 50% relative humidity for 48 hours, and three were subjected to wet cyclic conditioning according to the following modification of AITC Test 110. The three wet specimens were placed in an autoclave filled with water (65° to 80°F) so that the specimens were completely submerged. A vacuum (20 to 25 inches of mercury) was then applied and held for about 30 minutes. The vacuum was then released and air pressure of about 75 psi was immediately applied and held for about 120 minutes. The specimens were then placed in a forced air drier at 150°F and 8 to 10% relative humidity until they had returned to within 15% of the original weight (about 20 to 22 hours).

All conditioned specimens were sheared at 0.05 inch per minute load rate (applied to one of the offset ends); the maximum (breaking) load (psi) for each specimen was determined and separately averaged for dry and wet specimens. The shear strengths were then calculated [shear strength (psi) = maximum load in pounds per 3.0 square inch]. The percent wood failure along the glue line was estimated for each specimen and averaged for dry and wet specimens. The metal powders used in the following example were purified grades and consisted of substantially spherical particles of less than 40 mesh.

Example: (1) – A 4-liter reactor was charged with the following: 1,690.0 grams phenol (90.4%), 401.6 grams paraformaldehyde flakes (94%), 10.8 grams water and 16.8 grams calcium acetate. The temperature was adjusted to 25°C. Heating and agitation were then applied bringing the mixture to reflux (110°C) uniformly over a 60-minute period. The reflux was maintained for 2 hours, and the mixture was cooled slightly (100°C), whereupon there was added 1,880.8 grams m-hydroxy aniline hydrochloride (48.5%).

From 68°C the mixture was heated to reflux (105°C), the reflux was maintained for 2.5 hours and the resulting resin was then cooled to 25°C. The resin was water-washable, had a pH of 2.3 and a Gardner viscosity of WX (between W and X) at 25°C. Varying the pH of this resin substantially can cause some precipitation to occur. Therefore, it should be used with a neutral or preferably acidic hardener. (It is anticipated that by combining this resin with a hardener of substantially the same pH, a good primerless adhesive will be produced; see second footnote of Table 2.)

m-Hydroxy aniline hydrochloride (sometimes referred to as m-HA·HCl) utilized in the procedures of this example was prepared according to the following procedure. To 31.7 parts water were added 32.0 parts hydrochloric acid (38%) and to the resulting solution was added 36.3 parts m-hydroxy aniline with stirring. Stirring was continued until all the m-hydroxy aniline was dissolved.

(2) – A 5-gallon reactor was charged with the following: 6,569.7 grams phenol (90.4%), 2,238.6 grams water, 6,034.3 grams formaldehyde (50.2%) and 500.2 grams methanol. Temperature was adjusted to 35°C and there was added 383.9 grams sodium hydroxide (50.1%). The temperature of the reaction mixture was allowed to rise from 41° to 65°C in 70 minutes at a uniform rate; no external heating was required during the heat-up. The reaction mixture was maintained

at 65°C for 105 minutes and was then adjusted to pH 8.4 with 124.0 grams formic acid (90.7%). The reaction mixture was maintained at 65°C until 3 hours had elapsed from the start of pH adjustment.

There was then added 6,952.2 grams formaldehyde (50.2%) and the reaction mixture was mixed thoroughly while cooling to 25°C. A 5-gallon reactor was charged with 6,008 grams of the resulting base resin. The temperature was raised from 25° to 60°C and 8,013 grams resorcinol was added. Heating was immediately reapplied bringing the temperature from 44° to 71°C in 5 minutes at which point a mild exotherm was noted. No external heating was applied from 71° to 104°C where the exotherm subsided.

Heating was again applied and the reaction mixture was brought to reflux (105°C) and the reflux was maintained for 5 minutes. The mixture was cooled to 100°C, and 7,945 grams m-HA·HCl (48.5%) was added. From 76°C the mixture was heated to reflux (105°C) and the reflux was held for 2 hours. The resin was then cooled to 45°C and 714 grams of pyrogenic silica (Cab-O-Sil M-5, Cabot Corporation) was added with mixing; the pyrogenic silica renders the composition thixotropic. The resulting mixture was water-washable and had a pH of 2.60.

(3) – The procedure of (2) was repeated with the following modifications. The amounts of base resin, resorcinol and m-HA·HCl (48.5%) employed were 271 grams, 362 grams and 359 grams, respectively. After the addition of the m-HA·HCl, the reaction mixture was stirred thoroughly at 75°C and then there was added 8.5 grams of a hydroxyethyl cellulose having a 0.9 to 1.0 degree of substitution and a 1.6 to 2.0 molar substitution of ethylene oxide, and a 2% solution LVT Brookfield viscosity of 70 to 100 cp at 25°C in the L range (Cellosize QP 40 high, Union Carbide Corporation); no pyrogenic silica was added.

The reaction mixture was heated from 75°C to reflux with vigorous agitation to avoid lumping of hydroxyethyl cellulose. The reflux temperature (104°C) was maintained for 2 hours. The resin was then cooled to 25°C. The hydrophilic resin had a pH of 2.60 and a Gardner viscosity of MN at 25°C.

(4) – Liquid hardener formulations containing varying amounts of aluminum powder were prepared by mixing together the ingredients shown in Table 1 in the order listed. One part of the hardeners was individually combined with 2 parts of the final resins produced in (1), (2) and (3). Films of the resulting adhesives were cast, allowed to cure and then powdered; the stable pH values of aqueous slurries of the powders were determined as described previously and are recorded in Table 2. The results of Block Shear Tests on the adhesives are also shown in Table 2.

TABLE 1

Ingredients	Hardeners (Weight Percent) I	II	III
"U.F. 85"	60.0	60.0	56.0
Water	7.5	7.5	7.5
"Dow Latex 612"	25.0	24.0	23.5
Calcium Carbonate	1.0	1.0	1.0
Aluminum Powder	3.5	5.0	10.0
"Cab-O-Sil M-5" Pyrogenic Silica	3.0	2.5	2.0

UF 85 is a concentrated urea-formaldehyde resin precursor containing an excess of free formaldehyde and about 15% water (Allied Chemical Corporation). Dow latex 612 is a styrene-butadiene polymer (Dow Chemical Corporation); the polymer was added to enhance the impact strength of adhesives into which the hardeners are incorporated. Calcium carbonate causes evolution of CO_2 giving a desirable gap filling quality to the adhesives.

TABLE 2

Adhesive			Block Shear Tests* — Shear Strength (psi) % Wood Failure	
Resin	Hardeners (Table 1)	pH of Cured Adhesive	Dry	Wet
Example (1)	I	4.30	1280/55	0/0**
Example (1)	III	4.41	1760/25	0/0**
Example (2)	I	4.15	1300/95	790/85
Example (2)	II	4.20	2020/82	320/23
Example (2)	III	4.35	1970/92	370/52
Example (3)	I	4.28	1550/82	650/67
Example (3)	II	4.35	1500/100	460/80
Example (3)	III	4.49	1550/90	860/43

* Specimens were clamped at 75 p.s.i. for 30 min. at room temperature.

** These specimens delaminated upon wet-conditioning; this is believed to be due to inadequate wetting of the wood by the adhesive. These two adhesives give excellent bonding if the wood is first primed for example with a phenol-formaldehyde resin modified with resorcinol (see U.S. Patent Re. 26,881), diluted with methanol and cured with a 55% formaldehyde solution in methanol and water ("Methyl Formcel" — available from Celanese Corporation) as described in U.S. Pat. application Ser. No. 822,009, filed May 5, 1969.

UREA-FORMALDEHYDE RESINS FOR PARTICLE BOARD

Latent Catalyst

D.K. Sausman; U.S. Patent 3,893,963; July 8, 1975; assigned to Commercial Solvents Corporation has found that an improved aqueous urea-formaldehyde solution is obtained by incorporating therein catalytic amounts of a latent catalyst. The latent catalyst useful for the practice of this process is a mixture of the hydrochloride salt of an amino alkanediol corresponding to the formula:

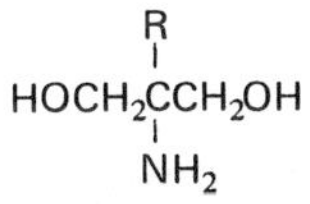

where R is methyl or ethyl, and an alkanolamine corresponding to the formula:

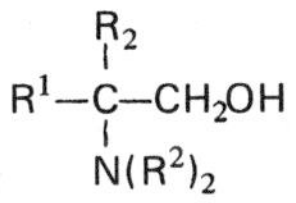

where R^1 is hydrogen or alkyl of 1 or 2 carbon atoms and R^2 is hydrogen or methyl, in a mol ratio of about 1.5 to 2.0 of the amino alkanediol per mol of alkanolamine.

The composition is particularly suitable for use as the adhesive in the manufacture of particle board and laminated wood articles and eliminates the resin cook heating step heretofore required. At the time of use, sufficient urea to provide a desirable mol ratio to formaldehyde, which ratio is known to those skilled in the art, is dissolved in the solution. During the manufacture of laminates, the resulting composition is applied directly without prior heating, to the surfaces of the components to be laminated or is mixed in suitable amounts with particles or chips. The components are then assembled, as is known, and heated in a press at an elevated temperature and pressure and for a length of time sufficient to produce the laminate.

For the manufacture of particle or chip board, the particles or chips are thoroughly mixed with the adhesive in effective amounts in accordance with known methods. The mixture is then heated in a press at a temperature and at a pressure and for a length of time effective to produce the board.

A suitable and convenient catalyst which is available commercially is Adamac Catalyst 20 (McWhorter Chemical Co.). This product is a nearly nonaqueous solution at about 20% by weight catalyst. When used in the process, it is employed at about 4 to 5% based on the weight of the urea-formaldehyde solution, thereby providing the preferred concentration of 0.8 to 1.0%.

Example 1: To 100 parts of an aqueous solution containing 51 parts of formaldehyde and 20 parts of urea there were added increments of Adamac 20 of 1, 2, 3, 4 and 5% by weight, respectively. Adamac 20 has the following composition: 111 parts 2-amino-2-ethyl-1,3-propanediol hydrochloride, 35 parts 2-amino-2-methyl-1-propanol, 567.5 parts isobutyl alcohol and 44 parts water.

At 20% latent catalyst solids, the Adamac 20 provided catalyst concentrations of 0.2, 0.4, 0.6, 0.8 and 1.0%, respectively. The mixtures were divided into two portions each. One set was permitted to stand for 7 weeks at ambient temperatures and the other set was permitted to stand for 4 weeks at 50°C. No polymer formation occurred, nor any discoloration. Control samples without latent catalyst developed yellow discoloration. The urea-formaldehyde solutions containing latent catalyst were thus determined to be suitable for handling, shipping and storage without risk of undue polymerization.

Example 2: To two portions of 100 parts each of the urea-formaldehyde solution of Example 1 there were added 5 and 10 parts, respectively, of Adamac Catalyst 20, thus providing latent catalyst concentrations of approximately 1 and 2%, respectively. Another portion of urea-formaldehyde solution without catalyst was used as a control. These portions were designated 2, 3 and 1, respectively.

Each of the above samples was divided into three portions which were designated 1a, 1b, 1c; 2a, 2b, 2c; and 3a, 3b and 3c, respectively. To each of the a samples there was added sufficient urea to provide a mol ratio of formaldehyde:urea of 0.91; to each of the b samples was added sufficient urea to provide a mol ratio of 1.22; and to each of the c samples was added sufficient urea to provide a mol ratio of 2.

To a portion, 20 parts, of each of the above 9 samples was added 6 parts of wood flour and a slurry was formed. The slurry was cast into ¼-inch plaques which were then baked at 275°F for 30 minutes. All of the catalyzed samples cured well. At a mol ratio of formaldehyde:urea of 0.91 the most rigid plaque was produced whereas at a mol ratio of 2, the plaques were crumbly and semisoft.

Example 3: The 9 mixtures prepared in Example 2 were further tested for preparing fir laminates. Strips of fir 5 inches by 1 inch by ⅛ inch were prepared and 3 strips were laminated by placing 1 milliliter of each urea-formaldehyde solution between each contacting surface. Each of the assembled articles was sealed in foil and laminated in a press at 250°F and 200 psig for 30 minutes. After 1 hour, the laminate was tested for satisfactory cure and water resistance.

The laminates were prepared in duplicate. One set was placed in boiling water for 1 hour after removing from the press, and the other set was allowed to age for 1 hour at room temperature. The strength of the laminate was then tested by wedging a scalpel at the fir strip interfaces and attempting to pry apart. Wood failure was reported as pass. Adhesive failure was reported as approximate surface area which failed. The results are tabulated in the following table.

Sample No.	Catalyst % Wt.	Mole Ratio HCHO/Urea	Stability Hours	Laminate 1 Hour	1 Hour in H_2O at 100°C
1a	0	0.91	96	Pass	—
1b	0	1.22	—	50% Fail	—
1c	0	2	—	60% Fail	—
2a	1	0.91	2	Pass	30% Fail
2b	1	1.22	3	Pass	Pass
2c	1	2	168	50% Fail	Pass
3a	2	0.91	1¾	Pass	Pass
3b	2	1.22	2¼	Pass	Pass
3c	2	2	168	Pass	70% Fail

Alkali Metal Salts

A process described by *W.R. Black; U.S. Patent 3,905,847; September 16, 1975; assigned to Borden, Inc.* relates to the method of bonding water-penetrable cellulosic substrates with urea-aldehyde resin adhesives as illustrated by bonding of wood flakes into particle boards, and in particular to a method of improving the efficiency of the adhesives, which comprises replacing from about 1 to 35% by weight of the resin solids by a salt selected from the group consisting of alkali and alkaline earth halides and mixtures.

In addition to the advantage that suitable choice of the replacing salt may effect substantial savings in material costs without loss of adhesive strength, it has been found that tensile strength in many cases is substantially increased at the same level of total solids add-on and without changing the temperature and pressure of curing. It has furthermore been found that high and uniform strength may be obtained even when the wood furnish comprises up to 50% by weight of waste material such as plywood trim.

Because of their relative cheapness, sodium chloride and magnesium chloride are currently preferred, but salts which function according to this process include, for example, potassium chloride, lithium chloride, sodium bromide, potassium iodide, magnesium bromide, calcium chloride, strontium chloride and barium chloride.

The salt may be first mixed with the urea-aldehyde resin, as above illustrated, and applied simultaneously with the resin to the material to be bonded. Alternatively, an aqueous solution of the salt may be separately applied to the substrate either before or after the resin. Thus, for example, the salt may be incorporated in a wax dispersion applied to impart water resistance.

Example 1: A furnish of 75% Ponderosa pine and 25% Douglas fir was reduced to particles in a Bauer attrition mill. A furnish of 100% white fir was reduced to chips in a Pallmann pulverizer. A 50-50 blend of these two furnishes was used in the experiments of Examples 1, 2, 3 and 4. An aliquot of these flakes was treated with a catalyzed urea-formaldehyde resin, (Casco Resin WS-114-79) in an amount corresponding to 6 parts by weight resin solids per 100 parts oven-dry wood, using a horizontal rotating drum type blender with internally mounted spray system. There was then added 0.5 part of a paraffin wax emulsion, care being taken to attain maximum homogeneity.

Square samples of board 14.5 inches square and 0.838 inch thick were preformed by mat consolidation at a pressure of 250 psi prior to the final pressing and curing at 310°F for 5 minutes. During this pressing cycle the pressure varied from about 650 to 50 psi. This example illustrates a treatment not including the halide of this process and tests on the product thereof are included for comparison in the first table on the following page together with corresponding data on the products of Examples 2, 3 and 4.

Example 2: The procedure of Example 1 was followed except that 3 parts by weight sodium chloride (NaCl) were added to 100 parts of liquid Casco Resin WS-114-79 and this mixture was used to treat the wood flakes in an amount corresponding to 6 parts by weight total solids per 100 parts of oven-dry wood. Since the solids content of Casco Resin WS-114-79 was 60%, this treatment differed from Example 1 in that 4.8% of the resin solids were replaced by sodium chloride.

Example 3: The procedure of Example 1 was followed except that 6 parts by weight sodium chloride were added to 100 parts of liquid Casco Resin WS-114-79 and this mixture was used to treat the wood flakes in an amount corresponding to 6 parts by weight total solids per 100 parts of oven-dry wood. This treatment differed from Example 1 in that 9% of the resin solids were replaced by sodium chloride.

Example 4: The procedure of Example 1 was followed except that 3 parts by weight magnesium chloride ($MgCl_2$) were added to 100 parts of liquid Casco Resin WS-114-79 and this mixture was used to treat the wood flakes in an amount corresponding to 6 parts by weight total solids per 100 parts of oven-dry wood. Since the solids content of Casco Resin WS-114-79 was 60%, this treatment differed from Example 1 in that 4.8% of the resin solids were replaced by magnesium chloride.

Test results on the particle board products of Examples 1 through 4 are summarized in the first table on the following page. These results show significant improvement in the internal bond strength achieved by the incorporation of the respective alkali halides in place of part of the resin solids.

The effect of alkali halides on UF resin efficiency using Casco Resin WS-114-79 (60% solids) and wood chip blend containing 37.5% Ponderosa pine, 12.5% Douglas fir and 50% white fir is shown in the first table on the following page. All treatments are at 6% total solids on the basis of oven-dried wood weight.

EXAMPLE Alkali halide	1 None	2 NaCl	3 NaCl	4 $MgCl_2$
Parts by weight of U-F				
Resin solids replaced	—	3%	6%	3%
Thickness (inches)	.838	.827	.820	.827
Density (lbs./cu.ft.)	41.5	42.2	42.4	42.3
Internal bond (psi.)	78	126	142	95

Example 5: A furnish of 100% white fir was reduced to chips in a Pallmann pulverizer. Using the blender and spray procedure of Example 1, aliquots of the pulverized furnish were treated with the respective compositions given in the following table in amounts corresponding to 6% total solids based on the weight of the oven-dried wood, Casco WS-138-43 being a noncatalyzed urea-formaldehyde resin of commerce and containing 68% solids. Boards having the stated thickness and density were prepared by pressing at 330°F for 5 minutes at 50 to 650 psi.

The test results show that the use of the process as exemplified in Experiment C in which 8.1% of the resin solids were replaced by sodium chloride resulted in superior modulus of rupture. Furthermore, although the treatment of Experiment C did not include addition of conventional ammonium sulfate catalyst, the products were comparable in regard to both internal strength and water absorption. The following table shows the effect on adhesive efficiency of UF resin when 6% sodium chloride (based on liquid resin) is used instead of catalyst in amount of ammonium sulfate.

EXPERIMENT	A	B	C
$(NH_4)_2SO_4$ based on o.d. wood weight	.25%	.5%	None
NaCl based on liquid resin	none	none	6%
Board thickness, inches	.814	.808	.814
Density lbs/cu.ft.	40.1	40.5	40.6
Internal bond strength, psi.	92	122	116
Modulus of rupture	2260	2346	2466

Examples 6 through 8: A furnish composed entirely of Ponderosa pine was reduced to flakes and aliquots were treated respectively as indicated in the table on the following page. The procedure was essentially that of Example 1, except that the UF resin used was Casco Resin WS-138-44 containing 65% solids and the combined resin and halide solids applied on an oven-dry basis was 5.5%. A paraffin wax emulsion was applied at 0.5% level on oven-dry basis. Treatment was carried out in a rotating drum blender equipped with internal spray.

Particle board samples were pressed for 4.5 minutes at 325°F and under 50 to 650 psi. Example 6 was a control sample prepared without any alkali halide. Examples 7 and 8 illustrate the process and differed from each other in that the sodium chloride in Example 7 was first mixed with the resin before spraying, whereas in Example 8 an aqueous solution of sodium chloride was first sprayed on the furnish, the furnish was then redried to the initial mixture content and the resin was then applied.

The results of corresponding tests, as summarized in the table on the following page, show that internal bond strength was substantially increased by the replacement of 8.25% of resin solids by sodium chloride both when the sodium chloride was applied simultaneously with the resin and also when the resin and sodium chloride were applied separately. The table below shows the effect of mode of addition of sodium chloride on adhesive efficiency of a UF resin.

EXAMPLE Procedure	6 No NaCl	7 NaCl Added With Resin	8 NaCl Added Separately
Board Thickness, Inches	.825	.814	.820
Density, lbs./cu.ft.	42.4	42.9	42.6
Water Absorption %	12.6	12.4	12.3
Internal bond, psi.	93	158	150

Three-Stage Resin Production Process

F. Brunnmueller, H. Henkel and J. Lenz; U.S. Patent 3,896,087; July 22, 1975; assigned to Badische Anilin- & Soda-Fabrik AG, Germany describe a process for the manufacture of impregnating urea-formaldehyde resins, particularly for use in the surface treatment of wood materials, which comprises the following steps:

(a) condensing urea and formaldehyde in aqueous acid solution in a molar ratio of from 1:1.5 to 1:2.5 in known manner at least to such an extent that a sample of the condensate solution causes turbidity when added dropwise to cold 50% magnesium sulfate solution, but not to an extent exceeding the point at which the resin still remains soluble in the solution at temperatures above 60°C;

(b) adding at least 0.2 mol of formaldehyde per mol of urea to obtain a molar ratio of urea to formaldehyde of from 1:2.5 to 1:4 followed by condensing at a temperature of between 40°C and the boiling point of the solution and at a pH of between 3.5 and 9 until the resin solution is capable of being diluted with at least 20 volumes of water at room temperature to form a clear solution; and

(c) adding urea if necessary to give a final molar ratio of from 1:1.5 to 1:2.5 and effecting postcondensation of the mixture at a pH of between 4.0 and 6.5 until all free urea has disappeared.

The process is illustrated with reference to the following examples, particularly as regards the interrelation of molar ratios, after-treatment times and amounts of substances used for the after-treatment. The impregnating time is determined by causing a flat sheet of light gray decorative paper weighing 100 grams per square meter to float on the surface of a resin solution which has been poured into a wide dish and measuring the time for the color of the sheet of paper to turn a uniform dark gray. With some practice this method of determination can be carried out to a reproducible accuracy of less than one-half of a second.

To demonstrate various methods of measuring the degree of condensation, urea and formaldehyde are condensed at the boil starting with molar ratio of urea to formaldehyde of 1:2.0 in approximately 50% solution. The pH is adjusted to 5.5 at the commencement of condensation and drops to 4.5 in the course of the condensation process. After 50 minutes a further amount of urea is added in the form of a concentrated aqueous solution until the molar ratio is 1:1.8.

Condensation Time (min)		Compatibility with Epsom Salts (g of $MgSO_4 \cdot 12H_2O$ per 100 ml of solution)	Compatibility with Water (1:5) (°C, expressed at temperature at which turbidity occurs)	Water Compatibility at Reaction Temperature
0	M.R. 1:2.0	-	-	-
10		-	-	-
20		-	-	-
30		-	-	-
40		50	-	-
45		40	5	-
50+	M.R. 1:1.8	35	12	-
55		30	13	-
60		25	18	-
65		20	20	-
70		15	25	1:8*
75		10	27	1:6**/1:4.5*
80		6	29	1:4**

+ = urea added M.R. = molar ratio *milky **precipitation

Example: A 50% aqueous solution of a urea-formaldehyde condensate having a molar ratio of urea to formaldehyde of 1:1.7 is produced at a pH of 4.8 in a number of separate experiments giving products of different degrees of condensation. The degree of condensation is indicated by diluting with 8 volumes of hot water and cooling. In the following table, the expression turbidity temp. means the temperature at which the dilute sample begins to show turbidity when cooled in a water bath, as measured with a submerged thermometer having a solid stem. This value is characteristic and reproducible to within about 1 to 2 degrees centigrade. Following the after-treatment with formaldehyde and urea, the resin again has a molar ratio of 1:1.7.

The table shows that to achieve a minimum impregnating time the amount of formaldehyde required for the after-treatment is less the lower the degree of precondensation of the resin.

It is also readily seen from the following table not only that the after-treatment with formaldehyde lowers the viscosity of the resin solution but also that the impregnating times are improved to an extent which is considerably more than to be expected from the mere change in viscosity (compare test No. 1 with tests No. 2 and No. 3 or test No. 6 with tests No. 7 and No. 11). Thus it follows that the after-treatment with formaldehyde not only lowers the degree of condensation but also causes a basic change in the structure of the condensate.

From the data it is seen that the ratio of urea to formaldehyde during the after-treatment with formaldehyde should not fall below 1:2.5, since at molar ratios below this value the viscosity no longer sinks to the desired extent but on the contrary rises and the impregnating times are lengthened (compare tests No. 2, No. 3 and No. 4 with test No. 5 or tests No. 10, No. 11 and No. 12 with test No. 13). In the case of resins showing a high degree of precondensation, the molar ratio of 1:2.5 to 1:2.7 is hardly adequate for the conversion of the high molecular weight resins to low molecular weight resins (compare test No. 15 with tests No. 16 and No. 17). It is particularly characteristic of the process that the after-treatment with formaldehyde does not impair but distinctly improves the solidifying time, i.e., the curing rate of the resin solutions when processed. This fact is particularly surprising bearing in mind that a reduction in viscosity previously meant prolongation of the curing process.

Test No.	Degree of condensation turbidity temp. on dilution with 8 vols of water (°C)	M.R. urea formaldehyde during after-treatment with formaldehyde	After-treatment time/temp. [min/°C]	Solidifying time of 50% soln. on adding 10% of 0.5% ammonium chloride soln. [sec]	Impregnating time [sec]	Flow time of 50% soln. from Ford Cup No. 3 at 20°C [sec]
1	10	untreated control		480	9	35
2	10	1 : 3.10		140	1.8	28.1
3	10	1 : 2.7		165	4	32.3
4	10	1 : 2.5		165	10	39.2
5	10	1 : 2.3		180	60	75.7
6	20	untreated control	60/90	595	22.8	43.1
7	20	1 : 4.0		150	1.2	25.7
8	20	1 : 3.5		160	1.6	27.1
9	20	1 : 3.3		180	1.8	28.5
10	20	1 : 3.1		195	2.3	28.5
11	20	1 : 2.7		200	7.0	37
12	20	1 : 2.5		198	15.4	55.5
13	20	1 : 2.3		207	55.5	100.6
14	40	untreated control	60/90	275	81.8	75.8
15	40	1 : 3.1	60/90	113	4.0	33.7
16	40	1 : 2.7	–/90	gelled when cooled to 60°C after treatment with formaldehyde		
17	40	1 : 2.5	–/90	gelled during after-treatment after 45 minutes		
18	10,20,40	1 : 2.3	–/90	gelled during after-treatment after 15 minutes		

OTHER WOOD BONDING PROCESSES

Acid-Modified Olefin Polymer, Polyvalent Metal Salt and Diepoxide

O. Ohara and N. Wakabayashi; U.S. Patent 3,900,440; August 19, 1975; assigned to Kuraray Co., Ltd., Japan describe an adhesive composition comprising: (a) a member selected from the group consisting of (1) a copolymer of an olefin and a carboxylic acid and (2) a salt of the copolymer; (b) a salt of a polyvalent metal and (c) a compound having at least two epoxy groups per molecule.

The water-resistant adhesion obtained by an adhesive composition comprising such mixture is fit for formation of ordinary plywoods and ordinary water-resistant plywoods [plywoods of Classes III and II defined according to Japanese Agricultural Standard (JAS)]; however, the adhesion does not reach the level required for highly water-resistant plywoods for outer decoration (plywoods of Class I defined according to JAS). The following examples illustrate the process; all parts and percentages are by weight unless otherwise specified.

Example 1: 25 grams of an isobutylene-maleic anhydride copolymer (mol ratio is 1:1) having a degree of polymerization of 1,100 was mixed with 63.2 grams of water and 11.8 grams of 28% aqueous ammonia, and the mixture was heated at 80°C to form a solution.

Separately, a latex was prepared by emulsion polymerization of styrene, butadiene and acrylic acid as unsaturated monomers. The water content in the latex was 51% and the total solids content was 49%. The latex-constituting polymer contained 44% styrene, 53% butadiene and 3% acrylic acid. Sodium dodecylbenzene sulfonate was used as a surfactant in an amount of 5 parts by weight per 100 parts by weight of the total monomers.

200 grams of the above latex, 125 grams of gypsum (calcium sulfate) and 75

grams of water were added successively to the above aqueous solution of the isobutylene-maleic anhydride copolymer and the mixture was agitated to form a white suspension. Then 10 grams of glycerin diglycidyl ether (WPE = 145.6) as a water-soluble epoxy compound was added to the solution to obtain an adhesive composition (pH 8.1). WPE means weight per epoxy equivalent.

After about 60 minutes had elapsed after addition of the epoxy compound, this adhesive composition was coated on red lauan veneers in an amount of 30 grams per 900 square centimeters and 3 veneers, each having a thickness of 1 millimeter, were piled to form a plywood. The plywood was allowed to stand still in the piled state for 20 minutes and clamping (cold pressing) was performed under application of a pressure of 10 kilograms per square centimeter for 20 minutes, following which hot pressing was carried out under application of a pressure of 10 kilograms per square centimeter at 120°C for 1 minute.

The obtained adhesive strength was determined according to the method specified in JAS. As a result, it was found that the adhesive strength according to the dry bonding test was 26.5 kilograms per square centimeter, the adhesive strength according to the hot and cold soaking test was 20.6 kilograms per square centimeter and the adhesive strength according to the cyclic boiling test was 16.2 kilograms per square centimeter.

Example 2: The procedures employed in Example 1 were repeated under the same conditions except that a latex composed of a polymer containing 49.5% styrene, 48% isoprene and 2.5% acrylic acid, which was prepared in the same manner as in Example 1 by using isoprene instead of butadiene, was employed. The adhesive strength according to the dry bonding test was 24.3 kilograms per square centimeter, the adhesive strength according to the hot and cold soaking test was 18.1 kilograms per square centimeter and the adhesive strength according to the cyclic boiling test was 13.4 kilograms per square centimeter.

Example 3: 25 grams of an isobutylene-maleic anhydride copolymer (mol ratio is 1:1) was mixed with 55.3 grams of water and 19.7 grams of 28% aqueous ammonia, and the mixture was heated at 80°C to form a solution. To the solution were successively added 67 grams of an aqueous emulsion of an ethylene-vinyl acetate copolymer (PANFLEX, Kuraray Co., Ltd., solids concentration of 56.1%), 87.5 grams of calcium carbonate and 45.5 grams of water. Then the mixture was agitated to obtain a white suspension. Then 8 grams of a glycidyl-etherified product of sorbitol (WPE = 165.8) was added to the solution to obtain an adhesive composition (pH 6.8).

The adhesive strength was determined with respect to the so obtained adhesive composition in the same manner as described in Example 1. It was found that the adhesive strength according to the dry bonding test was 10.1 kilograms per square centimeter, the strength according to the hot and cold soaking test was 9.2 kilograms per square centimeter and the adhesive strength according to the cyclic boiling test was 7.4 kilograms per square centimeter.

Example 4: 30 grams of an isobutylene-maleic anhydride copolymer (mol ratio is 1:1) having a degree of polymerization of 300 was mixed with 10 grams of sodium hydroxide and 225 grams of water, and the mixture was heated at 80°C to form a solution. Then 50 grams of calcium carbonate, 140 grams of the same latex used in Example 1 and 15 grams of diglycidyl ether of glycerin was added with stirring to the so formed solution to obtain an adhesive composition (pH, 8.5). The viscosity of this composition was about 500 cp at 25°C. The so formed adhesive composition (280 grams) was sprayed onto lauan chips (1,000 grams) and thereafter the chips were placed into a mold (300 mm wide by 300 mm long by 15 mm high) and then clamping (cold pressing) was performed under application of a pressure of 10 kilograms per square centimeter for 10 minutes, following which hot pressing was carried out under application of a pressure of 20 kilograms per square centimeter at 150°C for 15 minutes to obtain a single layer particle board.

This particle board was tested according to Japan Industrial Standard A 5908 and the results are: thickness, 15.03 mm; density, 0.68; water content, 6.7%; bonding strength, 201 kilograms per square centimeter; nail holding, 46 kilograms.

Polymerization Accelerators for Cyanoacrylate Compositions

A process described by *J.J. Delahunty; U.S. Patent 3,836,377; September 17, 1974* concerns polymerization accelerators or surface activators for increasing the rate of cure of cyanoacrylate adhesive compositions. The active ingredients of the activators are organic compounds containing a:

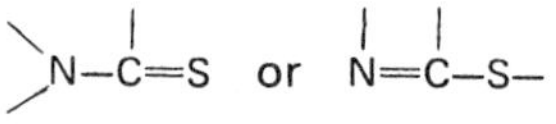

group. Examples of three specific classes of compounds which have been found excellent for purposes of this process are as follows.

(a) N-disubstituted metallic dithiocarbamates having the general formula $R_1R_2N{\cdot}C{\cdot}(S){\cdot}S_xM_y$ where each of R_1 and R_2, which may be the same or different, represents a saturated or an ethylenically or aromatically unsaturated group, preferably a hydrocarbon group, having a maximum of 10 carbon atoms, and R_1 and R_2 together with the nitrogen atom to which they are attached, may additionally represent a heterocyclic group having a maximum of 10 carbon atoms; M is any metal which forms dithiocarbamates, which metal is commonly any of the transition metals; and x and y are integers which depend in well known manner upon the valence of the metal, and are generally from 1 to about 4.

(b) N-disubstituted benzothiazole-2-sulfenamides having the general formula:

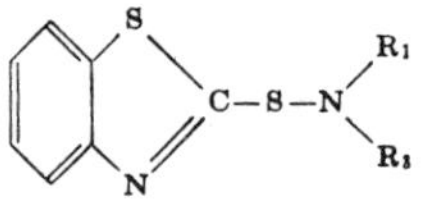

where R_1 is defined in (a) above and R_3 is either R_1 or hydrogen.

(c) N,N-tetrasubstituted thiuram mono- and oligosulfides having the general formula $R_1R_2N{\cdot}C(S){\cdot}S_zC(S){\cdot}NR_1R_2$, where each of R_1 and R_2, which may be the same or different, has the meaning given in (a) above and z is an integer not exceeding 5.

The polymerization accelerator generally is used in a volatile organic solvent, and applied as a solution to the substrate. Any solvent capable of dissolving the polymerization accelerator and which will evaporate from a thin film at standard conditions of temperature and pressure within about 1 minute can be used. Examples of suitable solvents are halogenated hydrocarbons, such as methylene chloride, trichloroethylene, trichloroethane and methyl chloroform. Other solvents such as alcohols, hydrocarbons and laquer type solvents can also be used. The polymerization accelerator may be used within broad ranges in the solvent, e.g., 0.01 to 20% by weight.

Example 1: Two pieces of oak wood were bonded together with an adhesive composition comprising methyl 2-cyanoacrylate containing 0.5% propane sultone and 0.01% hydroquinone as stabilizers. The pieces were clamped together for 15 minutes, after which it was found that no effective bond had been formed.

The above procedure was then repeated with the following variation according to this process. Both of the wood surfaces to be bonded were moistened with a 0.1% by weight solution of bismuth dimethyldithiocarbamate in 1,1,1-trichloroethane/chloroform (a 50/50 mixture by volume). The solvent was allowed to evaporate. The adhesive was applied to one of the surfaces, the pieces were joined and a firm bond resulted within 1 minute.

The first part of this example illustrates a disadvantage of adhesive compositions of the prior art. The second part illustrates the effectiveness of a solution of bismuth dimethyldithiocarbamate in promoting the bonding of wood, when used according to the process.

Example 2: Two pieces of oak wood were moistened with a 0.1% by weight solution of N-oxydiethylene benzothiazole-2-sulfenamide in 1,1,1-trichloroethane. The solvent was allowed to evaporate, and the adhesive of Example 1 was applied to one of the surfaces. The pieces were then joined together and a firm bond became established within 1 minute.

Wood Fibers Impregnated with Complex of Styrene and Maleic Anhydride

N.G. Gaylord; U.S. Patent 3,894,975; July 15, 1975; assigned to Champion International Corporation describes the preparation of polymer composites with porous cellulosic materials, such as wood. The process for forming such a composite involves impregnating the wood with a polymerizable complex of a monomer combination such as a complex of maleic anhydride and styrene. The polymerization is uncatalyzed, i.e., it is conducted without employing conventional means for inducing polymerization of monomers in wood, such as high energy radiation or the thermal decomposition of a chemical free radical precursor.

By the practice of this process, a wood-polymer composite may be prepared in which the original weight of the wood is increased by from 10 to 150% due to the incorporation therein of the polymer. Such a composite may be regarded as a plastic reinforced with the naturally oriented strands of the wood, the capillaries and interstices of the wood being filled with polymer. Thus, the wood is the continuous phase and the polymer is present as an ordered dispersion in intimate contact with the structure of the wood. Moreover, when the complex from which the polymer is formed contains a carboxyl group or an anhydride group which is reactive with hydroxyl groups on the cellulose substrate, chemical bonds are formed between the polymer and the wood, thus increasing the homogeneity of the system.

The polymer-wood composites of this process are dimensionally stable, water repellant, resistant to solvents and have improved mechanical properties, as compared to untreated wood, such as hardness, toughness, impact strength, tensile strength, flexural strength and compression strength. The following examples illustrate the process. Unless otherwise stated, the styrene used in these examples contains about 15 ppm of t-butyl pyrocatechol inhibitor.

Example 1: A mixture of maleic anhydride (9.8 grams; 0.1 mol) and styrene (10.4 grams; 0.1 mol) are heated to 83°C to form a homogeneous solution. Upon cooling, crystallization of the solution starts at about 35°C and the solution is transformed into a solid mass at room temperature. When reheated in a water bath, the solid crystals melt at 35°C. The polymerizable complex solution is easy to handle at temperatures between 40° and 50°C.

Example 2: The process of Example 1 is repeated using uninhibited styrene and adding 0.04 gram of t-butyl catechol. On cooling to 25°C, a stable solid mass is obtained which remains stable for several days. The mass has a melting point of 35°C. Upon the addition of 0.4 gram of benzoyl peroxide, the solution remains stable at 50°C for 6 hours.

Example 3: Maleic anhydride (93 grams; 1 mol), styrene containing no inhibitor (104 grams; 1 mol) and t-butyl catechol (0.4 gram) are heated together at 70°C until a clear solution is obtained. The resultant polymerizable complex is cooled to 35°C and benzoyl peroxide (0.4 gram) is added.

Pieces of pine wood which have been dried at 105°C to constant weight are immersed in the polymerizable complex and the immersed wood is subjected to a vacuum which is gradually increased to about 8 to 10 mm of Hg for about 15 minutes. While the wood is still immersed in the polymerizable complex, it is transferred to an autoclave heated to a temperature of 50°C. The autoclave is then flushed with nitrogen and pressurized with nitrogen to about 1,000 to 1,200 psi. This pressure is maintained for about 15 minutes and then slowly released.

The impregnated wood pieces are removed from the polymerizable complex, dried with a paper towel and weighed. The wood is then cured by placing the pieces in contact with plates heated to 100°C. Polymerization starts from the heated surface and propagates to the interior of the wood. The apparent density of the wood increases from an initial value of 0.5 gram per cubic centimeter to 1.0 gram per cubic centimeter. The Shore D hardness of the treated wood is 63 to 84. The wood increases in weight by 100%.

In the following examples, the weight increase, the density of the wood before and after impregnation and curing and the hardness of the final product are set forth in the table at the end of the examples.

Example 4: Pine wood is impregnated as described in Example 3 with the composition of that example. The impregnated wood is wrapped in aluminum foil and placed in an oven at 70°C to cure the polymerizable complex.

Example 5: Example 4 is repeated except that 0.1 mol of the maleic anhydride is replaced with 0.1 mol (23.3 grams) of di-n-butyl maleate and the benzoyl peroxide is omitted. The impregnated wood is cured in an oven at 80°C. The polymerization reaction starts after an induction period of about 40 minutes. The temperature in the wood reaches a maximum of 190°C.

Example 6: A polymerizable complex is prepared comprising uninhibited styrene (104.0 grams, 1 mol), maleic anhydride (88.0 grams, 0.9 mol), di-n-butyl maleate (23.2 grams, 0.1 mol) and t-butyl catechol (0.1 gram). Fir plywood, ¾ inch thick containing 5 layers, is placed in the complex so that only about the bottom layer is immersed. Impregnation is carried out applying vacuum but no pressure. The plywood is then heated in an oven at 120°C. The whole piece is impregnated through with an effectiveness of about 75%.

Example 7: A polymerizable complex is prepared by heating to about 70°C uninhibited styrene (93.6 grams, 0.9 mol), vinyl isobutyl ether (10.0 grams, 0.1 mol), maleic anhydride (98.0 grams, 1 mol) and hydroquinone (0.01 gram). The complex is cooled to about 40°C and t-butyl perbenzoate (0.05 gram) is added. Pine wood is impregnated with the complex as described in Example 3 and the wood is cured in an oven at 80°C.

Example 8: A composition is prepared by adding uninhibited styrene (93.6 grams, 0.9 mol), vinyl acetate (8.6 grams, 0.1 mol) and hydroquinone (0.01 gram) to molten maleic anhydride (98.0 grams, 1 mol). Pine wood is impregnated with the complex as described in Example 3 and polymerization is initiated by heating the wood in an oven for 70 minutes at 60°C. A maximum temperature of 202°C is reached after 2.5 hours. The copolymer with which the wood is impregnated is soluble in acetone and methyl ethyl ketone. Therefore, the treated wood may be surface polished by rubbing with these solvents.

Example 9: A polymerizable complex is prepared by heating uninhibited styrene (83.2 grams, 0.8 mol), vinyl acetate (17.2 grams, 0.2 mol) and maleic anhydride (98.0 grams, 1 mol). Pine wood is impregnated with the complex as described in Example 3 and polymerization is initiated by heating the impregnated wood in an oven at 70°C for about 15 minutes. The treated wood is much more resistant to water absorption than is untreated wood.

Example 10: The polymerizable complex described in Example 9 is used to impregnate pine wood. The wood is kept at 25°C for 48 hours and then is heated in an oven at 50°C for 5 hours and finally at 100°C for 2 hours. The peak temperature reached in the wood is 120°C. This example illustrates that by multistage curing process, the wood may be cured at a maximum peak internal temperature of 120°C.

EXAMPLE	APPARENT DENSITY (g./cc.) Before	After	WEIGHT INCREASE	SHORE D Hardness
4	0.51	1.05	98%	73–84
5	0.54	1.14	109%	80–87
6	0.48	0.85	71%	74–80
7	0.54	1.04	93%	70–76
8	0.54	1.14	111%	80–90
9	0.54	1.13	110%	85–88
10	0.60	1.17	133%	82–84

STARCH ADHESIVES FOR PAPER BONDING

Grinding Procedure for Starch-Clay Mixtures

H.H. Morris, R.J. Drexel, Jr. and K.L. Turner; U.S. Patent 3,804,653; April 16, 1974; assigned to Freeport Minerals Company describe a process for preparing a polymeric adhesive material. The process comprises subjecting an aqueous mixture of starch or protein, or at least two insoluble substances, and grinding media to an agitation in the presence of a pigment, if desired, and then removing the grinding media. The treated starch or protein may be further reacted with a polymerizable monomer or preformed polymer to form the polymeric material. A superior adhesive, which finds particular application in coating base materials, such as paper, is obtained. Stable emulsions formed are particularly useful in pharmaceutical and food applications.

Example 1: (a) – A slurry was prepared which contained 600 grams clay, 99 grams of Australian Casein, 5.94 grams sodium hydroxide, 1.0 gram tetrasodium pyrophosphate, 1,000 grams 1/10 by 1/10 inch diameter nylon cylinders and enough water to give 35% solids (based on clay and casein). The clay used was a coarse filler grade of clay having about 25% of its particles below 2 microns in equivalent spherical diameter. Such a clay is generally considered unsuitable for coating compositions.

The slurry was placed in a 1-gallon plastic container and agitated for 3 hours by means of a stirrer operating at about 1,600 rpm. After this treatment the nylon pellets were removed by screening the mixture through an 80 mesh screen. The coating color so obtained was used to coat paper sheets. The following table (Sample B) shows the results obtained.

	Calendered Sheet Properties	
Sample	Gloss**	G. E.*
A	35.5	69.8
B	67	74.3
C	57	74.0
D	71	75.0

*"G.E." brightness values are determined according to standard TAPPI (Technical Association of the Pulp and Paper Industry) Method T646 m-54, as reported in the Testing Methods-Recommended Practices-Specifications of the Technical Association of the Pulp and Paper Industry.

**Gloss values are determined according to TAPPI specification T480 ts-65, entitled "Specular Gloss of Paper and Paperboard at 75°."

Sheets were coated on one side by applying, through the use of a blade coater, 13 pounds (air dry basis) of coating per 3,300 square feet on a 54-pound-per-ream paper raw stock. The coated sheets were air dried at constant temperature and constant humidity (73°F, 50% relative humidity) before being tested and calendered. The conditioned sheets were calendered 8 nips at 1,600 pounds per lineal inch (pli) pressure through the use of a three roll (steel-cotton-steel) calender stack.

(b) – For comparative purposes a conventional coating was prepared by heating a 16% solids slurry of casein (17 parts), sodium hydroxide (1 part), and ammonium hydroxide (1.8 parts, added as a 28% solution) at 70°C for 20 minutes, until the casein had been solubilized. The casein solution was then mixed carefully with a previously prepared, dispersed slurry of the coarse filler clay (100 parts) used in (a). Final solids of the coating colors was 35%. The resulting coating color was then homogenized, carefully screened, and used to coat paper sheets as described in (a). The results obtained are shown in the preceding table (Sample A).

(c) – Another paper coating color was prepared in the conventional manner, as just described, but a commercial paper coating clay (Columbia Coating Clay, Freeport Kaolin Company, a commercial paper coating clay having 80% of its particles finer than 2 microns) was used in place of the coarse filler clay. Data obtained on paper coated with this coating color are listed in the preceding table also (Sample C).

(d) – A coating color was prepared as described in (a) except that a No. 2 grade coating clay (Columbia Coating Clay, Freeport Kaolin Company) was substituted for the coarse filler clay. The results obtained from use of this coating color are shown in the preceding table (Sample D). Sample C in the table shows that results obtained when a coating color of the same composition but prepared by conventional means, is used.

It can be seen from a comparison of the above data for samples A and B that the process for preparing coating colors gave a gloss improvement of better than 30 points and a 4.5 point improvement in GE brightness over a coating color prepared with a coarse filler by conventional means. Furthermore, a comparison of the above data for samples C and D shows that the coating color prepared by this process gave results superior to a coating color prepared in the conventional manner using a standard No. 2 grade of coating clay (Sample C).

By use of this process, a coating color prepared using a cheap filler clay has been found to give results better than those obtained when a more expensive coating clay prepared in the conventional manner is used. This process does not require the expense of separate solubilization of the adhesive and the cost of separate mixing and milling to blend properly the adhesive and pigment.

Example 2: In order to demonstrate the effect of this process on the adhesive portion of a coating color, the following experiment was performed. An 18% solids casein slurry was prepared by suspending 285 grams of Australian Casein and 17.1 grams of sodium hydroxide in 1,300 grams of water. Nylon pellets (755 grams, 1/10 by 1/10 inch diameter) were then added and the mixture was agitated for 3 hours through the use of a stirrer operating at about 1,600 rpm. The nylon pellets were then removed by screening and the casein dispersion was mixed with a previously prepared, delaminated clay slurry. The coating color so obtained was used to coat paper sheets as described in Example 1(a) and the data determined on these sheets are shown in the following table (Sample B).

	Calendered Sheet Properties	
Sample	Gloss	G.E.
A	60	77.8
B	71	78.5

A control coating color was prepared by conventional means, using the same ratio of casein and clay starting materials, and the results obtained on paper coated with the composition are shown in the preceding table (Sample A). The delaminated clay used in both preparations is a commercial product (NUCLAY, Freeport Kaolin Company). The data indicate that the treatment of the adhesive gives a coating color with an 11 point gloss improvement and 1 point improvement in sheet brightness over the best values obtainable by conventional means. In addition, a reduction in coating color viscosity was obtained. This is an important factor to many papermakers.

Example 3: (a) – A pearl starch adhesive was converted by mixing 1,500 grams styrene-divinyl benzene copolymer beads, 1,200 grams water, 300 grams pearl starch and 3 grams sodium carbonate with rapid agitation for 2 hours. The final temperature of the mixture was 88°C. The beads were removed from the converted starch solution by screening.

(b) – A second sample of pearl starch adhesive, used as a control for the above sample, was enzyme converted with 0.02% of an enzyme (Vanzyme No. 31, R.T. Vanderbilt Co., Inc.) at 20% solids. The formula for this mixture was 1,200.0 grams water, 300.0 grams pearl starch, 3.0 grams sodium carbonate and 0.06 gram Vanzyme No. 30.

The above mixture was held at 170°F for 20 minutes, then at 200°F for 10 minutes to deactivate the enzyme. The above described adhesives were at the same percent solids and at approximately the same consistency. The adhesives were used to bond wood to wood samples together and allowed to cure or dry under constant pressure and temperature for 12 hours prior to testing. Wood-adhesive-wood samples from both of the above described starch adhesives were tested on a Dillon Model L.W. Tester designed for bond or strength tests and the results are shown in the following table.

		lbs./in.^{2} required to pull blocks apart
a.	styrene-divinyl benzene copolymer treated starch	52.7
b.	enzyme converted starch	49.3

Phenol, Alkaline Dispersed Starch and an Oxazolidine

J.E. Voigt and E.M. Bovier; U.S. Patent 3,802,897; April 9, 1974; assigned to Anheuser-Busch Incorporated have found that an adhesive composition comprising a combination of a phenol, alkaline dispersed starch, and an oxazolidine represented by the formula:

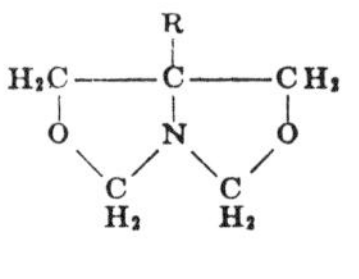

where R is an ethyl or hydroxymethyl group, possesses water resistant characteristics. When R is an ethyl group, the oxazolidine is 5-ethyl-1-aza-3,7-dioxabicyclo[3.3.0] octane, which will be referred to as oxazolidine E; when R is a hydroxymethyl group, the oxazolidine is 5-hydroxymethyl-1-aza-3,7-dioxabicyclo-[3.3.0] octane, which will be referred to as oxazolidine T.

This adhesive is made similarly to the phenol-aldehyde alkaline curing starch base adhesives. The critical difference is the substitution of an oxazolidine for an aldehyde. The concentration of the oxazolidine is related by way of a mol ratio to the phenolic compound, which should be greater than about 0.5 to 1 and preferably 1 to 1. The amount of oxazolidine may be as much as 3 to 1.

The bicyclic oxazolidines used in the adhesives may be prepared according to the method of M. Senkus, *J. Am. Chem. Soc.,* 67, pages 1515 to 1519 (1945) or that of W.B. Johnston, U.S. Patent 2,448,890. Oxazolidine E is preferred, being more reactive than oxazolidine T.

There are two procedures to be used in practicing the process, a two-batch preparation and a one-batch preparation, each of which depends upon the needs of the manufacturer. Example 1 illustrates the two-batch procedure and Example 2 illustrates the one-batch procedure.

A preferred procedure is to mix a part of the starch and the phenolic compound in water under agitation and heat to a temperature of 125° to 200°F, preferably 140° to 160°F, for about 10 minutes. At this time the swelling of the starch begins, and alkali is added to give the slurry a pH of about 10 to 11.5. Complete dispersion of the starch has not yet occurred and is not desired at this time.

More water, the remainder of the starch, and the oxazolidine is added. The temperature is lowered to 90° to 120°F and within 1 to 2 hours an adhesive composition will be obtained which is essentially odorless, has a stabilized viscosity of 15 to 120 sec measured by the Brass Cup method, and a satisfactory pot life of 24 to 48 hours or longer. The preferred stabilized viscosity is 20 to 60 sec. The product can be used in the manufacture of corrugated or solid paperboard and possesses a waterproof bond on drying.

The final product involves a polymerization reaction involving starch, the phenolic compound and the oxazolidine compound. This results in a number of complex polymers with varying molecular weights with no single resultant formula. The following examples illustrate the process; all parts of ingredients are by weight.

Example 1: A mixture of 100 parts pearl cornstarch and 45 parts of resorcinol was slurried in 600 parts of water. 26 parts of a 25% NaOH solution was added to give a pH of 10. The slurry mixture was heated to 150°F and maintained at that temperature for 10 minutes to paste the starch. It was cooled by the addition of 625 parts water and stirred for 10 minutes.

In a second batch, 500 parts of pearl cornstarch and 60 parts of oxazolidine E were slurried in 1,045 parts of water. The primary mixture was slowly mixed into the secondary. The resultant mixture was kept at 105° to 115°F for 1 to 2 hours and was ready for use on corrugated board and gave a water resistant bond comparable to the phenol-aldehyde resin adhesives. It is not necessary to wait 1 to 2 hours before using; however, better water resistance is obtained by doing so. This gives the starch, the phenolic compound and the oxazolidine a chance to polymerize.

Another adhesive was prepared according to the procedure used for this example with the exception that 25 parts of paraformaldehyde were added instead of 60 parts of oxazolidine E. This adhesive, when used to glue corrugated board 1 hour after preparation, gave excellent water resistant bonds. However, it had a very strong, irritating odor and its viscosity significantly increased with time. In fact, after 24 hours it was so viscous that it could not be used to glue corrugated board.

On the other hand, the adhesive prepared with oxazolidine E, as set forth in this example, was essentially odorless and after 24 hours the viscosity remained unchanged. Corrugated board glued with the oxazolidine E adhesive 24 hours after preparation still gave excellent water resistant bonds. This illustrates that the adhesive prepared according to this process has improved viscosity stability and improved pot life over a similarly prepared adhesive made with formaldehyde.

In this procedure, the unpasted starch of the second batch is held by the pasted carrier starch of the first batch. The hot rolls of the corrugating machine paste the unpasted starch in the usual manner and firm bonds are produced. But this formulation is also effective in cold laminating applications where the laminating rolls are not heated. The presence of the unpasted starch in the formulation provides for effective waterproofing.

This is significant since only a limited amount of pasted starch can be used due to its high viscosity and a much larger amount of unpasted starch may be used without increasing the viscosity of the mixture. This is an unexpected result in the improvement in waterproofness, for the unpasted starch would normally be expected to act as a diluent and weaken the adhesive bond.

Example 2: A mixture of 40.7 parts of pearl starch and 8.7 parts of resorcinol were slurried in 275 parts of water with agitation. The slurry mixture was heated to 140°F for 10 minutes. 21.2 parts of 25% NaOH solution was added and mixed for 10 minutes. Approximately 284 parts of water was added followed by 164 parts of pearl cornstarch. Finally 11.6 parts of oxazolidine E was added. This mixture was kept at a temperature of 105° to 115°F for 1 to 2 hours, and upon drying after application to corrugated board in the usual manner, gave a water resistant bond comparable to the phenol-aldehyde resin adhesives.

The second portion of water is added to lower the temperature of the pasted starch so that when the unpasted starch is incorporated into the mixture it does not gelatinize.

Water-Resistant, Short Tack Starch Adhesive

G.F. Bramel; U.S. Patent 3,844,807; October 29, 1974; assigned to A.E. Staley Manufacturing Company describes a preblend adhesive system for tailoring adhesive systems to particular paper products bonding applications requirements. The preblend system comprises two premix blends of acid hydrolyzed starch, hydroxyethylated starch, buffering salts and a specific type starch modifier, more particularly polyoxyethylene laurate.

The salts provide a desirable pH for good storage stability of the resulting adhesive paste after the above preblends and mixtures thereof are slurried with water and resin added for water resistance and the mixture is cooked to form a short, water-resistant liquid adhesive paste, particularly useful in the manufacture of multiwall envelope bags having adhesive bonded bottom and side seams.

In the adhesive systems of this process, only about one-half the amount of resin previously used is required to obtain a water-resistant bond, and the water resistance develops in a considerably shorter time after fabricating the adhesive bonded product, thereby shortening the time period required for aging newly bonded products prior to using them in an environment requiring full water-resistant properties. The dry preblends are formulated as follows.

	Preblend, percent by weight (as is)	
Ingredient	**1**	**2**
Hydroxyethylated cornstarch (2% reacted with ethylene oxide)	4.7	17
Acid hydrolyzed cornstarch	89.3*	77**
Granular zinc chloride	0.6	0.4
Predispersed HT clay No. 2	3	3
Potassium chloride	2	1.4
Polyoxyethylene laurate	0.4	1
Powdered aluminum sulfate or Na_2CO_3	Sufficient for pH adjustment	

*Thin, 100 to 200 cp, Brookfield viscosity, 100°F, 100 rpm, No. 1 spindle using 80 grams starch, 320 grams water and cooked

**Thick, 11,500 to 12,500 cp, Brookfield viscosity, 15% solids, 20 rpm, No. 5 spindle

The above described preblend No. 1 has been found most useful for bonding the side seams of envelope type multiwall paper bags. It has the following advantages:

- good tack, quickly forms a tight, water-resistant bond;
- reduces penetration of bag plies, thus eliminating inner bonding;
- produces excellent bonds on glueable polyethylene coated papers and other specialty papers;
- add only resin and water to make pastes, no other ingredients, no pH adjustment required;
- no pH buffering;
- reduces resin usage by 50% (using UF);
- excellent viscosity stability during storage, can be held 2 to 3 weeks before using;
- performs well in machine operations, remaining fluid in pans during prolonged runs; and
- is acceptable as a food packaging adhesive under FDA regulations.

When preblend No. 1 is used, the amount of urea-formaldehyde resin typically used in a seam paste can be reduced by approximately 50% without sacrificing desired water-resistant properties. This reduced ingredient requirement is a significant cost savings factor. Preblends No. 1's pH buffering system eliminates the need to manually adjust the pH of the paste after make-up, even though the liquid paste may be stored 2 weeks or more before using.

A typical adhesive formulation using preblend No. 1 is 100 pounds preblend No. 1, 27 gallons water and 7 pounds urea-formaldehyde resin (65% solids). The above is mixed at room temperature, then heated to 190°F for 20 minutes. It is then cooled to application temperature. The prepared paste has the following properties: color, off-white; solids, 25 to 27%; pH, 5.5; and Brookfield viscosity (80°F, No. 3 spindle, 20 rpm), 400 to 800 cp.

Preblend No. 2 has a much thicker acid hydrolyzed starch ingredient (subjected to less acid hydrolysis), and is specifically useful for bottom seals of multiwall paper bags. The amount of emulsifier-surfactant present (polyoxyethylene laurate) is more than double that present in preblend No. 1, and the balance between the hydroxyethylated starch/acid hydrolyzed starch has shifted to about three times the amount of hydroxyethylated starch. The zinc chloride is less, as is the potassium chloride. Both preblends are designed to produce stable liquid resin-containing paste adhesives at acid pH levels (3 to 5), which are still useful after 2 to 3 weeks of storage in liquid form.

As with preblend No. 1, preblend No. 2 is blended with water and a resin such as urea-formaldehyde (to give water-resistant bonds). The higher viscosity of the resulting adhesive paste is more suitable to the applicator technique used for bag bottoms. The amounts of water and urea-formaldehyde resin used to make the bottom seam adhesive paste are substantially as set forth above for preblend No. 1.

The urea-formaldehyde cannot normally be stored with the preblends because it tends to crosslink the starch too soon. Other earlier adhesives using this type of resin normally used about twice the amount, and this may have been the reason they could only be stored for very short periods in the liquid form. The preblends of this process are believed to be stable because of the combined action of the emulsifier-surfactant (polyoxyethylene laurate), zinc chloride and potassium chloride. The presence of the minor amount of ethylated starch improves the rheology tack and storage properties. The clay reduces penetration of the adhesive.

The above preblends 1 and 2 can be blended in any desired proportions to give an adhesive blend anywhere in between the two preblends in any desired intermediate viscosity and at any desired intermediate solids level from about 15 to 30% solids. The two preblends are compatible at any combination, and give completely predictable properties with a long liquid storage shelf life potential. Only about one-half the amount of the normally required urea-formaldehyde resins is needed to impart good water resistance to preblends 1 and 2.

The nonsoap emulsifier surfactant seems particularly important in the realization of the lower resin requirement and the long liquid storage stability. Apparently resin/starch reactivity is increased by the polyoxyethylene laurate, but not while in liquid form. Another possible emulsifier-surfactant which might replace the presently used one is the reaction product of tridecyl alcohol with 4 mols of ethylene oxide. This product is believed to have generally the same properties which make the polyoxyethylene laurate effective for this purpose.

Continuous Production of Starch-Based Adhesive

R.J. Hampton and B.P. Bonnell; U.S. Patent 3,836,376; September 17, 1974;

assigned to The Ogilvie Flour Mills Company, Limited, Canada describe a process for preparing and continuously delivering a finished starch-based adhesive composition, containing a gelatinized starch portion and a raw starch protein, to a site of use, such as a corrugating machine. The process includes continuously delivering formula components, including raw starch slurry, water and caustic, to an adjustable metering device by which means the volume of each component is precisely controlled to attain relative proportions of such components corresponding to a desired formula.

Controlled volumes of raw starch slurry, water and caustic are delivered to a first mixing zone where the ingredients are heated to produce a gelatinized starch mass of relatively stable viscosity. The gelatinized mass is passed to a second mixing zone. A further volume of water is simultaneously forwarded to the second mixing zone. The gelatinized starch and water in the second mixing zone is blended to form a gelatinized starch medium. The gelatinized starch medium is passed from the second mixing to a third mixing zone. Simultaneously, a further volume of raw starch slurry is forwarded to the third mixing zone.

The raw starch slurry is commingled with the gelatinized medium so as to intimately combine the two starch fractions; the gelatinized starch becomes a carrier to hold raw starch slurry in suspension thereby forming the finished starch-based adhesive composition. The finished starch-based adhesive is delivered to a container for temporary storage. When necessary there is further adjusting of the volumes through the metering device of one or more of the adhesive components relative to the other volumes thereby altering prior to use the relative proportions in the finished starch-based adhesive composition, to produce a correct formula. The finished starch-based adhesive composition is passed from the container to the site of use.

OTHER WATER-SOLUBLE ADHESIVES

Solid Water-Soluble Product Containing Stearoyl Isethionate Ester

A process described by *D.D. Hutson and C.A. Rader; U.S. Patent 3,838,084; September 24, 1974; assigned to The Gillette Company* relates to a coherent, self-sustaining adhesive composition which will resist slumping and change in form at temperatures up to as high as about 50°C and being adapted to be readily spread upon paper-like surfaces to form an adhesive film.

The adhesive composition is a skeleton structure of interlocked elongated crystals of a matrix builder which is virtually water-insoluble, at room temperature, and is the salt of a compound having the formula $RCOO{-}R'{-}SO_3H$, where R is aliphatic having from 7 to 35 carbon atoms and R' is lower alkylene, the skeleton structure forming a matrix for an aqueous medium containing a water-soluble synthetic resin adhesive. The method for producing the adhesive composition described above includes the step of forming a mixture of an aqueous solution of a water-soluble adhesive and a matrix builder; heating the mixture to a temperature between that at which a significant amount of the matrix builder dissolves and the boiling point of the compositions; slowly mixing the composition to form a homogeneous slurry or paste without formation of a significant amount of air bubbles; cooling the composition to about a temperature at which the matrix builder is substantially insoluble in water; and curing the matrix builder at between the insolubility temperature and a temperature 15°C lower than the insolubility temperature.

A preferred composition of the process would contain 13 to 20% of matrix builder, 20 to 35% of water-soluble adhesive, 10 to 20% of humectant and/or plasticizer, and 20 or 25% to 55 or 60% water, all percentages being by weight based on the entire weight of the composition. It is generally more preferred if the amount of matrix builder varies from 14 to 16% by weight, based on the weight of the entire composition. In the following examples all parts and percentages are by weight.

Example 1: Fifteen parts by weight of sodium stearoyl isethionate, 25 parts by weight of polyvinylpyrrolidone (average molecular weight about 500,000), 15 parts by weight of glycerol, 0.1 part by weight of methyl Parasept, and 45 parts by weight of water were mixed together at room temperature. After all the components were wetted, the mixture was heated to between 93° and 97°C. Mixing was continued for approximately 2 to 3 hours at a slow rate so as to avoid formation of air bubbles. After that time the resulting composition was placed in suitable dispensers and maintained in an oven for about 20 hours at 50° to 60°C at which time the composition was allowed to cool to room temperature. The resulting composition met all the criteria of an adequate solid composition given above.

Example 2: Forty parts by weight of water, 32 parts by weight of polyvinylpyrrolidone, 16 parts by weight of sodium stearoyl isethionate, and 12 parts by weight of glycerol were mixed together and then heated with mixing until the boiling point was reached and the mixture refluxed, with stirring, until all of the air bubbles disappeared. Thereafter, the composition was poured into dispensers (each dispenser held about 8 grams) and held at about 60°C for 15 hours.

Example 3: 49.9 parts by weight of water, 24.0 parts by weight of polyvinylpyrrolidone, 16.0 parts by weight of sodium palmitoyl isethionate, 14.0 parts by weight of glycerol and 0.1 part by weight of methyl Parasept were mixed together and then heated with mixing until the boiling point was reached and the mixture refluxed, with stirring, until all of the air bubbles disappeared. Thereafter the composition was poured into dispensers and held at about 60°C for 23 hours.

Grafted Water-Soluble Cellulose Ether

S. Lindenfors and N.J. Lundberg; U.S. Patent 3,882,057; May 6, 1975; assigned to Mo och Domsjo, Sweden have found that adhesive compositions which contain as a binder a special type of graft polymer based on water-soluble cellulose ethers have good application properties, a high degree of adhesion and a good initial tack. The adhesive composition is mainly characterized by the fact that the binder comprises a graft polymer of (a) an unsaturated carboxylic acid of the general formula:

$$R'{-}CH{=}\underset{\displaystyle R''}{\underset{|}{C}}{-}COOH$$

where R' and R", independently of each other, are hydrogen or an alkyl group having 1 to 2 carbon atoms; (b) an alkyl acrylate having 1 to 8 carbon atoms in the alkyl portion; and (c) vinyl acetate on a water-soluble cellulose ether. Tests have shown that all three monomer groups are necessary in order for the adhesive composition to obtain satisfactory properties. Thus, the alkyl acrylate contributes primarily to an excellent initial tack, the vinyl acetate to good application properties, while the unsaturated carboxylic acid provides a high degree of adhesion.

The adhesive composition can be used to bond a large number of different materials, such as wood, paper, different types of textiles, leather, rubber, plastics, and metals, together with different aggregations and laminates of these materials. The adhesive composition is primarily intended for gluing different types of wall coverings in dry surroundings to supporting surfaces, for example, wallpaper, wall fabric, etc. Other conventional uses to which the composition can be put include the bonding of paper to form, for instance, paper bags, plastic foil, regenerated cellulose or different combinations of these materials.

Example 1: To a glass synthesis vessel there were charged under a nitrogen atmosphere 42 grams of ethyl hydroxyethyl cellulose having a viscosity of 300 cp determined in a 2% aqueous solution at 20°C according to Brookfield, 21 grams of cellulose powder, 3 grams of sodium bicarbonate, 20 grams of an emulsifier comprising mainly addition products between nonylphenol and ethylene oxide, and 760 grams of water. The total charge was stirred for 5 minutes until a homogeneous mix was obtained, whereafter the mix was allowed to stand until all cellulose ethers had dissolved.

The mix was then heated to 35°C and an initiator system consisting of 1.13 grams of ammonium persulfate and 0.90 gram of sodium pyrosulfite dissolved in a small amount of water was added. Subsequent to agitating the mix until the initiator system was homogeneously dispersed, there was successively added a monomer mixture consisting of 60 grams of vinyl acetate, 67.4 grams of ethyl acrylate and 21 grams of acrylic acid over 25 minutes. The reaction temperature was maintained at about 55°C and the reaction time was 2 hours.

A buffer system consisting of 5.0 grams of sodium acetate and 2.6 grams of acetic acid was added at the end of the reaction time. To enable comparisons to be made, other compositions were prepared which did not contain one of the monomers vinyl acetate, ethyl acrylate and acrylic acid. With the comparison compositions, the following ingredients were used.

Adhesive Composition Not Containing Vinyl Acetate	Grams
Ethyl hydroxyethyl cellulose	52.5
Sodium bicarbonate	7.5
Emulsifier	20.0
Ammonium persulfate	1.13
Sodium pyrosulfite	0.9
Ethyl acrylate	60.0
Acrylic acid	90.0
Sodium acetate	5.0
Acetic acid	2.6
Water	740.0

Adhesive Composition Not Containing Ethyl Acrylate	Grams
Ethyl hydroxyethyl cellulose	75.0
Cellulose powder	15.0
Sodium bicarbonate	3.0
Emulsifier	20.0
Ammonium persulfate	2.25
Sodium pyrosulfite	1.8
Vinyl acetate	75.0
Acrylic acid	75.0
Sodium acetate	5.0
Acetic acid	2.6
Water	580.0

Adhesive Composition Not Containing Acrylic Acid	Grams
Ethyl hydroxyethyl cellulose	52.5
Cellulose powder	21.0
Sodium bicarbonate	2.0
Emulsifier	20.0
Ammonium persulfate	1.13
Sodium pyrosulfite	0.90
Vinyl acetate	75.0
Ethyl acrylate	75.0
Sodium acetate	5.0
Acetic acid	2.6
Water	740.0

The adhesive compositions obtained were tested for adhesion, both with and without pregluing, application properties, initial tack and consistency. The adhesion was determined by gluing a 10 centimeter wide gauze strip to a fiber plate substrate. The adhesive was applied to a thickness of 200μ. The load, expressed in grams, required to remove the gauze strip from the substrate gave the adhesion. The greater the load required, the greater the adhesion. The term pregluing as used here implies that the fiber plate was first coated with adhesive, which was allowed to dry, and then furnished with a further quantity of adhesive before the gauze strip was applied and the degree of adhesion determined.

The initial tack was determined by applying a film of adhesive 700μ thick to a fiber plate. After 30 seconds a tack plate having a tight jute weave 5 centimeters by 5 centimeters was lowered against the adhesive film and allowed to lie for 10 seconds with its own weight of 264 grams as the load acting thereon. The surface of the adhesive was then subjected to a tensile load of 364 grams. The time taken to break the adhesion was noted. The adhesions which held for 2 minutes were loaded with a further 100 grams and the time taken to break the adhesion at this higher load was noted. Tests carried out on the adhesive compositions gave the following results.

Craft polymer monomers	Adhesion without gram/10 cm	Pregluing with gram/10 cm	Initial "tack"	Application properties	Consistency
vinyl acetate ethyl acrylate acryl acid	2800	3200	2 min. + 12 sec.	good	medium stiff
ethyl acrylate acrylic acid	3200	3200	2 min. + 15 sec.	poor	stiff tough
vinyl acetate ethyl acrylate	1400	2300	1 min. 50 sec.	good	medium stiff
vinyl acetate acryl acid	3500	4300	54 sec.	fairly good	medium stiff

As will be evident from the results, the adhesive composition according to the process gave good results throughout with regard to adhesion, initial tack, application properties and consistency. On the other hand the results obtained with the compositions based on only two of the three monomer groups were unsatisfactory. Thus, the absence of ethyl acrylate from the composition resulted in a much poorer initial tack, while the absence of acrylic acid gave poorer adhesion and a relatively poor initial tack. The absence of vinyl acetate greatly impaired the application properties and the adhesive obtained a stiff and tough consistency, which made it impossible to apply the adhesive by roller.

Example 2: In the same manner as that of Example 1, there was produced an adhesive composition based on carboxymethylcellulose having a viscosity of 330 cp in a 2% aqueous solution at 20°C, vinyl acetate, ethyl acrylate and acrylic acid. The following ingredients were included.

	Grams
Carboxymethylcellulose	35.0
Cellulose powder	28.5
Sodium bicarbonate	4.0
Emulsifier	20.0
Ammonium persulfate	1.13
Sodium pyrosulfite	0.90
Vinyl acetate	40.0
Ethyl acrylate	60.0
Acrylic acid	30.0
Sodium acetate	5.0
Acetic acid	2.6
Water	740.0

The adhesive composition obtained was tested with respect to adhesion, initial tack and application properties and was found to meet the requirements placed on an adhesive composition. The following results were obtained. Adhesion not preglued was 2,400 grams per 10 centimeters; adhesion preglued was 2,900 grams per 10 centimeters; initial tack time was 2 minutes and 7 seconds; application properties were good and consistency was somewhat thin.

Pullulan Ester

According to a process described by *H. Hijiya and M. Shiosaka; U.S. Patent 3,873,333; March 25, 1975; assigned to Hayashibara Biochemical Laboratories, Incorporated, Japan* adhesives or pastes are prepared by dissolving or dispersing uniformly a pullulan ester and/or pullulan ether in water or in a mixture of water and acetone in an amount between 5 and 40% of the solvent. The adhesives are moistureproof and have better water retention, stability and adhesiveness than conventional starch pastes or glue. They are also nontoxic and resistant to the growth of fungi.

Pullulan, the starting material for preparing adhesives and pastes, is obtained by submerged culture of *Pullularia pullulans* IFO 6353 or *Dematium pullulans* IFO 4464 in a medium containing sucrose, glucose, date extract, invert sugar, or starch hydrolyzate as a carbon source. The molecular weight of pullulan is freely adjustable between 10,000 and 5,000,000 by varying the fermentation conditions. Generally, a pullulan with a molecular weight of 10,000 to 500,000 is used in this process. The pullulan recovered from the culture broth as a white powder by precipitation with an organic solvent has extremely high water solubility. The process is illustrated by the following examples in which all parts and percentages are by weight unless stated otherwise.

Example 1: (a) – Pullulan acetic esters were prepared by dissolving 50 grams pullulan (molecular weight 450,000) in a mixture of 500 grams dimethylformamide and 100 grams pyridine at 60°C, adding thereto dropwise either 0.35 or 1.0 mol of acetic anhydride per one glucose residue of pullulan over 1 hour with stirring, and holding the mixture for 1½ hours at 60°C. After the reaction, ethanol was added, and the precipitate formed was recovered and purified by washing with ethanol. The pullulan acetic esters were obtained in a yield of about 95%, their DS values were 0.31 and 0.95, respectively, and they were a white, water-soluble powder.

(b) – Pullulan propionic acid esters were prepared as in (a), except that either 0.2 or 0.7 mol propionic anhydride was added instead of acetic anhydride. The yield was about 90%, and the DS values were 0.18 and 0.6, respectively.

(c) – Pullulan ethyl ethers having DS values of 0.07 and 0.1, respectively, were prepared by dissolving pullulan of molecular weight 300,000 in 300 milliliters of 10% aqueous sodium hydroxide solution, adding diethyl sulfate dropwise in 20% excess, holding the mixtures at 60°C for 2 hours with stirring, neutralizing the mixture, precipitating the ethers with methanol, filtering them, washing them with methanol, and drying them.

(d) – Pullulan carboxymethyl ether was prepared by dissolving 50 grams of pullulan of molecular weight 50,000 in 250 milliliters of a 5% aqueous sodium hydroxide solution, adding dropwise 0.1 mol monochloroacetic acid per glucose residue of the pullulan, holding the mixture at 40°C with stirring, neutralizing the mixture, recovering the carboxymethyl ether after precipitation by ethanol, washing the precipitate with ethanol and drying. Thus, 47 grams carboxymethyl ether having a DS of 0.07 was obtained and was readily water-soluble.

(e) – Pullulan hydroxypropyl ether was prepared by dissolving 50 grams pullulan of molecular weight 300,000 in 300 milliliters of a 2.5% aqueous sodium hydroxide solution, adding dropwise 1 mol propylene oxide per glucose residue of the pullulan with stirring at 40°C for 3 hours, neutralizing the reaction solution with acetic acid, precipitating the ether with ethanol, filtering the precipitate, and washing it with ethanol. Thus, 40 grams pullulan hydroxypropyl ether powder of high water solubility and having a DS value of 0.08 was obtained.

(f) – The pullulan acetic acid ester of DS 0.31 obtained in (a) was made into a 15% paste with warm water. A control was prepared by dissolving a commercial dextrin product in water with heating in a concentration of 15%. Respective end portions (30 millimeters by 3 millimeters) of fine kraft paper strips (30 millimeters by 60 millimeters) were coated with the pastes in an amount of 30 grams of paste per square meter, the paper strips were allowed to stand for 5, 30, 120 or 240 seconds, whereupon the strip ends carrying the paste were pressed against the bare ends of the same paper strips, and the bonded strips were allowed to stand for additional 10, 20 or 40 seconds before the bonds were tested on a tensile strength tester. The test results are listed in the table on the following page in kilograms per square millimeter.

Interval between coating and bonding	Bond strength, kg/mm² After 10 seconds P. ester	Dextrin	After 20 seconds P. ester	Dextrin	After 40 seconds P. ester	Dextrin
5 sec.	12.5	10.5	21.1	10.5	29.1	19.1
30 sec.	16.1	10.9	23.5	10.1	30.5	20.5
120 sec.	13.9	3.1	15.5	4.2	21.1	7.1
240 sec.	7.5	0.5	8.1	0	15.1	0.2

As is apparent from the table, even when the interval between coating and bonding is longer for the pullulan ether paste than for the dextrin paste, the former exhibits higher adhesiveness, which is probably due to better moisture retention. In addition, the pullulan acetic acid ester paste has higher adhesiveness after bonding, and the increase in bond strength after bonding is much greater than for the control.

Example 2: Comparison tests were carried out between a pullulan propionic acid ester paste, DS 0.18, and the dextrin control paste. The results were almost equal to those of the pullulan acetic acid ester paste of the preceding example.

Example 3: Pullulan ethyl ether, DS 0.07, was dissolved completely in warm water to form a 10% paste. The paste was tested as in Example 1 (f) against the same dextrin paste as a control. The pullulan ether paste retained adhesiveness for approximately the same period after coating as the pullulan acetic acid ester paste of Example 1. In addition, the adhesiveness increased rapidly after bonding as with the pullulan acetic acid ester paste, and the increase was much greater than that of the control.

Example 4: A 10% paste prepared by dissolving pullulan acetic acid ester of DS 0.31 in warm water was spread on one end of a kraft paper strip in an amount of 40 grams per square meter. The paper strip was allowed to dry at 50°C overnight and thereafter for 1 hour at 70°C. After 1 day at a relative humidity of 60%, the portion of the paper strip (30 millimeters by 3 millimeters) on which the paste was spread was slightly moistened and was bonded under manual pressure to the bare end of another paper strip of the same quality. The bonded papers were tested and compared with control paper strips prepared with dextrin paste as described in Example 1 (f).

The pullulan ester paste had a higher water absorption rate, developed its adhesiveness immediately after moistening, and was ready for use upon bonding. Thus, the paste is suitable for labeling containers on continuous labeling machines. In addition, the pullulan ester paste retained its adhesiveness at low humidity without delamination of the bonded materials. While the pullulan ester paste developed immediate adhesiveness upon absorption of moisture, the dissolution loss upon remoistening is small due to the relatively low water solubility of the gel formed by the paste.

Example 5: Paste layers prepared from pullulan acetic acid and propionic acid esters, pullulan ethyl, carboxymethyl, hydroxypropyl ethers and dextrin were tested for support of fungus growth in a room kept at a relative humidity of 80 to 90%. On the dextrin paste, a considerable number of fungus colonies was noted, whereas paste prepared with the pullulan derivatives did not show such growth.

Example 6: A 40% pullulan acetic acid ester (DS 0.9) was prepared by dissolving the ester in a water-ethylene glycol mixture (1:3) with stirring and heating. The paste had satisfactory adhesiveness when spread on paper or tape, and showed hardly any change in adhesiveness after an extended period.

TEXTILE ADHESIVES AND PLASTICS BONDING

Adhesive compositions are becoming increasingly used, for example in the garment manufacturing trade, for bonding fabrics together. In the garment manufacturing trade it is known to bond two fabrics together using a hot melt adhesive composition which is capable of being activated by heat and pressure in a pressing operation. However, in many cases the physical properties of the adhesive composition have to be quite specific in order to retain a satisfactory adhesive bond throughout the lifetime of a garment.

For example, an adhesive composition for use in the garment manufacturing trade may be required to retain its bond strength when a garment is being washed in warm soapy water and also when the garment is being dry cleaned in cleaning solvents such as perchloroethylene, trichloroethylene or alcoholic spirits. In addition to the requirement of resistance to dry-cleaning solvents and to washing in warm soapy water, it is important that adhesives have a relatively wide softening point range to enable their use effectively in industrial adhesive bonding of fabric sheet, for example, in the known procedures for depositing spaced dots of adhesive from powdered resin adhesive.

It is well known that the adhesive bonding of synthetic plastic materials presents difficult problems and that such bonding of polyvinyl chloride either to itself or to other substrates is particularly difficult, especially when the polyvinyl chloride is plasticized. Flexible, plasticized polyvinyl chloride (PVC) is difficult to bond for a number of reasons. First, it tends to concentrate peel stresses unevenly in the glue line of the adhesively bonded article. Also, inasmuch as the plasticizer tends to migrate out of flexible PVC on aging, it may enter the adhesive layer and adversely affect its performance, or may simply accumulate at the PVC-adhesive interface in a structurally weak layer.

Adhesives used to bond flexible PVC articles must, of course, be flexible themselves to withstand the repeated flexing of the article in use. Of no lesser importance is the fact that a suitable adhesive for bonding synthetic resin substrates must be able to withstand the environmental conditions of humidity, heat and cold to which the bonded article will be subjected in use. A suitable adhesive for plasticized polyvinyl chloride should retain its flexibility and bonding integrity in the presence of variable amounts of moisture and humidity at temperatures about -40°F up to 210°F. Developments in adhesive systems for textile bonding, PVC lamination and glass reinforced polyesters are described.

TEXTILE ADHESIVES

Diacetone Acrylamide and Aminoplasts

A process described by *J.W. Forsberg; U.S. Patent 3,879,325; April 22, 1975; assigned to The Lubrizol Corporation* relates to water-soluble compositions prepared by reacting, in the presence of an alkaline reagent, an aliphatic aldehyde containing not more than 4 carbon atoms, or a reversible polymer thereof, with an N-3-oxohydrocarbon-substituted acrylamide having the formula

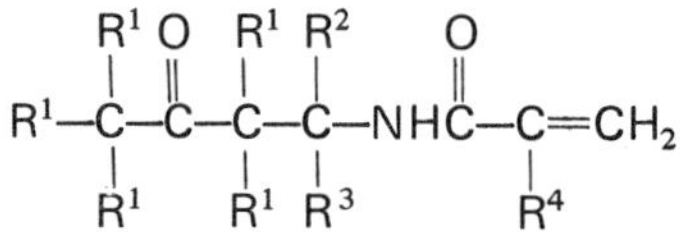

where each R^1 is individually hydrogen or a hydrocarbon or substituted hydrocarbon radical, at least one R^1 being hydrogen; each of R^2 and R^3 is hydrogen or a hydrocarbon or substituted hydrocarbon radical; and R^4 is hydrogen, halogen or a lower alkyl or substituted lower alkyl radical; the reaction being effected in a diluent comprising (1) water, or (2) an organic liquid which is a solvent for the reactants or the product or both, or (3) a mixture of diluents 1 and 2.

For preparation of the preferred compositions of this process, from one to three and preferably all five of the R^1 radicals in the N-3-oxohydrocarbon-substituted acrylamide reagent are hydrogen; R^2 and R^3 are lower alkyl radicals; and R^4 is hydrogen or methyl. Suitable N-3-oxohydrocarbon-substituted acrylamides are described in U.S. Patents 3,277,056 and 3,425,942.

Examples are N-(1,1-dimethyl-3-oxobutyl)acrylamide, referred to as diacetone acrylamide, and N-(1,3-diphenyl-1-methyl-3-oxopropyl)acrylamide, referred to as diacetophenone acrylamide. Because diacetone acrylamide is preferred and is most readily available, it will frequently be referred to in this process. The preparation of the water-soluble compositions of this process is illustrated by the following examples. All parts and percentages are by weight unless otherwise indicated.

Example 1: A solution of 338 parts (2 mols) of diacetone acrylamide in 1,000 parts of water is heated to 50°C and 45 parts of a 3% aqueous solution of trisodium phosphate is added. Dropwise addition of a 37% aqueous solution of formaldehyde is then begun and is continued for one hour, a total of 650 grams (8 mols of formaldehyde) being added. During the formaldehyde addition, three further portions of trisodium phosphate solution, one of 14 parts and two of 15 parts, are added. Heating is continued for five hours after formaldehyde addition is complete, and during that time additional increments of trisodium phosphate solution are added, care being taken that the pH of the mixture never increases above 10, until a total of 299 parts have been introduced (0.16 mol or 2.0

mol percent based on formaldehyde). The solution is cooled and filtered, and volatile materials are removed by heating under vacuum at 55°C. The product is a 61% aqueous solution of the desired water-soluble composition.

Example 2: Following the procedure of Example 1, a similar reaction product is prepared from diacetone acrylamide and acetaldehyde.

Example 3: Following the procedure of Example 1, a similar reaction product is prepared from diacetone acrylamide and n-butyraldehyde.

Example 4: A solution of 280 parts (1.66 mols) of diacetone acrylamide in 296 parts of distilled water is heated to 33°C, and 164 parts of paraformaldehyde (5 mols based on monomeric formaldehyde) is added over 20 minutes. The solution is then heated to 47°C and 8.3 parts of a 10% aqueous solution of potassium hydroxide is added over 10 minutes. The reaction mixture is stirred and heated to 52°C over about ½ hour, at which time an exothermic reaction begins; stirring is continued and the temperature is kept at about 50°C by passing cooling water through a jacket on the reaction vessel.

At the end of the 2-hour stirring period, an additional 8.3 parts of the potassium hydroxide solution is added, and a final 8.3 parts is added after a further 2-hour stirring period (total 0.9 mol percent based on formaldehyde). The mixture is stirred for an additional 2 hours, cooled to 24°C and filtered. The product, a 55% aqueous solution of the desired water-soluble composition, contains 3.04% nitrogen.

Example 5: To a solution of 644 parts (3.81 mols) of diacetone acrylamide in 681 parts of water, at 31°C, is added, with stirring, 372 parts of paraformaldehyde (11.3 mols based on monomeric formaldehyde). The mixture is heated to 43°C and 19 parts of a 10% aqueous solution of potassium hydroxide is added. The mixture is further heated, with stirring, to 48° to 55°C and maintained at this temperature for 2 hours.

An additional 10 parts of potassium hydroxide solution is then added and stirring is continued for 2 hours, followed by addition of a third 19 ml portion of potassium hydroxide solution (total 0.87 mol percent based on formaldehyde) and stirring for 2 more hours. The solution is then cooled to 29°C, 0.044 part of methylhydroquinone is added and the mixture is filtered. The product, a 53% aqueous solution of the desired water-soluble composition, contains 3.04% nitrogen.

Example 6: A mixture of 1,352 parts (8 mols) of diacetone acrylamide, 1,136 parts of methanol and 480 parts of a solution comprising 55% formaldehyde (8.8 mols of formaldehyde), 35% methanol and 10% water is heated to 44°C, and 10 parts of a 10% solution of potassium hydroxide in methanol (0.2 mol percent of potassium hydroxide based on formaldehyde) is added. The mixture is heated at 44° to 47°C for 7 hours, with stirring, and is then stripped of volatile materials by distillation at 48°C/4 torr. The water-soluble product contains 7.08% nitrogen.

Example 7: Following the procedure of Example 6, a water-soluble product containing 6.16% nitrogen is obtained from 1,014 parts (6 mols of diacetone acrylamide, 1,706 parts of methanol, 491 parts (9.0 mols of formaldehyde) of formaldehyde-methanol-water solution, and 10 parts of methanolic potassium hydroxide (0.2 mol percent potassium hydroxide based on formaldehyde).

Example 8: Following the procedure of Example 6, a water-soluble product containing 6.38% nitrogen is obtained from 1,014 parts (6 mols) of diacetone acrylamide, 1,767 parts of methanol, 654 parts (12 mols of formaldehyde of formaldehyde-methanol-water solution, and 10 parts of methanolic potassium hydroxide (0.15 mol percent potassium hydroxide based on formaldehyde).

Example 9: A mixture of 291 parts of ethyl acrylate, 16 parts of the product of Example 4, 596 parts of water and 90 parts of a 21% aqueous solution of a sodium alkaryl polyether sulfate anionic emulsifier sold under the trade name Triton X-301 is purged with nitrogen for 45 minutes, after which a solution of 1.5 parts of ammonium persulfate in 10 parts of water and a solution of 0.3 part of sodium formaldehyde sulfoxylate in 15 parts of water are added. The reaction vessel is cooled as the mixture is stirred and an exothermic reaction takes place which causes the temperature to rise to 59°C. After 10 minutes an additional solution of 0.3 part of sodium formaldehyde sulfoxylate in three parts of water is added and the mixture is filtered, yielding the desired copolymer latex.

Example 10: A mixture of 87 grams of diacetone acrylamide, 30 grams of the product of Example 6, 180 grams of butyl acrylate, 3 grams of acrylic acid, 418 grams of water and 34 grams of an anionic emulsifier sold under the trade name Abex 18S is purged with nitrogen. To the mixture is added 5 ml of a 10% aqueous solution of sulfuric acid, 1.0 gram of sodium formaldehyde sulfoxylate and 1.5 grams of ammonium persulfate. The mixture is stirred under nitrogen for about 15 minutes and then heated to 72°C, after which 1 ml of t-butyl hydroperoxide is added. Heating is continued at about 60° to 65°C for 2 hours. The desired tetrapolymer latex is then filtered through cheese cloth and neutralized with 10% aqueous ammonia solution.

Example 11: Following the procedure of Example 10, a tetrapolymer latex is prepared from 30 parts of diacetone acrylamide, 30 parts of the product of Example 8, 120 parts of butyl acrylate, 20 parts of methacrylic acid and 283 parts of water, and is then neutralized with 10% aqueous ammonia solution.

Example 12: Following the procedure of Example 10, a terpolymer latex is prepared from 70 parts of the product of Example 7, 110 parts of butyl acrylate, 20 parts of methacrylic acid and 280 parts of water, and is subsequently neutralized with 28% aqueous ammonia solution.

Example 13: A pressure bottle is charged with 25 parts of styrene, 70 parts of butadiene, 9.6 parts of the product of Example 5, 200 parts of butadiene, 0.2 part of potassium persulfate, 4 parts of diamyl sodium sulfosuccinate, 0.25 part of sodium naphthalene sulfonate emulsifier sold under the trade name Tamol N, 0.5 part of cumene hydroperoxide, 0.5 part of t-dodecyl mercaptan and 0.2 part of sodium formaldehyde sulfoxylate. The bottle is sealed and agitated for 5 hours at 50°C; it is then opened and 2 parts of an alkylphenyl polyethoxyethanol nonionic emulsifier sold under the trade name Triton X-405, 7.8 parts of water and 0.2 part of sodium diethyl dithiocarbamate are added. Volatile materials are removed by steam stripping, yielding the desired terpolymer latex.

Example 14: Following the procedure of Example 13, a tetrapolymer latex is prepared from 15 parts of styrene, 70 parts of butadiene, 10 parts of diacetone acrylamide, 9.6 parts of the product of Example 5 and 200 parts of water.

Example 15: A reaction flask is charged with 500 parts of water, 36 parts of styrene, 36 parts of butyl acrylate, 4 parts of the product of Example 4, 1 part of acrylic acid, 5 parts each of an oxyethylated alkylphenol water-soluble emulsifier sold under the trade name Igepal CO-710 and a similar oil-soluble emulsifier sold under the trade name Igepal CO-520, and 2.5 parts of sodium lauryl sulfate. The mixture is purged with nitrogen and stirred as a solution of 0.2 part of ammonium persulfate in 5 parts of water and a solution of 0.2 part of sodium formaldehyde sulfoxylate in 5 parts of water are added.

Stirring is continued as the mixture is heated to 68°C, and there are simultaneously added (1) a mixture of 410 parts of water, 286 parts of styrene, 286 parts of butyl acrylate, 36 parts of the product of Example 4, 6 parts of acrylic acid, 5 parts each of Igepal CO-710 and Igepal CO-520, and 1 part of sodium lauryl sulfate; (2) a solution of 1.4 parts of ammonium persulfate in 50 parts of water; and (3) a solution of 1.4 parts of sodium formaldehyde sulfoxylate in 50 parts of water. The addition takes place over 1 hour at 68° to 72°C. Stirring is continued at 72° to 80°C as additional portions of ammonium persulfate and sodium formaldehyde sulfoxylate are added, followed by portions of t-butyl hydroperoxide and sodium formaldehyde sulfoxylate.

The mixture is cooled to room temperature and filtered through cheesecloth, a solution of Igepal CO-710 in 25 parts of water and a solution of Triton X-405 in 7 parts of water are added, and the material is stirred briefly. The product is the desired tetrapolymer latex.

Example 16: A tetrapolymer latex is prepared by the method of Example 15, containing 4.1 parts of the product of Example 4, 63.8 parts of butyl acrylate, 31.1 parts of acrylonitrile and 1 part of acrylic acid. The emulsifiers are Triton X-301 and Triton X-405.

Example 17: Following the procedure of Example 15, an interpolymer latex is prepared containing 5.8 parts of the product of Example 4, 68 parts of ethyl acrylate, 14.5 parts of butyl acrylate, 10.7 parts of acrylonitrile and 1.0 part acrylic acid. The emulsifiers are Triton X-301 and Triton X-405. Following its preparation, the latex is neutralized with aqueous ammonia solution.

Example 18: A solution of 77 parts of vinyl acetate, 20 parts of butyl acrylate, 5.4 parts of the product of Example 4, 0.75 part of hydroxyethylcellulose, 15 parts of Triton X-405, 10 parts of sodium lauryl sulfate and 1.5 part of sodium bicarbonate in 700 parts of water is purged with nitrogen, and 0.4 part each of sodium formaldehyde sulfoxylate and ammonium persulfate, each dissolved in 10 parts of water, are added.

The solution is heated to 72°C and there are simultaneously added, over 3 hours, (1) a mixture of 1,078 parts of vinyl acetate, 280 parts of butyl acrylate, 76.4 parts of the product of Example 4, 15 parts of Triton X-405, 5 parts of sodium lauryl sulfate, 875 parts of water and about 3 parts of t-butyl hydroperoxide, this solution being added via a subsurface addition tube; (2) a solution of 2.8 parts of ammonium persulfate in 50 parts of water; and (3) a solution of

2.8 parts sodium formaldehyde sulfoxylate in 50 parts of water. The mixture is maintained at 70° to 75°C for ½ hour after the addition of the above described solutions is complete, and is then cooled and filtered. The product is the desired terpolymer latex.

The addition polymers of this process, particularly latexes thereof, have self-crosslinking properties which make them particularly useful as adhesives. Because of these self-crosslinking properties, the polymers may be cured under acidic (preferably) or alkaline conditions at temperatures from room temperature to about 200°C to form adherent films with a high percentage of acetone insolubles. Such polymers are especially useful as laminating adhesives and binders for textiles and nonwoven fabrics. It is possible to formulate an adhesive in which the only ingredient present in substantial amounts in the polymer latex of this process, with the optional presence of such substances as thickeners, antifoam agents, protective colloids, pigments and the like which are known in the art. Particularly useful for this purpose are polymers containing units derived from an acidic compound such as acrylic acid.

For example, the polymer of Example 10, 11 and 12 may be combined with a 5% aqueous hydroxyethylcellulose solution (as a protective colloid) in an amount to provide 4.5 parts of hydroxyethylcellulose per 100 parts of polymer to form a composition suitable for use as an adhesive. However, such adhesives ordinarily cure only at undesirably high temperatures, frequently above 150°C. The preferred adhesive compositions additionally contain at least one substance selected from the group consisting of alkaline reagents (e.g., sodium hydroxide, potassium hydroxide) and the aminoplast compositions previously described herein. Both of these substances may be present, but it is usually satisfactory to use only in an amount of about 0.5 to 10.0 parts per 100 parts of the polymer of this process. It is particularly preferred to use hexamethoxymethylmelamine. Typical adhesive compositions containing the addition polymers are listed below.

	Adhesive, Parts by Weight				
	F	G	H	J	K
Product of Example 9	-	-	-	100	-
Product of Example 16*	100	100	-	-	-
Product of Example 17	-	-	-	-	100
Product of Example 18	-	-	100	-	-
Hexamethoxy-methylmelamine	-	3.4***	2.3†	2.3†	3.6†
Potassium hydroxide**	1.5	-	-	-	-
Vinyl acetate copolymer thickener	8	8.5	-	-	-

*Neutralized with aqueous ammonia solution.
**Added as a 10% aqueous solution.
***Added as a 50% solution in a 2:3 ethanol-water mixture.
†Added as a 30% aqueous solution containing 30% of a non-ionic emulsifier.

The utility of the adhesive compositions containing the polymers of this process is illustrated by a procedure in which the adhesive is applied in a checkered pattern of pyramidal dots from an engraved roll onto a strip of cotton cloth resting on a piece of soft rubber. A layer of tricot is placed over the adhesive and a cylindrical weight is rolled over the laminate to assure bonding. The laminate is then placed in a forced-air oven, tricot side down, at 150°C for 3 minutes. The cured laminates are found to have excellent peel resistance, dry and wet strength and "hand". Latexes of the addition polymers containing units derived from styrene and butadiene (e.g., the products of Examples 13 and 14) may be used as glass-to-rubber adhesives.

Lactam Copolyamides

F.R. Halas and C. Rossitto; U.S. Patent 3,850,887; November 26, 1974; assigned to USM Corporation have found that acceptable bond strength in fabric to fabric bonds and good resistance to washing in warm soapy water and to dry cleaning may be obtained by use of an adhesive composition comprising a polyamide formed from a mixture in controlled proportions of at least one lactam, a mixture of aliphatic dicarboxylic acids and at least one diamine. The components are selected to give polymer molecules having a degree of irregularity which interferes with crystallization and contribute to a substantial degree amorphous characteristics giving a wide melting point range.

In addition, the compounds selected for copolymerization include components having a relatively large number of carbon atoms in chains in repeating units of the copolyamide molecular chain to impart improved wash resistance to the composition together with components giving improved resistance to attack by dry-cleaning solvents. Also, by balancing the range of relative proportions and the conditions of polymerization to control the molecular weight and softening point range of the resulting copolyamide material having physical properties enabling it when reduced to powdered condition to retain this condition without undue lumping while at the same time having a relatively wide range of softening or melting temperatures particularly avoiding critical temperature control and other handling problems encountered in use of the material for the hot melt adhesive between, for example, layers of fabric.

Example: A series of copolyamides were prepared by the procedure set forth below using reagent mixtures and forming products having the properties listed in the following table.

Mols per 100 mols Copolymer	Number							
	2	3	4	5	6	7	8	9
Caprolactam	66.7	66.7	55	65	65	65	66.7	66.7
Adipic	10.6	10.6	–	–	–	–	–	10.6
Azelaic	–	–	–	–	–	–	10.6	–
Sebacic	–	–	18	14	14	14	–	–
Dodecanedioic	22.7	22.7	27	21	21	21	22.7	22.7
Hexamethylene diamine	30.0	33.3	33.3	33.3	29.75	32.3	33.3	33.3
Trimethyl hexamethylene diamine	3.3	–	–	–	5.25	2.7	–	–
Melting point, °C	135-140	138-142	144-148	128-137	121-135	122-140	125-137	130-135
Viscosity, cp 180°C	400,000	392,000	352,000	350,000	570,000	500,000	420,000	400,000
Printing character	(*)	(**)	(**)	(***)	(*)	(*)	(*)	(**)
Laundering	(**)	(**)	(***)	(***)	(***)	(***)	(***)	(**)
Dry cleaning	(***)	(***)	(***)	(***)	(***)	(***)	(***)	(***)
Incipient melting, °C	89	117	128	98	75	100	85	100
Endotherm peak, °C	128	136	142	118	111	142	113,142	129
Difference, °C	39	19	14	20	36	42	–	129

*Good **Fair ***Excellent

The procedure used for making the copolyamide involved melting the lactam and the dicarboxylic acid components together in a reaction vessel. Thereafter, diamine was added and the temperature allowed to rise to from 100° to 120°C. The reaction mixture was heated under reflux conditions until salt formation was completed and at this point the temperature was raised to 200°C to eliminate water. The temperature of 200°C was maintained for 2 hours and then a vacuum was applied to the reaction mixture and heating continued for a further hour at 200°C. At this point, polymerization was complete and the resinous polymeric materials was poured out into a casting tray. The resinous material was cryogenically ground and screened to form a uniform powder having a particle size of from 60 to 210 microns. The copolyamide powders were used to bond pieces of fabric and the bonds between the pieces of fabric were tested by the following procedure.

The powder was disposed on the surface of a silicone rubber mat formed on that surface with dimples about 1 mm in diameter and spaced about 2.5 mm apart and the mat was scraped leaving powder in the dimples. A 7" x 9" piece of fabric lining material having a fiber content with 65% polyester and 35% cotton was disposed on the surface of the mat and heated and pressed with an iron set at a temperature of about 300°F.

The heat caused the powder in the dimples to sinter together and to adhere to the fabric. The fabric with the adhered sintered powder was stripped from the dimpled mat and examined. A substantial proportion of the powder that had been in the dimples adhered to the fabric. The fabric with the dots thereon was heated under an infrared heater for 10 seconds to completely fuse the powder. A 7" x 9" piece of suiting fabric with a fiber composition of 65% polyester and 35% cotton was assembled against the surface of the first piece of fabric and the assembly was pressed in an electric press at a temperature of 325°F.

After aging for one day, the sample was subjected to a laundering test involving subjecting the assembly to five 20 minute cycles of washing in a washing machine using wash water detergent containing 1% by weight at a temperature of 140°F with drying of the assembly between wash cycles. The bond between the pieces of fabric was inspected after each drying cycle and the results noted in terms of resistance to separation of the pieces of fabric. Washing resistance listed as fair means some separation of the pieces of fabric; good means some weakening of the bond and excellent means no significant separation or weakening of the bond. Identical samples were subjected to 5 cycles of dry cleaning; the copolyamides withstood the test with no weakening of bond.

It can be seen by comparison of copolyamides 1, 5 and 6 with the other copolyamides, that the inclusion of a hindered diamine, that is, trimethyl hexamethylene diamine, increases the temperature range over which softening occurs so that the resin powder is picked up more satisfactorily from the dimples in the silicone rubber mat, i.e., prints more satisfactorily, than do the copolyamides in which the hindered diamine is not used. It will also be observed that resistance to laundering is better where sebacic acid having 10 carbon atoms is used as in copolyamides 3 to 6 than where adipic acid having only 6 carbon atoms is used as in copolyamides 1, 2 and 8 and also that azelaic acid having 9 carbon atoms is also more resistant to laundering than the copolyamides containing adipic acid.

Polyurethane

According to a process descrsibed by *K.H. Remley; U.S. Patent 3,835,081; September 10, 1974; assigned to American Cyanamid Company* polyurethane textile adhesive compositions in the form of aqueous dispersions which furnish films of superior wet strength are obtained by (1) forming an isocyanate-terminated prepolymer by reacting tolylene diisocyanate with a glycol mixture of polypropylene glycol having a molecular weight between about 900 and 1,200 and 2,2'-bis(hydroxymethyl)propionic acid wherein the diisocyanate to total glycol molar ratio is between 1.6 and 1.75 and the prepolymer reaction is terminated when the free NCO content of the prepolymer is between 6.19 and 7.4% by weight based on the weight of the prepolymers (2) adding the prepolymer to water containing a tertiary aliphatic amine and allowing chain extension to take place and (3) adding a thickening agent to the aqueous dispersion.

The adhesive of the process can be used with any of the fabrics and plastic sheeting or foams normally used in the laminating processes. These include woven knit and nonwoven fabrics of natural and synthetic fibers, and a variety of plastic sheeting and foams, but particularly those made of poly(vinyl chloride) or polyurethane. In the flocking processes the adhesive can be used with any of the substrates normally used in the flocking process, including woven and nonwoven fabrics of natural and synthetic fibers, paper, plastic coated fabrics, plastic films and sheets, wood, metal, rubber, glass, etc. Any of the flocking materials may be used. These include both synthetic and natural fibers.

Example 1: To a suitable reactor there was charged 3,550 parts of polypropylene glycol (MW 1025) and 273 parts of 2,2'-bis(hydroxymethyl)propionic acid. The mixture was heated to 55°C, whereupon 1,650 parts of tolylene diisocyanate (80/20 mixture of 2,4- and 2,6-isomers) was added and the temperature was maintained at 60°C until analysis indicated that the free NCO content by weight was 7.2% of the reaction mixture. (If allowed to go to completion, the reaction would contain about 6% of free NCO.) At this point the prepolymer was added to a vigorously stirred solution of 210 parts of triethylamine in 9.275 parts of water at a temperature of 12° to 16°C. When the addition was finished, stirring was continued at 30°C for five hours. The aqueous dispersion was thickened to a viscosity suitable for a textile adhesive composition by the addition of ammonia and a thickener of the acrylic carboxy-containing polymer type.

Example 2: (A) a series of prepolymer samples was prepared by reacting together 450 parts (0.4390 mol) of polypropylene glycol (MW 1025), 34.5 parts (0.258 mol) of 2,2'-bis(hydroxymethyl)propionic acid, and 210 parts (1.21 mol) of tolylene diisocyanate (80/20 mixture of 2,4- and 2,6-isomers). The ratio of diisocyanate to total glycol was 1.73:1.0. The mixture was stirred at about 70°C. At hourly intervals, the mixture was analyzed for free NCO and a 50-part sample was removed. Each sample was immediately added to a vigorously stirred mixture of 93 parts of ice and water and 1.87 parts of trimethylamine. Sufficient water (about 12 parts), ammonium hydroxide (about 0.3 parts) and commercial thickening agent of the polyacrylic acid type, Acrysol ASE-60 (Rohm & Haas). (about 3.0 parts) were added to provide the desired viscosity.

The wet strengths of the bonding films provided by the adhesive compositions were measured by (1) placing a swatch of cellulose acetate taffeta on a flat surface, (2) covering the taffeta with a perforated template, (3) spreading the adhesive composition evenly and cleanly across the template thereby applying a pattern of small dots of adhesive to the taffeta, (4) removing the template, (5) placing a swatch of cellulose acetate tricot on the taffeta, (6) pressing the two pieces of fabric together for 30 seconds by the weight of a glass plate, and (7) heating the fabric assembly in an oven at 133°C for 1½ minutes. The bonded fabric assemblies were cut into 1 inch strips. Three strips of each fabric assembly were placed in boiling water with stirring for five minutes and then in cool water. Peel strengths of the bonds of the wet strips were measured by a spring scale. The average force in ounces required to peel one fabric from the other of each assembly is shown in the second column "A" of the following table.

(B) For comparison, an adhesive composition was prepared by the same general procedure using a polypropylene glycol of molecular weight 2,025 and a ratio of diisocyanate to total glycol of 1.9:1.0. The results are shown in second column "B" of the table below.

Prepolymer reaction time, hr.	Percent free NCO		Wet peel strength, oz.	
	A[1]	B	A	B
1	7.82	7.35	4.17	1.3
2	7.22	6.70	7.35	3.5
3	6.58	6.34	6.85	4.0
4	6.33	6.06	6.0	4.2
5	6.19	5.98	5.67	4.0

[1] The free NCO content when the reaction is carried to completion is about 6.1%.

This example demonstrates that adhesive compositions which provide films of superior wet strength can be obtained by using a polypropylene glycol of molecular weight about 1,000 as compared with a polypropylene glycol of molecular weight about 2,000. It also demonstrates that when a polypropylene glycol of molecular weight about 1,000 is employed and the prepolymer reaction is terminated before the reaction is finished, an adhesive composition is obtained which provides a film with superior wet strength.

Thixotropic Polyurethane Adhesive for Carpets

The tufting method which is now generally employed for the manufacture of carpets comprises looping pile fibers of natural or synthetic material through a relatively inexpensive woven or nonwoven textile base, known as the primary fabric backing material. Short loops of the long pile fibers are pushed through the primary fabric backing material such that one single continuous length of fiber constituting a complete row of pile in the carpet is formed. The elongated loops extending from the base (the top side) of the primary fabric backing material can remain connected or severed, depending on whether a loop pile or a cut pile carpet is desired. The loops on the bottom side of the carpet are not cut. These pile loops or tufts are not securely fastened in the tufting process.

Without additional anchorage these fibers or the tufts can be pulled from the primary fabric or otherwise disarrayed. Necessary anchorage is provided by applying an adhesive material in liquid form to the underside of the carpet. The adhesive applied to the underside of the carpet is accordingly of major importance to the quality and performance of the carpet. It retains the pile fibers or tufts in place, secures the individual fibers of the yarn, prevents pilling of the yarn and controls dimensional stability.

Without additional backing material applied to the adhesive material on the underside of the carpet, the carpet has a unitary backing. Carpets having a unitary backing are used principally as commercial carpeting. If a sponge-like material (or foam) is applied after the adhesive material is applied, the adhesive coating is referred to as a precoat composition. Uncured foam material can be applied directly to the precoated carpet back and cured in place, or it can be cured as a separate sheet and then laminated to the back of the carpet by means of the precoat or use of another adhesive. The precoat provides good tuft lock, while the foam material, such as polyvinyl chloride or styrene butadiene copolymer, serves as a cushion back for the carpet.

For the standard double back carpet, the adhesive layer is referred to as a laminating adhesive. After the laminating adhesive is applied to the underside of the primary fabric backing material of a double back carpet, a further backing layer of secondary fabric material, known as the scrim, is applied to the coated underside of the carpet. The scrim serves to improve dimensional stability, appearance of the carpet and also enhance the tuft lock, i.e., the strength with which the fibers are retained in the primary fabric backing. The laminating adhesive for double back carpets serves not only to anchor the pile fibers or tufts, but also to adhere the scrim to the carpet. Conventionally, in the process of making double back tufted carpets the laminating adhesive is applied in liquid form to the underside of the carpet and the scrim is applied to the same side while the laminating adhesive is still wet and uncured. The carpet is then passed through an oven to dry and cure the laminating adhesive.

According to a process described by *R.H. Sheffler, R. Koncos and K.A. Riches; U.S. Patent 3,895,149; July 15, 1975; assigned to Atlantic Richfield Company* a carpet backing is formed by applying a thixotropic polyurethane adhesive composition as either a laminating adhesive, a precoat adhesive or a unitary backing to the underside of the primary fabric of tufted carpet.

The thixotropic adhesive composition, comprising a liquid, hydroxyl-terminated diene polymer, a low molecular weight reinforcing polyol, an isocyanate, a filler, and a catalyst, provides substantially complete bundle wrap of each fiber tuft without penetration through the primary fabric backing material to the top side or face of the carpet. The thixotropic polyurethane adhesive composition is applied to the underside of the primary fabric backing material in measured quantity and cured by the application of heat with or without a secondary fabric being applied prior to curing.

The adhesive compositions comprise a mixture of a liquid hydroxyl-terminated diene homopolymer or copolymer, a polyol having an equivalent weight of between about 50 and about 300 present in an amount up to 5 times the equivalents of diene polymer; a polyol having an equivalent weight of between about 50 and about 300 present in an amount between about 1.5 and about 8 times the equivalents of diene polymer; an isocyanate material having a functionality of between 2 and 3 present in an amount to provide a NCO/OH equivalents ratio of between 0.95:1 and 1.5:1.

It also comprises at least one filler present in an amount of between about 40 and about 800 parts per 100 parts by weight of diene polymer; an oil extending hydrocarbon liquid present in an amount up to 200 parts per 100 parts by weight of the diene polymer; water present in an amount up to 10 parts per 100 parts by weight of the diene polymer; and at least one catalyst capable of accelerating the cure time of the composition, present in an amount between about 0.02 and 4 parts per hundred parts by weight of the diene polymer. The resulting composition has an initial Brookfield viscosity at 5 rpm using a number 5 spindle of between about 3,000 and about 100,000 and the ratio of the viscosity at

1 rpm and 20 rpm for the thixotropic composition is between about 1.3:1 and 10:1. Conventional additives such as oxidation inhibitors, pot life inhibitors, stabilizers, pigments and the like can be incorporated in the adhesive compositions for improved characteristics.

Example 1: A thixotropic adhesive composition was prepared by blending 100 parts by weight of a liquid, hydroxyl-terminated polybutadiene (Resin R-45HT, ARCO Chemical) having equivalent weight of 1180, a hydroxyl content of 0.85 milliequivalents per gram, and 0.05 weight percent moisture; with 24 parts by weight of bisisopropanol isophthalate having an equivalent weight of 141 grams; 196 parts by weight of naphthenic process oil (Acme S-60 oil); 310 parts by weight of calcined kaolin (Glomax H.E.); 0.02 parts by weight of dibutylin dilaurate (T-12, MeT Chemical Co); and 43.8 parts by weight of diphenyl methane 4,4'-diisocyanate (Isonate 143-L), having an equivalent weight of 144 grams. This formulation can be used as an adhesive composition for rug and carpet backing applications.

Example 2: A thixotropic adhesive composition was prepared by blending 100 parts by weight of R-45-HT resin (identified in Example 1) with 17.7 parts by weight of bisisopropanol aniline (Isonol C-100) having an equivalent weight of 105 grams; 118 parts by weight of saturated naphthenic process oil (Tufflo 6024 Oil): 141 parts by weight of talc (Mistron Vapor); 94 parts by weight of dry ground fatty acid treated calcium carbonate (Quincy-2-Electro); 0.25 parts by weight of dibutyltin dilaurate; and 40.4 parts by weight of diphenyl methane 4,4'-diisocyanate. The resulting adhesive formulation gave good adhesion of jute backing to carpet at a 15 oz/yd^2 rate of application.

It was found that calcium carbonate (whiting) can be introduced at higher levels than clay (Example 1) or talc without causing excessive viscosity increases and thus can be used to contribute to lower cost. The use of higher levels of calcium carbonate does not impart thixotropy and contributes little to reinforcement; hence a balance between the filler and calcium carbonate loadings must be achieved.

Example 3: An adhesive formulation was prepared by blending 100 parts by weight of R-45HT resin (identified in Example 1); 16.6 parts by weight of bisisopropanol aniline; 100 parts by weight of a low viscosity saturated naphthenic process oil (Tufflo 6004 Oil); 100 parts by weight of dry ground calcium carbonate; 200 parts by weight of calcined kaolin; 1.06 parts by weight of water; 0.07 parts by weight of dibutyltin dilaurate; and 70.8 parts by weight of diphenyl methane, 4,4'-diisocyanate.

The resulting adhesive composition had a NCO/OH ratio of 1.36. When applied at an application rate of 28 oz/yd^2, tuft lock equaled 12.7 lb (average) and 15 lb (maximum). In addition, adhesion of the backing was excellent; jute was destroyed when attempting to delaminate double backed carpet. Addition of water to the formulation of this example prior to isocyanate cure was found to be very beneficial for the following reasons:

(a) The thixotropy of the uncured mix was greatly increased, presumably due to flocculation of kaolin clay.

(b) Tuft lock was increased. The same formulation without added water and the equivalent amount of diisocyanate had a maximum tuft lock of only 8.5 lb. The water reaction contributes urea linkages to the polymer which should provide additional reinforcement.

(c) Water reaction with isocyanate results in a chemical flow (foam formation) which appears to be beneficial since it increases the volume of the mix and allows lower application rates, provides continued blow after application which helps prevent voids between the carpet tufts, thereby resulting in more effective use of the adhesive, particularly if the foam is crushed at the proper stage of tackiness, and chemical blow brings the adhesive out of the tufts into contact with the secondary backing thereby assuring good adhesion of the latter to the carpet even if initial penetration of the mix is excessive.

POLYVINYL CHLORIDE BONDING

Ethylene-Vinyl Acetate

A process described by *S. Ishihara and S. Kobayashi; U.S. Patent 3,886,234; May 27, 1975; assigned to Showa Denka Kabushiki Kaisha, Japan* relates to a resin composition having excellent adhesiveness which comprises (A) 100 parts by weight of (1) an ethylene-vinyl acetate copolymer or (2) a combination of an ethylene-vinyl acetate copolymer and polyethylene, the vinyl acetate content in (A) being 1 to 35% by weight based on the total weight of (A); (B) 0.05 to 200 parts by weight of unsaturated carboxylic acids which are solid at normal temperature; (C) 0.005 to 10.0 parts by weight of organic peroxides; and, optionally, (D) 0.05 to 5.0 parts by weight of at least one phenol compound selected from the group consisting of monophenol compounds, bisphenol compounds, trisphenol compounds and tetraphenol compounds.

The resin composition can be used as an adhesive for various synthetic resins, rubbery materials, fibrous materials, including paper and wood, and moldings prepared from inorganic compounds. In particular, the resin composition has been found to be useful for bonding plasticized polyvinyl chloride or polyolefins such as polyethylene and polypropylene which generally exhibit poor adhesiveness to other articles with adhesives commonly used.

In the examples, the adhesiveness of samples was determined basically according to the criteria given in ASTM standards D 903-49, i.e., by measuring the peel strength of the sample in a direction of 180° at tensile rate of 100 mm/min. The gel content in adhesive composition was determined by dissolving 1 gram of the composition in 100 ml of toluene as a solvent at a temperature of 90°C for 3 hours and filtering the solution while hot through a screen having a 325 mesh size. The undissolved portion on the filter is referred to as the gel content.

Example 1: 100 parts by weight of an ethylene-vinyl acetate copolymer containing 20% by weight of vinyl (Ultrathene UE 631), 10 parts by weight of fumaric acid and 0.5 part by weight of dicumyl peroxide were mixed for 5 minutes in a hot roll at a temperature of 85°C to prepare an adhesive composition. The resulting composition was then pressed into a film having a thickness of 0.1 mm by a hot press at a temperature of 100°C. The film thus obtained was placed between an aluminum foil having a thickness of 0.1 mm previously degreased with trichloroethylene and a 0.2 mm thick film of high density polyethylene having a high load melt index of 5.1 grams per 10 minutes as (determined under a load of 21.6 kg and at 190°C) and a density of 0.950 g/cc (Sholex) to prepare a laminate. The resulting laminate was then pressed for 5 minutes under a gauge pressure of 10 kg/cm^2 using a hot press at a temperature of 160°C. The resulting bonded laminate showed a peel strength of 18.90 kg/2.5 cm width.

As described above, a film of the adhesive composition and a bonded laminate were prepared with the exception that fumaric acid was not used in the adhesive composition. The resulting bonded laminate showed a peel strength of 5.10 kg/2.5 cm width. In further comparison, a film of the adhesive composition and a bonded laminate were prepared with the exception that dicumyl peroxide was not used in the adhesive composition. The resulting bonded laminate showed a peel strength of 5.80 kg/2.5 cm width.

Example 2: In the same manner as described in Example 1, a film of the adhesive composition and a bonded laminate were prepared with the exception that 2-methylene glutaric acid was used in place of fumaric acid and that pressure was applied to the laminate at a temperature of 190°C. The resulting bonded laminate showed a peel strength of 20.10 kg/2.5 cm width.

Example 3: 100 parts by weight of an ethylene-vinyl acetate copolymer containing 24% by weight of vinyl acetate (Evatate K2010), 5 parts by weight of itaconic acid and 1.0 part by weight of 2,5-dimethyl-2,5-dibenzoyl peroxyhexane were mixed for 5 minutes in a hot roll at a temperature of 80°C to prepare an adhesive composition. The resulting composition was then pressed into a film having a thickness of 0.1 mm by a hot press at a temperature of 100°C.

The film thus obtained was placed between an aluminum foil having a thickness of 0.1 mm previously degreased with trichloroethylene and a 0.2 mm thick film of crystalline polypropylene having an intrinsic viscosity of 3.3 dl/g in decalin at 135°C (Shoallomer). The resulting laminate was pressed for 3 minutes under a gauge pressure of 10 kg/cm^3 using a hot press at a temperature of 200°C. The bonded laminate thus obtained showed a peel strength of 17.90 kg/2.5 cm width.

Example 4: An adhesive film having a thickness of 0.1 mm prepared in accordance with the procedure described in Example 1 consisting of ethylene-vinyl acetate copolymer, fumaric acid and dicumyl peroxide was placed on both sides of a 0.2 mm thick film of high density polyethylene having a density of 0.96 g/cc (Sholex) to prepare a laminate comprising adhesive-polyethylene-adhesive layers.

A stainless steel sheet (SUS 27) having a thickness of 1 mm previously degreased with trichloroethylene was then placed on one side of the laminate and a 0.1 mm thick aluminum foil previously degreased with trichloroethylene was placed in the other side of the laminate. The resulting laminate (five layers) was then pressed for 5 minutes under a gauge pressure of 10 kg/cm^2 using a hot press at a temperature of 190°C. The bonded laminate thus obtained showed a peel strength of 21.30 kg/2.5 cm width as determined by a peeling off test of the aluminum foil.

Example 5: A bonded laminate was prepared in the same manner as described in Example 4 with the exception that a 0.2 mm thick galvanized iron sheet previously degreased with trichloroethylene was used in place of the stainless steel sheet. The laminate (five layers) thus obtained showed a peel strength of 20.70 kg/2.5 cm width.

Example 6: A bonded laminate was prepared in the same manner as described in Example 4 with the exception that the polyethylene sheet used in Example 4 was replaced by a sheet having a thickness of 5 mm prepared from a mixture of 30 parts by weight of a medium density polyethylene having a density of 0.935 g/cc (Sholex) and 70 parts by weight of aluminum hydroxide using a hot press and the resulting laminate was pressed at a temperature of 160°C. The bonded laminate thus obtained showed a peel strength of 19.20 kg/2.5 cm width.

In related work, *S. Ishihara and S. Kobayashi; U.S. Patent 3,904,571; September 9, 1975; assigned to Showa Denko Kabushiki Kaisha, Japan* describe a resin composition which comprises (A) 1 to 99 parts by weight of ethylene-vinyl acetate copolymer containing 5 to 50 parts by weight of vinyl acetate and (B) 99 to 1 part by weight of a mixture consisting of 1 to 99 mol percent of unsaturated dicarboxylic anhydride and 99 to 1 mol percent of esters of carboxylic acid and, if necessary, (C) 5.0 parts by weight at most of organic peroxide.

Hydroxyethyl Methacrylate Capped Urethanes

A process described by *W.J. Owston and D.D. Howard; U.S. Patent 3,873,640; March 25, 1975; assigned to Lord Corporation* concerns the adhesive bonding of synthetic resin substrates and particularly polyvinyl chloride. The process employs shelf-stable, catalytically curable compositions including (A) at least one polymerizable monomer such as styrene, methyl methacrylate, ethyl methacrylate, etc., including mixtures of these monomers and partially polymerized syrups of one or more of them; (B) an organic polyisocyanate, preferably a diisocyanate or isocyanate functional urethane capped with hydroxy functional monomer from the group including hydroxyethyl methacrylate, β-hydroxypropyl methacrylate, allyl alcohol, vinyl alcohol and the like; (C) the reducing component of a redox catalyst polymerization system such as a tertiary amine, a metallic salt or preferably, N,N-diisopropanol-p-toluidine and, optionally, (D), a polymerizable monomer or polymer.

The proportions of (A), (B), (C) and (D) in the composition should be about 10 to 90% (A), 90 to 10% (B), 0.04 to 4% (C) and, 0 to 20% of (D). The composition is adapted to cure rapidly at room temperature on contact pressure alone to produce an environmentally stable bond between synthetic substrates by the addition of the oxidizing component of the redox catalyst system, such as benzoyl peroxide, to the catalytically curable composition or by the application of that composition over a coating of a lacquer containing the oxidizing component previously applied to the substrate. The following examples illustrate the process in which all percentages are by weight of the total composition.

Example 1: To a suitable reaction vessel equipped with stirrer, thermometer, dry air purge, and Dean Stark trap and condenser, were charged 1.0 mol of polycaprolactone triol, molecular weight average of 540, (PCP-0300), 0.65 mol of polycaprolactone diol, average molecular weight of 2,000, (PCP-0240) and sufficient methyl methacrylate monomer to give a nonvolatile content of 80% after all ingredients had been added. A known excess of methyl methacrylate monomer was added and the contents of the reaction vessel heated to boiling with stirring and dry air purge. The excess methyl methacrylate monomer was removed along with any water present in the reaction vessel, as methyl methacrylate and water form a low boiling azeotrope. The temperature was lowered to 60°C and 4.3 mol of toluene diisocyanate added to the reaction vessel. The temperature was maintained at 70° to 80°C until all -OH groups had been reacted (about 2 hours), yielding -NCO terminated urethane dissolved in the methyl methacrylate monomer, 4.3 mols of hydroxy ethyl methacrylate (HEMA) were then added to the reaction vessel and the temperature held at 70° to 80°C until all -NCO had been reacted (about 2 hours). A slight excess (about 10%) HEMA may be added to facilitate completion. Dibutyltin dilaurate or stannous octoate catalyst may also be used.

An 80% solution of HEMA capped polyurethane resulted, dissolved in methyl methacrylate monomer. This HEMA capped urethane was cooled and used in making the following adhesive:

	Ingredient	Parts by Weight
(B)	HEMA capped urethane	10
(A)	Methyl methacrylate monomer	58
(A)	Butyl methacrylate monomer	15
(D)	Methacrylic acid monomer	5
(D)	Medium molecular weight polymethyl methacrylate polymer, (Acryloid A-11)	10
(D)	Fumed silica, (Cab-o-Sil H-5)	3
(C)	Dimethyl-p-toluidine	0.4
	Hydroquinone	0.01

After blending and agitating the ingredients until a complete, clear solution was obtained, the adhesive was used to bond nylon fabric-reinforced plasticized vinyl strips to steel and polycarbonate plastic sheeting. Low carbon steel coupons 0.50 mm thick were cut to 1 x 2¾ inch sections. The polycarbonate sheet stock was ⅛ inch thick and was cut into 1 x 3 inch coupons. Such stock is available as mold-extruded sheet made from such trademarked plastics as Lexan or Merlon. Prior to bonding, the vinyl surfaces were wiped with a cloth saturated with isopropyl alcohol to remove residual oils and plasticizers. The adhesive was catalyzed with a 55% benzoyl peroxide paste dispersion in butyl benzylphthalate. The amount used to catalyze was 4% by weight, admixed with 100 parts of adhesive resin. The catalyzed adhesive was brushed onto the metal and polycarbonate bond surfaces with similar sized strips of reinforced vinyl pressed onto the adhesive to complete the test composite.

The glue line thickness was measured and found to range between 1 and 45 mils in thickness. After storage at 72°F for 30 minutes, during which time the adhesive fully hardened, the test assemblies were destructively delaminated using a commercially available tester machine. The 180° peeling force required to delaminate the vinyl from the steel strip averages 18 pounds per linear inch (pli), resulting in a failure between the vinyl and the reinforcing nylon fabric core. A similar test with the vinyl to steel assembly yielded 28 pli. A layer of vinyl was left on the adhesive surface after failure of the assembly occurred.

Example 2: The procedure of Example 1 was followed except 2.0 mols of polyoxypropylene glycol, molecular weight average of 1,000, (PPG-1025) were used instead of 1.0 mol of polycaprolactone triol, molecular weight average of 540 (PCP-0300) and 0.65 mol polycaprolactone diol, molecular weight average of 2,000, (PCP-0240), as in Example 1, and 3 mols of TDI, (an 80:20% mixture of 2,4-ditolylene diisocyanate to 2,6-ditolylene diisocyanate) were added to the reaction vessel after removal of the azeotrope, instead of 4.3 mols. When all -OH had reacted, 2.0 mols of HEMA were added and the cook continued until all -NCO was reacted.

	Ingredient	Parts by Weight
(B)	Acrylic ester urethane*	90
(A)	Methyl methacrylate monomer	8.5
(D)	Diethylene glycol dimethacrylate	1.0
(C)	Diisopropanol-p-toluidine	0.5
	Hydroquinone	0.001

*Compensated for the prepared concentration of 80% in methyl methacrylate monomer.

The above ingredients were blended together in a vessel until a clear solution was obtained. The adhesive thus prepared was then catalyzed with 2% of a benzoyl peroxide paste. The catalyzed adhesive was poured onto and spread evenly over 1 inch square bond areas of polycarbonate sections 1 x 3 x ⅛ inch in size and polysulfonate coupons ½ x 4 x ¼ inch in size. Equal size sections of the plastics were pressed onto the adhesive coated bond areas forming an overlap assembly with an overlap area of one inch. The glue lines were measured and found to range between 0.05 and 75 mils in thickness.

One specific composite was prepared using a glue line of 200 mils in thickness using polycarbonate sections. After storage for one hour, the bonded assemblies were destructively tested by shearing. All plastic substrates failed outside the glue line area, leaving a thin layer of plastic on the hardened glue surfaces. The strengths ranged from 480 to 850 psi, averaging 530 psi for the polycarbonate and 650 psi for the polysulfone stock. The adhesive had hardened to yield a structural bond within 6 minutes at 72°F after catalyzing.

Vinyl Acetate-Polyvinyl Alcohol-Monocarboxylic Acid

L. Krutzel; U.S. Patent 3,906,135; September 16, 1975; assigned to Air Products and Chemicals, Inc. describes an adhesive composition that, when cured, provides a bond having excellent water resistance. The adhesive comprises an aqueous emulsion of: (A) a mixture of (1) an addition polymer of about 50 to 100 weight percent vinyl acetate and a balance, if any, consisting essentially of one or more of dibutyl maleate, dibutyl fumarate, ethylene, butyl acrylate, and 2-ethylhexylacrylate and (2) polyvinyl alcohol; (B) a monocarboxylic acid which can esterify polyvinyl alcohol, e.g., benzoic acid; and (C) water.

Thus, it has been found that vinyl acetate polymer/polyvinyl alcohol aqueous emulsion adhesive can yield cured bonds having substantially improved water resistance by including in the adhesive a monocarboxylic acid which can esterify polyvinyl alcohol. Moreover, this increased water resistance can be attained without loss of emulsion stability and with an acceptably slight decrease in dry bond strength. The monocarboxylic acid ingredient of the composition can be aliphatic or aromatic, for example alkanoic of up to 10 carbon atoms or monocyclic aromatic of up to 10 carbon atoms. Specifically suitable are formic, acetic, propionic, benzoic, and ortho-toluic acids. Uses for the composition include the preparation of laminates containing two or more superposed layers of solid material, e.g., wood, paper, resinous film or cloth, and are especially suitable for manufacturing water-resistant cloth-to-cloth or cloth-to unsupported polyvinyl chloride film laminates. In the latter two applications, suggested cure times will vary with pH, choice of catalyst, solids content, etc., but it is generally advantageous to cure the adhesive at about 250° to 350°F for about 0.5 to 5 minutes.

Example: Three adhesive compositions were prepared according to the formulations presented in the following table.

RUN:	1	2	3
Ingredients (parts by weight)			
Polymer Emulsion	100	100	100
o-Toluic Acid in Solox (25% Solution)		8	8
Solox	8		
Aerosol O. T. (75% Aqueous Solution)	0.3	0.3	0.3
H_3PO_4 (75%)			3 drops
pH	5.55	5.02	3.3
Viscosities (centipoises)			
Initially:			
At 12 RPM	1200	1590	1660
At 60 RPM	680	880	926
After 6 days:			
At 12 RPM			1800
At 60 RPM			960
Peel Strengths (lbs./linear in.)			
Cure and Test Conditions:			
No cure, dry	2.8 ± .2	2.4 ± .2	3.5 ± .2
No cure, wet	0.7 ± .1	1.0 ± .2	0.9 ± .1
250°, dry	3.2 ± .2	2.8 ± .2	2.5 ± .1
250°, wet	1.4 ± .1	1.8 ± .1	1.9 ± .1
300°, dry	2.7 ± .1	2.6 ± .2	2.5 ± .1
300°, wet	1.5 ± .2	1.7 ± .1	3.0 ± .2
350°, dry	2.3 ± .1	2.4 ± .1	2.5 ± .1
350°, wet	1.2 ± .1	1.7 ± .1	2.0 ± .1

First there was prepared an aqueous emulsion of vinyl acetate/ethylene copolymer and polyvinyl alcohol. The solids portion of the emulsion contained about 4 weight percent of the polyvinyl alcohol and the water content of the emulsion was about 45 weight percent. The ethylene/vinyl acetate copolymer contained about 80 weight percent of polymerized vinyl acetate, and about 40% of the copolymer was insoluble in benzene. The benzene-soluble portion of the copolymer exhibited an intrinsic viscosity of about 1.5 to 1.8.

The polyvinyl alcohol had a molecular weight of about 25,000 and was the product of about 88% hydrolysis of polyvinyl acetate. The other ingredients listed in the above table were mixed with the polymer emulsion in the amounts indicated. Solox is the trademark for a commercially available solvent reported as consisting of 100 parts by weight of denatured ethanol, 1 part ethyl acetate, 1 part aviation gasoline, and 2 parts denaturing grade methanol. Aerosol O.T. is a 75 weight percent aqueous solution of sodium dioctylsulfosuccinate.

Each adhesive composition was tested for effectiveness as a lamination adhesive in laminating cotton poplin cloth to polyvinyl chloride film. The lamination procedure involved coating the polyvinyl chloride film with the adhesive, using a No. 40 wire-wound rod, immediately placing the cloth over the adhesive and rolling the resulting laminate once with a rubber roller. Each test lamination was prepared on the first day following the preparation of the adhesive composition. The laminations were then dried for 72 hours under ambient conditions. Eight test laminates, divided into four pairs, were prepared from each adhesive. One pair was tested without being subjected to high temperature curing; another pair was cured at 250°F for 5 minutes; a third pair was cured at 300°F for 5 minutes; and the fourth pair was cured at 350°F for 3 minutes. One laminate in each pair was tested for peel strength without further treatment, while the other laminate was first immersed for about one day in room temperature water and then tested for peel strength.

Peel strengths were measured on an Instron tester at a speed of 2 inches per minute. The results are reported in the previous table. In all of the peeled samples the adhesive remained mostly on the polyvinyl chloride film, indicating that the type of failure was adhesive rather than cohesive. Viscosities reported in the table were determined at 25°C with a Brookfield viscometer, using a No. 3 spindle.

The results reported in the table testify to the greater water resistance of those bonds (Runs 2 and 3) prepared with the monocarboxylic acid adjuvant, orthotoluic acid. Moreover, this increase in wet strength was attained substantially without sacrifice in dry strength, especially when cure temperature of 300°F or higher were employed, or a pH below 4.0. Viscosities were not significantly increased by the presence of the monocarboxylic acid, thus indicating that its use is compatible with the need for relatively long periods of pot stability.

Poly(Ethyl Acrylate), Ethylene-Acrylic Acid Copolymer and Epoxy Resins

R.M. Ziegert; U.S. Patent 3,905,931; September 16, 1975; assigned to American Can Company describes a water-based laminating adhesive which cures, upon aging, to a water-resistant, thermally stable material of high bond strength when used as a laminant for a variety of polymer films. The adhesive is applied as an emulsion composition containing, as essential components, a poly(ethyl acrylate), a copolymer of ethylene and acrylic acid, and an epoxy resin, each component being present within critical concentration limits.

A particularly satisfactory epoxy resin of the epichlorhydrin-bisphenol A type has an epoxy content of 180 to 195, a viscosity at 25°C of 100 to 160 poises, hydroxyl content equivalent (OH/100 grams of resin) of 0.06 and a refractive index of 1.570 to 1.575 at 25°C. The average molecular weight of the epoxide is approximately 380. About 85 grams of the resin will esterify one mol of an acid. The refractive index of the resin is 1.570 to 1.575 at 25°C. This resin is commercially available (Epon-828) and is emulsified at 40% by weight resin concentration in water using 2.5% of Elvanol 52-22, an emulsifying agent. This epoxy material has no value, by itself, as a laminating adhesive, due to complete lack of bond strength.

The poly(ethyl acrylate) component, referred to as PEA, has the very desirable property of high initial adhesion or green tack, as well as contributing substantially to the final bond strength of the laminated structure. In direct contrast, although the ethylene-acrylic acid copolymer, referred to as EAA, exhibits a very low bond strength in a peel type test procedure, when cross-linked with the third component, an epoxy resin, and in the presence of the PEA component, the composition containing EAA forms an adhesive bond which is initially quite satisfactory and which increases substantially in strength upon aging, reaching and maintaining maximum values after a period of about ten days. In addition to the cross-linking which certainly occurs between the EAA component and the epoxy resin over a period of time, it is believed that a minor, but possibly significant amount of cross-linking also occurs between the PEA component and the epoxy resin, since the ultimate bond strength achieved with a blend of the three components is greater than that achieved with any proportions of the EAA and epoxy resin without the inclusion of some PEA.

It is to be noted that, as the cross-linking type of cure proceeds, the bond formed between the plies of a laminar sheet construction not only increases in strength, but also becomes very substantially more resistant to the deteriorative effects of water, heat, abrasion, solvent vapor and the like. It has been found that substantial increases in bond strength are obtained over the first ten days of aging, after which the laminate remains relatively constant in bond strength for periods up to at least 4 to 6 months.

The preferred compositions also generally contain small but effective amounts of an aliphatic dihydric alcohol, preferably dipropylene glycol, and also of a low molecular weight alcohol such as methyl alcohol or ethyl alcohol. The inclusion of up to about 4% of the total composition weight of dipropylene glycol in the adhesive composition of this process results in an improved bond strength between a variety of flexible packaging film materials and so modifies the bond in such laminates that two film plies laminated by the adhesive will separate in a smooth, continuous motion upon the application of a separating force in excess of the bond strength, in contrast to some adhesive laminations which separate in a series of small staccato jerks under the influence of similar forces. In the preferred adhesive compositions, the amount of dipropylene glycol added is about 2% based on the total weight of a 25% solids composition, or about 8% based on the total solids weight of such a composition.

Example: The following table includes bond strength data for a lamination of polyvinyl chloride film to the saran of saran (polyvinyldinine chloride) side coated low density polyethylene film using, as a control, a laminating adhesive of 25% solids concentration, the solids portion of which contained, on a weight basis, 67% of PEA, 11% of EAA, 14% of the epoxy resin of the epichlorohydrin-bisphenol A type noted above as being particularly effective and 8% dipropylene glycol.

Effect of Glycol Component on Bond Strength

Adhesive	Bond Strength in gm./in., after				
	3 days	5 days	10 days	14 days	19 days
Control, containing dipropylene glycol as 8% of total solids	713	825	1003	Film* Tear	Film* Tear
Containing no glycol	487	572	495	399	334
Containing tetra ethylene glycol as 8% of total solids	620	710	842	856	981
Containing glycerine as 8% of total solids	390	457	832	900	718

*The term film tear indicates that a rupture of the base film occurred rather than failure of the adhesive bond.

The table gives comparative bond strength values for laminates including the above adhesive and for laminates in which the dipropylene glycol was omitted or replaced with other polyhydroxyalcohols as indicated in the table. In all cases, the base films were laminated with 2 lb, solids basis, per ream of the respective adhesive composition under a pressure of approximately 8 lb/in^2 and at a temperature of 170°F. In forming the lamination, the adhesive composition, in a 25% solids suspension, was applied to one of the base sheets by means of an etched cylinder and the aqueous suspension medium removed in an oven at about 250°F to leave a tacky coating of the resinous components on the base film. The second film was then pressed against the tacky adhesive surface by passage through a two roll pin under pressure at 170°F.

The bond strength values of the resulting laminates were obtained after varying storage periods by separating the two adhered layers on an Instron Testing Unit at a rate of 10 inches per minute pull speed. Force readings were integrated to give average bond strength in grams per inch of width of the sample.

It is evident from the data in the table, that the presence of a small amount of a polyhydric alcohol in the adhesive composition has a beneficial effect on bond strength and that the bond strength increases substantially as the bond is aged over a period of days. In some cases, the aged adhesive bond strength exceeded the tensile strength of the base film, as indicated by the designation film tear in the tabulated data.

Further tests have shown that, after a 10 day storage period, laminations made with the same adhesive composition as set forth above but with only 4% of dipropylene glycol (solids basis) displayed seal strengths about 20% lower than the preferred 8% glycol-containing composition, whereas a 12% glycol level produced seals 10% stronger than 8% glycol composition. The glycol composition may range up to about 16% of the total solids weight in the preferred compositions, substantially higher percentages of glycol tending to overplasticize the composition and cause a reduction in seal strength and of water resistance.

It should be noted that the bond strengths developed by the adhesive are somewhat influenced by the nip temperatures used in forming the pressure laminations. Temperatures above about 150°F are to be preferred, higher temperatures being more desirable, since the bonds formed exhibit greater strength. Temperatures above about 200°F should be avoided, however, since excessive heat tends to distort some of the films being laminated, with resultant puckering and poor functional properties of the laminates. The adhesive compositions display excellent bond strength to a substantial variety of sheet materials commonly utilized in the packaging of food products as well as many other commodities. The preferred adhesive compositions are particularly effective in laminating plies of polyvinyl chloride, polyvinylidine chloride, polyethylene, nylon and Surlyn (a metallic ionomeric derivative of polyethylene.)

FRP BONDING

Urethane Adduct Primer for FRP Bonding

Molded or laminated polyester compositions, especially fiber reinforced polyester compositions and thermoplastic compositions (referred to as FRP and plastic compositions, are widely utilized today to form assemblies which make up automobile body panels, truck cabs, boats and the like. The preferred method of joining molded FRP or plastics to form the desired assemblies is with a structural adhesive. It has been determined that such adhesives act to evenly distribute throughout the area of attachment stresses encountered when the assemblies are subjected to normal vibration, shocks and forces encountered when in use as part of an operating vehicle, for example. In contrast, in assemblies formed with conventional techniques such as rivets and screws, stresses encountered in use concentrate at the points of attachment, causing the holes required for the rivets and screws to crack or widen. Since the FRP or plastic is not as strong as the metal rivets or screws, holes in the FRP or plastic wear until the metal fasteners can fail by being pulled out on the seam holes.

Joining molded FRP or plastics into desired assemblies with structural adhesives is not without problems. The normally smooth surface of inert, molded FRP or plastic substrates does not foster the formation of a secure and enduring bond between the structural adhesive and the FRP or plastic substrate. Various techniques have been employed to promote adhesion between adhesives and FRP or plastic substrates.

A common practice, especially during early use of FRP and plastic assemblies, involved abrading of the bond area of the FRP or plastic substrate by grit or sandblasting to provide a surface more amenable to adhesive assembly. This technique has severe limitations and is falling into disuse. Surface abrasion by grit or sandblasting in the bond area weakens the FRP or plastic compositions in the precise areas where maximum strength is needed and renders the bond area susceptible to absorption of paint solvents which can weaken and attack the FRP or plastic.

Absorbed paint solvents are especially deleterious since they can reemerge against the structural adhesive interface causing deterioration of the adhesive bond during high temperature bake cycles used to cure the paint and primer applied to a finished assembly. Of course, abrading the bond area is costly since it involves an additional processing step and requires skilled labor to prepare the seams without destroying the cosmetic surface contour of the soon to be painted exterior automobile or truck body panels. Also, the fiberglass particles created during blasting of FRP compositions are a severe health hazard, requiring protective equipment for workers and expensive exhaust and removal equipment.

Also, the current widespread use of FRP compositions made with low profile polyester resins have made the use of surface abrasion techniques impossible due to unacceptable strength loss of the FRP. Low profile polyester resins are polyester resins filled or diluted with thermoplastic polymers such as polyvinyl chloride. The thermoplastic diluent prevents the polyester resin from shrinking away from the surface of the reinforcing glass fibers which are incorporated in the resin to give it strength.

The reduced shrinkage gives the polyester resin a low profile or very smooth regular surface as contrasted to the high or standard profile of usual FRP resins which have irregularly contoured surfaces as the result of exposed reinforcing glass fibers in the resin. Low profile FRP compositions are especially desirable in forming automotive assemblies since they yield better painted surfaces with little of the previously required sanding and preparation prior to painting. However, the low profile FRP compositions are somewhat lower in strength than standard FRP compositions due to the polyvinyl chloride diluent. Thus, any strength loss due to abrasion of the low profile FRP is unacceptable. With the recognition of the serious problems created with sandblasting, and the like, attempts have been made to develop alternative techniques for preparing FRP and plastic substrates for surface bonding with adhesives. Various treating agents or primers such as the tertiary amines described in U.S. Patent 3,647,513 have been applied to FRP surfaces prior to application of certain urethane adhesives. However, it was found that the amines and solvents do not adequately soften the FRP substrates in the area of attachment and cause severe weakening of plastic substrates.

A solution of polyisocyanate in trichloroethylene has also been utilized as a surface primer in joining FRP compositions. However, FRP compositions treated with the polyisocyanate solution had to be adhesively joined within the narrow period between the time the polyisocyanate primer initially hardened, usually after about 2 days of open air moisture cure, and the time several days after application of the solution when the polyisocyanate primer coating becomes so hardened and inert that adhesive bonds cannot be formed with it. Thus, it can be seen that the art is still searching for suitable agents for treating FRP and thermoplastic substrates to make them amenable to adhesive bonding.

W.J. Owston; U.S. Patent 3,838,093; September 24, 1974; assigned to Lord Corporation describes a method for improving the adhesion between an adhesive of the class of polymerizable liquid ethylenically unsaturated compounds containing a single terminal $>C{=}CH_2$ group and solid polyester or thermoplastic substrates comprising treating the solid polyester or thermoplastic substrate with a solution of a partial urethane adduct, applying the adhesive to the treated polyester or thermoplastic substrate and allowing the adhesive to cure. The solid polyester or thermoplastic composition is treated with a solution of from about 0.85 to about 40 parts, by weight, of a partial urethane adduct in from about 60 to about 100 parts of a low boiling, inert organic solvent, the partial urethane adduct being the reaction product of from about 0.1 to 0.7 molar equivalent weight of a hydroxy functional monomer and 1 molar equivalent weight of polyisocyanate. The solution thus prepared will be referred to as a treating agent or primer.

The partial urethane adduct is formed by combining a polyisocyanate of tri or higher functionality with an amount of a hydroxy functional monomer such as the hydroxy alkyl acrylates or methacrylates including hydroxyethyl methacrylate, (HEMA), hydroxypropyl methacrylate or other hydroxylated vinyl functional monomers such that the OH:NCO ratio is 0.1 to 0.7 and heating the mixture to above room temperature up to about 130°C. Mixtures having an OH:NCO ratio within this range may be obtained by combining from about 0.1 to about 0.7, preferably 0.2 to 0.6 molar equivalent weight of a hydroxy functional monomer with about 1 molar equivalent weight of polyisocyanate.

This reaction will generally take place in a nonreactive, compatible solvent such as styrene, methylmethacrylate, trichloroethylene, toluene or the like under a dry atmosphere and is continued until about 50% of the reactive NCO groups of the polyisocyanate are converted to a urethane linkage through reaction with the functional hydroxy moieties of the monomer. The resultant partial urethane adduct thus contains some free NCO moieties. Polyisocyanates of tri or higher functionality include the phosgenated mixture, polymethylene polyphenylisocyanate, obtained by phosgenation of the rearrangement product of the reaction of aniline and formaldehyde described in U.S. Patent 2,683,730. Polyisocyanates of this class are known to contain small quantities of a diisocyanate to improve the overall balance of physical properties thereof. A solution of the partial urethane adduct just described may be utilized alone as an adhesion promoting primer for solid poyester or thermoplastic compositions; however, it has also been found that solutions containing a mixture of this adduct and a partial urethane adduct formed by combining a polyisocyanate with hydroxylated butadiene polymers or copolymers may be especially useful for applications where flexibility and lack of brittleness in the primer film are desirable.

Example 1: 131 grams (1 equivalent weight) of polyisocyanate A and 13 grams (0.1 equivalent weight) of 2-hydroxyethyl methacrylate (HEMA), where polyisocyanate A is the phosgenated mixture obtained by phosgenation of the rearrangement product of the reaction of aniline and formaldehyde as described in U.S. Patent 2,683,730 together with 51 grams of anhydrous trichloro-ethylene solvent were added to a nitrogen purged reaction vessel equipped with a reflux condenser. The mixture was heated for 2 hours at 180°F while being stirred. After cooling to room temperature, a clear partial urethane solution was obtained ready for use. Like mixtures were prepared in the same manner utilizing 0.2, 0.3, 0.5, 0.6 and 0.7 equivalent weights of HEMA respectively.

15% solids solutions of these partial urethane adducts in trichloroethylene along with a like solution of unmodified polyisocyanate A, were applied to clean FRP sections cut from 0.125 mil sheet stock. The primer films were cured for 24 hours at 72°F and 55% relative humidity and post baked at 280°F for 30 minutes. The baked sections were cooled to room temperature and cut into 1" x 4" coupons.

The coupons were subjected to an impact force of 30 inch-lb using a Gardner Impact Tester. Coupons treated with a solution of polyisocyanate A alone were fractured by the impact, the cracked film being flaked off by scraping with a knife edge. Coupons treated with the solution containing 1 equivalent weight polyisocyanate A and 0.1 equivalent weight HEMA fractured on impact, but resisted flaking with a knife edge. Multiple impacts of the Gardner mandrel fractured this film so that it would flake. Treating coupons with the other adducts containing greater amounts of HEMA resulted in the formation of films which were only slightly cracked after multiple impacts with the Gardner mandrel and resisted all attempts to lift or be removed with the knife edge.

Example 2: 1" x 4" test coupons cut from ⅛" FRP test stock were dipped into 12% solids solutions in trichloroethylene of the urethane adducts of Example 1. The primed coupons were dried for 15 minutes at ambient, room temperature conditions and cured at room temperature, exposed to ambient moisture (curing method A), or cured by baking in an air convection oven at 280°F for 10 minutes (curing method B). Coupons cured in each manner were bonded together at varying times after being cured with the modified acrylic adhesive described in Example 7 of U.S. Patent 3,333,025 catalyzed with 3%, by weight, of a mixture of 55% benzoyl peroxide in butyl benzyl phthalate forming one inch square single overlap bond joints. The adhesive was allowed to cure at 75°F for 24 hours and the bonded assemblies were destructively tested by shearing with a tensile tester using a crosshead speed of 1" per minute.

The table on the following page contains the results of these tests and illustrates the quality of the adhesive bond formed between FRP coupons treated with the various primer solutions described therein. Ideally, the adhesive bond should have a strength equal to or in excess of that of the polyester or thermoplastic composition it is bonding together. For this reason the destructive tests carried out on the samples in the following table were expected to cause failure within the FRP composition itself rather than in the bonded joint. Failures occurring between the FRP compositions and the structural adhesive indicate that the adhesive has failed; such failures are unacceptable and are abbreviated as "AF" in the table.

Acceptable failures, those indicating integrity, or greater strength in the adhesive bond than in the FRP composition are called fiber tearing or partial fiber tearing failures, abbreviated as "FT" or "PFT" respectively. Thus in the table, fiber tearing failure (FT), is the standard of acceptability against which the usefulness of the bonding methods must be compared.

Failure Occurrence on Destructive Testing

Sample number	Urethane adduct solution: Polyisocyanate A[1]	Urethane adduct solution: HEMA[1]									
1	1	0.7	PFT	AF	PFT	PFT	PFT-AF	PFT	AF	PFT-PF	PF
2	1	0.6	PFT	PFT	PFT	FT	FT	FT	PFT	PFT	PF
3	1	0.5	FT	FT	FT	FT	FT	FT	FT	FT	PF
4	1	0.3	FT	FT	FT	FT	FT	FT	FT	FT	PF
5	1	0.2	FT	FT	FT	FT	FT	FT	FT	FT	PF
6	1	0.1	PFT	PFT	FT	PFT	PFT	PFT	PFT-AF	PFT	PF
7	1		PFT	PFT	FT	PFT	PFT	AF	AF	AF	PF
Curing method: A=room temperature cure; B=Bake			B	B	B	B	B	B	B	B	A
Layover between cure and adhesive application, (days)			(2)	1	5	10	15	20	30	40	(2)

[1] Molar equivalent weight of ingredients dissolved in trichloroethylene at 12% solids.
[2] 1 hour.

The above table illustrates that a primer solution containing polyisocyanate A alone loses its effectiveness with aging. Resistance to tearing is increased when HEMA is combined with the polyisocyanate forming a partial urethane primer, especially in samples 2 through 5. All samples cured at room temperature prior to adhesive application exhibited failure in the primer itself.

Polyurethane Adhesive

H.J. Fabris, D.R. Sponseller, H. Uelzmann and A.E. Schmucker; U.S. Patent 3,886,122; May 27, 1975; assigned to The General Tire & Rubber Company describe polyurethane-based adhesive compositions comprising an intimate blend of (A) an isocyanate-terminated prepolymer preferably containing from about 5 to 30 weight percent free isocyanate groups and (B) a cross-linking mixture of an organic polyhydroxy compound free of isocyanate groups and containing groups having active hydrogen atoms (OH or OH plus a small amount of NH) reactive with the isocyanate groups in the prepolymer and a urethane catalyst.

This process comprises the improvement wherein the ratio of the total isocyanate reactive hydrogen atoms to isocyanate groups is from about 1.1:1 to 1.9:1, the ratio of the calculated branching coefficient to the critical branching coefficient is at least 1.4:1 and the critical branching coefficient is the reciprocal of f-1 wherein f is the weighted average functionality of the reacting molecules in the composition which have functionalities greater than 2. The branching coefficient is calculated as outlined in *Principles of Polymer Chemistry,* Chapter IX, page 347 et seq. by P.J. Flory, Cornell University Press, 1953 based on the assumption that all isocyanate groups have reacted. The critical branching coefficient is the value above which formation of a continuous polymer network occurs. When the recited criteria are met, the composition is sufficiently beyond its gel point to develop adequate strength at elevated temperatures. The following examples illustrate the process. Unless noted otherwise, all quantitative measurements are parts by weight.

Example 1: In this example component (A) was an isocyanate-terminated polymer prepared by heating for 1 hour at 100° to 105°C a mixture of 211.4 parts of polypropylene ether glycol (1,000 average molecular weight), 61.95 parts of polypropylene ether glycol (400 average molecular weight), 170.8 parts of an 80/20 mixture of 2,4- and 2,6-tolylene diisocyanates (TDI) and 126.5 parts of talc, yielding a prepolymer having a viscosity of 24,000 cp at 25°C and a free isocyanate content of 11 weight percent corresponding to an equivalent weight of 384. Component (B), the curative or cross-linker, was a blend prepared by mixing to a homogeneous solution at 50° to 70°C the following ingredients:

Ingredient	Parts
a commercial polypropyleneether triol having 1000 average mol. wt.(OH No.=168)	20.0
N,N,N',N'-tetrakis-2-hydroxy propyl ethylenediamine	47.0
a commercial propylene oxide adduct of methyl glucoside having a hydroxyl number of 300, an equiv. wt. of 181, a hydroxyl functionality of four and a molecular weight of 724	22.0
2,6-di-t-butyl-4-methyl-phenol	1.0
trimethylolpropane	5.5
dibutyltin dilaurate	0.05
ethylenediamine	3.0
a commercial 1,4-dialkylamino-anthraquinone dye (Calico Oil Blue ZV, American Cyanamid)	0.025
triethylene diamine	1.0

Component (B) had a viscosity of 6,000 cp at 23.3°C and an average equivalent weight of 97. A number of 2.54 x 7.62 x 0.32 cm samples were cut from a molded commercial FRP sheet (glass fiber mat impregnated with polyester-styrene mixture and cured) and exposed to air for 20 hours at 75°C and 25% relative humidity. The sample surfaces to be bonded were wiped with tissue paper saturated with a 2% solution in methylene chloride of a commercial phosgenated aniline-formaldehyde condensation product having an average molecular weight of 350 and an average isocyanate functionality of 2.6 (Mobay Mondur MRS).

A series of adhesives were prepared by mixing components (A) and (B) in various ratios corresponding to various values of r (as shown below). Pairs of samples were bonded together in a 90° crosswise fashion with the center overlapping 6.45 cm^2 bonded by applying a 0.89 mm layer of each adhesive on the previously wiped surfaces and pressing the surfaces together with finger pressure. The bonded samples were cured by heating them for 6 minutes at 120° ± 5°C. After cooling for about an hour at room temperature, the samples were destructively tested on a cross-pull tester. The cross-pull tester holds the cross shaped sample in such a way as to permit application of tension stress normal to the adhesive layer. There may be adhesion failure at the interface between the adhesive and the substrates, cohesive (CFA) and cohesive failure in the substrates (CFS). The principal type of failure was visually estimated in each case. The results were as follows:

Run	I-1	I-2	I-3	I-4	I-5	I-6
Component (A) (parts)	45	40	35	30	25	20
Component (B) (parts)	10	10	10	10	10	10
r, [H_R/NCO]	0.89	1.0	1.14	1.33	1.6	1.89
$\frac{\alpha(\text{calc.})}{\alpha(\text{crit.})}$	—	2.75	2.36	1.98	1.6	1.34
Gel time, RT about 25°C. (min.)	5	3.5	3.75	3.25	2.5	2.5
Gel time, 120°C. (min.)	2.5	2.25	2.5	1.75	2.5	3.0
Break strength (kg.) (ca 25°C.)	23.1	31.3	40	39.5	42.6	42.6
Failure (type and %)	CFA100	CFA95	CFA75	CFA90	CFS50	CFS80

None of the examples exhibited failure at the interface.

Example 2: A second curative component (B-II) was prepared by blending into a homogeneous solution 80 parts of a commercial poly(propylene oxide)triol having an average molecular weight of about 4,700, 18 parts of the commercial polypropylene oxide adduct of methyl glucoside the same kind as used in Example 1, 1.5 parts of dibutyltin dilaurate and 1.33 parts of the TDI the same kind as used in Example 1 to increase viscosity. After stirring for 1 hour at room temperature (about 25°C) there were added 35.5 parts of N,N,N',N'-tetrakis-2-hydroxypropylethylenediamine and 2.0 parts of ethylenediamine after which the whole mixture was stirred at 40° to 50°C until homogeneous. The resulting component (B-II) had a viscosity of 6,000 cp at 24.4°C and an equivalent weight of 198. Adhesives were prepared by mixing component (B-II) with various amounts of prepolymer (A) from Example 1. Test specimens were then prepared, cured and tested as described in Example 1 with the following results.

Run	II-1	II-2	II-3
Component (A) (parts)	25.6	29.1	33
Component (B-II) (parts)	10	10	10
r, [H_R/NCO]	1.13	1.0	0.88
$\frac{\alpha(\text{calc.})}{\alpha(\text{crit.})}$	2.54	2.94	—
Gel time, 25°C. (min.)	2.75	2.75	3.25
Gel time, 120°C. (min.)	1.5	1.5	1.5
Break strength (kg.) (ca 25°C.)	43.5	43.1	40.8
Failure (type and %)	CFS100	CFS55	CFA80

Example 3: In this example a prepolymer was prepared by reacting 49.5 parts of a commercial polypropyleneether glycol (1,000 average molecular weight) with 26 parts of 80/20 TDI in admixture with 24.5 parts of talc at 100° to 105°C for approximately 1 hour. The resulting prepolymer had a free isocyanate content of 8.4% and a viscosity of 22,000 cp at 25°C. This prepolymer was blended with an equal weight of the same type of prepolymer used as component (A) in Example 1 resulting in a component (A-III) for this example having a free isocyanate content of 9.44%, a viscosity of 23,000 cp at 25°C and an average equivalent weight of 457. A curative component (B-III) was prepared by mixing until homogeneous the following ingredients in the parts by weight as indicated:

the commercial polypropyleneether glycol (1000 av. mol. wt.) as used above	20.0
N,N,N',N'-tetrakis-2-hydroxy propylethylenediamine	47.5
the commercial propylene oxide adduct of methyl glucoside the same as used in Example I	23.54
2,6-di-t-butyl-4-methyl-phenol	1.0
trimethylolpropane	5.5
ethylenediamine	2.25
dibutyltin dilaurate	0.05
the dye as used in Example I	0.02

The resulting component (B-III) mixture had a viscosity of 6,500 cp at 23.9°C and an average equivalent weight of 101. Test sample strips were cut as described in Example 1, conditioned at 60% and 80% relative humidity (RH) at 25°C for 24 hours prior to bonding and wiped as described in Example 1. Prepolymer component (A-III) and curative component (B-III) were then mixed together in the ratio r, H_{R}/NCO as shown in the table below, and the adhesive mixture was then used to coat the sample strips. Test samples were then prepared from these strips, cured and tested, all as described in Example 1 with the following results.

Run	III-1	III-2	III-3	III-4	III-5
RH (%)	60	60	80	80	80
r, $[H_R/NCO]$	1.1	1.2	1.2	1.3	1.4
α(calc.) / α(crit.)	2.47	2.24	2.24	2.05	1.89
Break strength (kg.) (ca 25°C.)	29.0	41.7	43.1+	43.1+	43.1+
Failure (type and %)	CFA100	CFS100	CFA90	CFS75	CFS100

RUBBER ADHESION

When fabric materials such as nylon, vinylon, rayon, or polyester synthetic fibers are used as a reinforcing member for rubber articles such as pneumatic tires, belts, air cushions, hoses, or rubber vibration insulators, it is necessary to firmly adhere rubber compounds to these synthetic fabric materials. The adhesive composition used for treating a large quantity of fabric materials is preferably noninflammable and nonpoisonous. From the above point of view, it is convenient and economical to use water as the medium for an adhesive composition. It is also desirable that the adhesive composition be durable and have a long shelf life.

As the conventional adhesive composition for this purpose, there may be mentioned a mixture of a rubber latex and a condensed product obtained by the reaction of resorcinol with an aldehyde, particularly formaldehyde, in the presence of a catalyst such as an alkaline material or acid material (generally called RFL). This RFL has been and is now widely used as an adhesive liquid suitable for mass production.

The amount of consumption of synthetic rubbers, particularly styrene-butadiene copolymer (SBR) has increased year by year. SBR is generally mixed with natural rubber (NR), but the compounding rate of SBR tends to increase with the progress of rubber processing techniques and, furthermore, compositions consisting of SBR alone are now increasing.

When a blended composition of SBR and NR or a composition of NR alone is to be reinforced by a synthetic fabric material, a synthetic fabric material treated with RFL, which is believed to be ideal, is very effective for reinforcing a composition of NR alone and a blended composition of SBR and NR containing a high ratio of NR. When said synthetic fabric material is treated by RFL, however, it is not so effective for reinforcing a blended composition of SBR and NR containing a high ratio of SBR because the adhesive strength between the synthetic fabric material and the blended composition of SBR and NR containing a high ratio of SBR is low. Since such a low adhesive strength results in the shortening of the life of a rubber product subjected to a large, periodic deformation, it is desirable to prepare an adhesive composition having improved adhesive strength between SBR and a synthetic fabric material.

In a conventional process for producing RFL, resorcinol and formaldehyde are reacted in a solvent such as water in the presence of an alkaline catalyst and then latex is added after the reaction product reaches a certain degree of condensation. In general, excess, preferably 1.25 to 2.5 mols, of formaldehyde is added to 1 mol of resorcinol. As the alkaline catalyst, there are used alkali metal hydroxides such as sodium hydroxide and potassium hydroxide or weak bases such as amines and, particularly, ammonium hydroxide; the resulting condensation product of resorcinol with formaldehyde belongs to the resol type. This area has received considerable attention in recent years as steel wire reinforcement and new aromatic polyamide type fibers are being used in tires.

Adhesive compositions for elastomer-to-metal represent a highly developed art and many commercial formulations are available. Adhesive compositions which are employed include mixtures of chlorinated rubber and at least one polyalkylene polyamine adhesion promoter; mixtures of halogenated ethylene-propylene copolymer and sulfur; mixtures of chlorosulfonated polyethylene, orthoalkoxy aryl diisocyanates and dinitrosobenzene; chlorinated rubber-epoxylated novolak epoxy resin curing agent admixtures; mixtures comprising chlorine-containing polymers, polyisocyanates, epoxylated novolaks, gamma-methacryloxypropyltrimethoxysilane and dinitrosobenzene; and the like. This work continues at a good pace as the performance demands for rubber adhesion to a variety of substrates continually increase.

TIRE CORD—FABRIC ADHESION

Resol-Type Resin

T. Honda, T. Ueki and Y. Fukuura; U.S. Patent 3,884,859; May 20, 1975; assigned to Bridgestone Tire Co., Ltd., Japan describe a process for the production of an adhesive composition for adhesion between fabric material and rubber which is characterized in that a monohydric phenol is reacted with an aldehyde in the presence of an alkali metal hydroxide as a catalyst, the mixture is dissolved in dilute ammonium hydroxide in the absence of free aldehyde and then mixed with a rubber latex.

The process is as follows: A monohydric phenol and aldehyde are reacted in a solvent such as water and organic solvent, for example, alcohols, ketones, esters, and aromatic hydrocarbons, in the presence of an alkali metal hydroxide such as sodium hydroxide, potassium hydroxide, and the like, and then an acidic material such as an organic acid, for example, acetic acid, formic acid, propionic acid, chloroacetic acid, glyoxylic acid, benzene sulfonic acid, p-toluene sulfonic acid, and an inorganic acid, for example, hydrochloric acid, carbonic acid, sulfuric acid, boric acid, nitric acid, and phosphonic acid, are added to separate the resulting resin.

The thus separated resin is dissolved in an aqueous solution of ammonium hydroxide, or an amine compound having a low boiling point, for example, methyl amine, dimethyl amine, trimethyl amine, ethyl amine, diethyl amine, triethyl amine, n-propyl amine, n-butyl amine, n-amyl amine and the like, in the absence of free aldehyde, and thereafter, rubber latex is added to prepare a liquid adhesive. It has been found that the adhesive composition produced by the above described process has excellent adhesive properties for rubber compounds containing a high ratio of SBR. The adhesive composition has, of course, good adhesion to the NR compounds.

An important point in the process is to produce the resol type resin first by using an alkali metal hydroxide as a catalyst for the reaction between the monohydric phenol and the aldehyde. (An acidic material should not be used because a novolac type resin is produced thereby.) Ammonium hydroxide or basic organic compounds such as amines should not be used as the alkaline catalyst in the process. When amines are used as catalysts, a complicated reaction occurs between the amines and the free aldehyde produced during the reaction; in addition, the amine will react with a phenol to produce a long chain, water insoluble resin.

It is also important that free aldehyde should not be present when the separated resinous material is dissolved in ammonium hydroxide. When free aldehyde is present, a long chain, water insoluble resin is produced as mentioned above. This is not desirable.

The advantage of the adhesive composition of the process is that the adhesive composition can be put in practical use just after being produced by dissolving the resinous material in dilute ammonium hydroxide and mixing with rubber latex. Even if the adhesive composition is allowed to stand for more than one month after being produced, the adhesive strength is not decreased. On the contrary, a conventional RFL can not be put into practical use unless RFL stands and ages for sixteen hours after being produced, and can not be used for as long a period as the adhesive composition of the process.

The resinous material is easily dissolved in ammonium hydroxide and a transparent solution is obtained. If the resinous material is dissolved in ammonium hydroxide in the presence of free aldehyde, the resinous material will react with the free aldehyde to produce a white precipitate which is not soluble in water. In this process, the resinous material is dissolved in ammonium hydroxide in the absence of free aldehyde.

Example: One hundred parts (1.1 mols) of phenol, 440 parts (5.4 mols) of 37% formalin and 20 parts (0.12 mol) of 6N caustic soda were reacted with stirring together in a flask fixed with a reflux condenser at 80°C for 1.5 hours. The solution was then cooled and neutralized with 6N acetic acid and a precipitate settled out of solution.

The precipitate was collected by a centrifugal separator, dissolved in 500 parts of acetone, evaporated and dried under reduced pressure, and crushed to a powder of resinous material. Forty parts of said resinous material was dissolved in dilute ammonium hydroxide (50 parts of 28% ammonium hydroxide was diluted with 550 parts of water) to obtain a uniform, clear solution. The resulting solution was mixed with 400 parts of 40% vinylpyridine styrene butadiene terpolymer latex, stirred sufficiently and immediately put in practical use as an adhesive solution.

The nylon 6 cord of 1260 D/2 was dipped in the adhesive solution and was dried at 190°C for 5 minutes. The rubber compositions A and B were prepared as shown in the following table. Composition A is a blended rubber composition of NR and SBR containing a high ratio of NR, and composition B is a blended rubber composition containing high ratio of SBR.

Rubber Composition

Composition	A	B
NR	80	20
SBR	20	80
Carbon black	40	40
Softner	10	10
Stearic acid	2	2
N-phenyl-β-Naphthylamine	1.5	1.5
2-benzothiazolyl disulfide	0.75	0.75
diphenyl guanidine	0.75	0.75
Sulfur	2.5	2.5
Adhesion strength between the cord treated with the adhesive solution, Kg/cord	2.8	2.5

The adhesive strength was measured as follows. The treated cords were set near the surface of the unvulcanized rubber composition sheet and were vulcanized under a pressure of 30 kg/cm^2 at 150°C for 20 minutes. One cord was separated from said vulcanized rubber sheet at the speed of 30 cm/min to measure separation resistance. The separation resistance per one cord was designated as the adhesive strength.

For comparison, RFL was produced as follows: Eleven parts (0.1 mol) of resorcinol was dissolved in 519 parts of water; 16.2 parts (0.2 mol) of 37% formalin was added thereto. Ten parts (0.025 mol) of 10% aqueous solution of caustic soda was added immediately, the mixture was allowed to stand for four hours; 244 parts of 40% vinylpyridine styrene butadiene terpolymer latex was added, and thereafter, the mixture was aged for sixteen hours.

The preparation of test pieces and tests with RFL were carried out by the same process as in the example. The rubber compositions A and B were the same as those of the example. The adhesive strength of composition A as 2.6 kg/cord; the adhesive strength of composition B was 1.3 kg/cord. From the example and the comparison, it is found that the adhesive composition of the process is not different from RFL for the rubber composition containing a low ratio of SBR, but is remarkably better than RFL for the rubber composition containing a high ratio of SBR.

Polyester Fiber-Rubber Bonding with Polynucleus Halophenol Compound

According to a process described by *T. Ozeki, K. Ito, Y. Maeda and T. Kudo; U.S. Patent 3,855,168; December 17, 1974; assigned to Mitsubishi Rayon, Ltd., Japan* the bondability of polyester fiber to rubber is markedly improved by incorporating a polynucleus halophenol compound (A) into a conventional adhesive composition consisting of a thermosetting resin (B) of resorcinol and formaldehyde and a rubber latex (C), in such a proportion that the solid weight ratio (A)/(B) is 10/1 to 1/10 and the solid weight ratio [(A) + (B)]/(C) is 2/1 to 1/10.

(A)

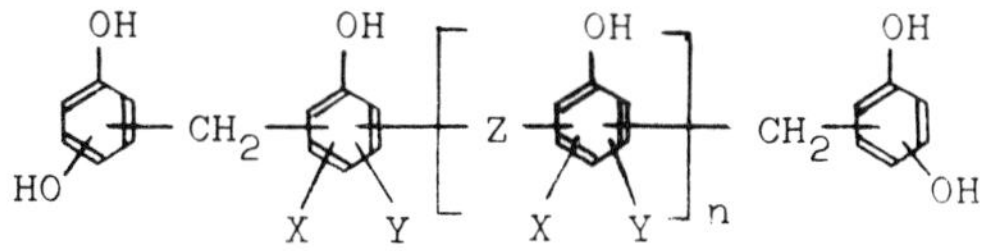

Compound (A) has the general formula above where X is chlorine or bromine, Y is chlorine, bromine, hydrogen, hydroxyl or an alkyl of 1 to 4 carbon atoms, Z is $-S_x-$ in which x is an integer of 1 to 4; $-S(O)_y-$ in which y is 1 or 2; or

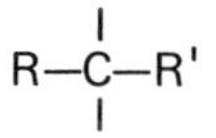

in which R and R' are hydrogen or $C_{1\text{-}8}$ alkyl, $C_{6\text{-}18}$ aryl, $C_{7\text{-}18}$ aralkyl, $C_{1\text{-}8}$ haloalkyl or furyl groups, and n is an integer of 1 to 6.

Compound (A) may be prepared in high yields by, for example, the following reaction scheme:

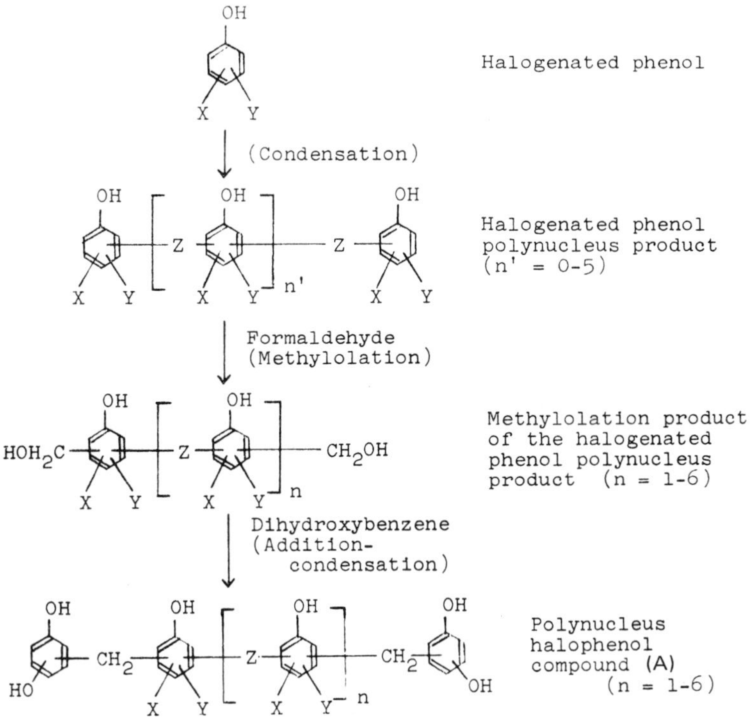

In the above formulas, X, Y and Z have the same meanings as defined above.

Examples of polynucleus halophenol compounds (A) for use in this process are as follows:

(A_1)

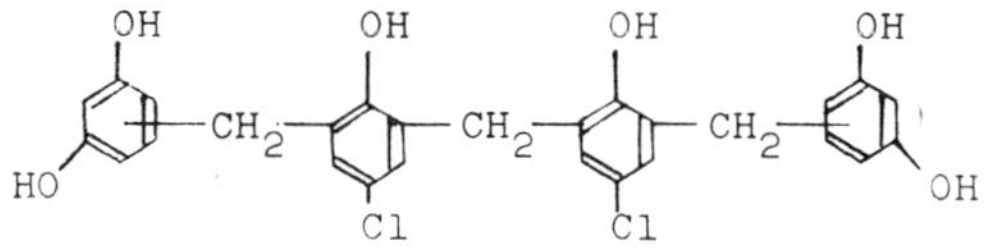

(A_2)

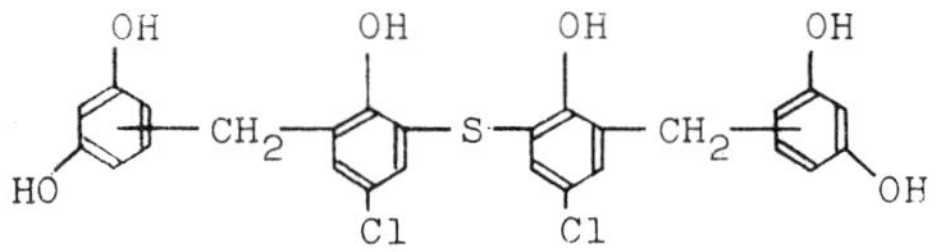

(A_3)

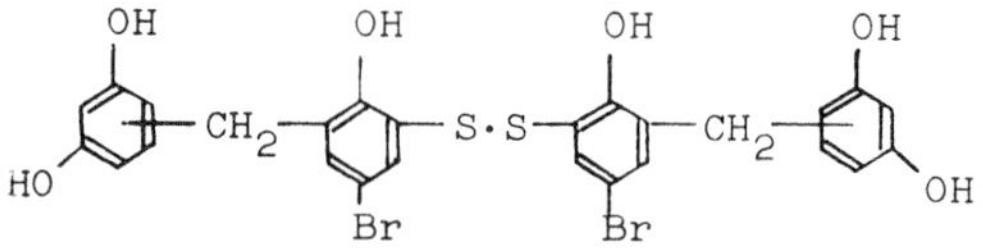

(A_4)

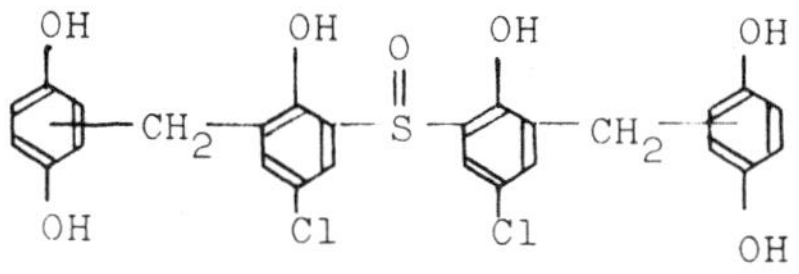

The adhesive composition has a very excellent effect as an adhesive for firmly bonding rubber to polyester fiber in the fields where the best use of the characteristic properties of polyester fiber is made, such as tires and conveyor beltings. The following examples illustrate the process. The bond strength was measured at a pulling speed of 500 mm/min according to the cord pull-out test (H-test) described in *India Rubber World,* 114, pages 213-217 (1946) and expressed in terms of kg/9 mm. In the examples, all parts are by weight.

Example 1: Preparation of Aqueous Solution of a Polynucleus Halophenol Compound — Five mols of formaldehyde was added to 1 mol of bis(5-chloro-2-hydroxyphenyl)methane and subjected to condensation in an aqueous alkaline solution. Subsequently, the reaction system was neutralized with acetic acid. The reaction product which was separated was recrystallized from aqueous methanol to obtain pure bis(5-chloro-2-hydroxy-3-hydroxymethylphenyl)methane. In 500 g of dioxane was dissolved 0.5 mol of the thus obtained compound, mixed with a solution of 2.5 mols of resorcinol in 500 g of water and allowed to react with heating under reflux for 4 hours.

Then, the reaction mixture was added dropwise under stirring into a large excess of water to precipitate a yellowish brown solid, the chief constituent of which was found on elementary analysis to be a polynucleus halophenol compound [Compound (A_1)] represented by the above-given formula (A_1). Sixty parts of the compound (A_1), 30 parts of 28%-ammonia and 210 parts of water were mixed to obtain a solution [solution (A_1)].

Preparation of Resorcinol-Formalin-Rubber Latex — 11.0 parts of resorcinol, 9.7 parts of 37% formalin, 3.0 parts of a 10% aqueous solution of sodium hydroxide and 116.3 parts of water (140.0 parts in total) were allowed to react with stirring at 25°C for about 6 hours to obtain an aqueous solution of thermosetting resorcinol-formaldehyde resin (B_1). 140 parts of the aqueous solution (B_1) was added to 280 parts of a vinylpyridine-styrene-butadiene copolymer rubber latex (C_1) of 40% solid content and left standing for 24 hours at 25°C to obtain a liquid mixture, (B_1) + (C_1) (hereinafter referred to as RFL-1).

300 parts of the solution (A_1) and 200 parts of water were added to 420 parts of RFL-1 to obtain an adhesive composition, (A_1) + (B_1) + (C_1), about 20% in solid content.

Two 1,000-denier 192-filament yarns of polyethylene terephthalate were twisted at 500 turns/m of the primary twist and at 500 turns/m of the final twist to obtain a cord. The cord was dipped into the above-obtained adhesive composition, then dried at 130°C for 1 minute, and heat-treated at 240°C for 1 minute. The amount of the adhesive applied was about 6.5%. The treated cord was embedded in an unvulcanized rubber compound mentioned below, and bonding was effected by vulcanizing at 140°C under a pressure of 20 kg/cm^2 for 40 minutes.

The adhesion of the polyester fiber to rubber in the above laminate was evaluated by means of the H-test. The mean pull-out force was found to be 13.6 kg/9 mm at room temperature. Observation of the fracture in test specimen revealed that the rupture took place not in the bond layer but, in most of the cases, in the rubber. On the other hand, the above laminate was maintained at a temperature of 120°C for 10 minutes and the H-test was carried out under these conditions to obtain a mean pullout force of 10.8 kg/9 mm, showing a favorable adherence retention at high temperatures.

In order to examine stability of the adhesive composition of this process, the above adhesive composition was kept at room temperature for 1 month and then applied on a cord in a manner similar to that mentioned above. On evaluating the adherence by the H-test, it was found that the pull-out force was 13.2 kg/9 mm, showing an excellent stock stability. The breaking strength of the polyester cord used in the present example was 14.5 kg. Composition of the unvulcanized rubber compound was as follows:

	Parts
Natural rubber RSS No. 3	100
Zinc oxide	5.0
Stearic acid	2.0
Carbon black, GPF	30.0
Antioxidant D (phenyl-β-naphthylamine)	1.0
Aromatic oil	7.0
Sulfur	2.25
Accelerator DM (dibenzothiazole disulfide)	1.0

Comparative Example 1: For comparison, a cord was treated in a manner similar to that in Example 1 with RFL-1 which was composed of (B_1) and (C_1), that is the composition obtainable by eliminating the component (A_1) from the adhesive composition, $(A_1) + (B_1) + (C_1)$ in Example 1. It was found that the mean pull-out force by the H-test was 5.5 kg per 9 mm at room temperature and 3.8 kg per 9 mm at 120°C, and rupture took place, in most of the cases, at the interface of polyester and the adhesive layer.

On the other hand, a cord was treated in a manner similar to that in Example 1 with an adhesive composition, $(A_1) + (B_1) + (C_1)$, in Example 1, in which (B_1) had been replaced by an aqueous sodium hydroxide solution containing formaldehyde, that is the component (B_1) without resorcinol. The mean pull-out force by the H-test was 5.7 kg per 9 mm at room temperature, and that rupture took place in the adhesive layer.

Comparative Example 2: For comparison, an adhesive composition was prepared by using a compound of the formula

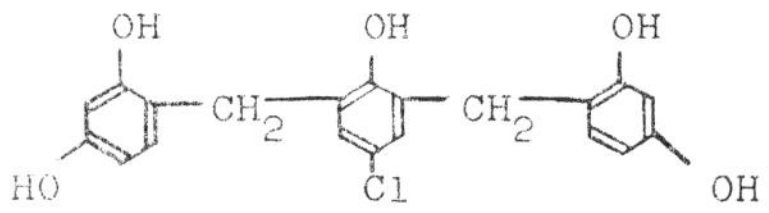

described in British Patent 1,140,528 in place of the component (A_1) in the adhesive composition, $(A_1) + (B_1) + (C_1)$, in Example 1. A cord was treated in a manner similar to that in Example 1 with the above composition. The mean pull-out force by the H-test was 13.5 kg per 9 mm at room temperature and 9.8 kg per mm at 120°C, showing a decreased adherence retention at high temperatures.

Example 2: 0.5 mol of bis(5-chloro-2-hydroxyphenyl) sulfide, 4 mols of formaldehyde (162 g of 37% formalin) and 1 mol of sodium hydroxide were dissolved in about 1 liter of dioxane, and allowed to react with stirring at 40°C for 48 hours. After completion of the reaction, the reaction mixture was neutralized with acetic acid to pH 6, freed from the excess formaldehyde and dioxane by distillation, and washed with water to obtain dimethylol derivative of bis(5-chloro-2-hydroxyphenyl) sulfide.

0.5 mol of the above dimethylol derivative was mixed with 2 mols of resorcinol and allowed to react under a nitrogen atmosphere while heating up to 150° to 170°C and removing the liberated water by distillation. The reaction mixture was freed from the excess resorcinol by extraction with warm water and the residue was dried to obtain a resinous matter melting at 150° to 170°C. On elementary analysis, the main constituent was found to be a polynucleus halophenol compound [compound (A_2)] represented by the above-given formula (A_2).

The above compound (A_2) was treated in a manner similar to that in Example 1 to obtain 300 parts of a 20% aqueous ammonia solution [solution (A_2)]. 300 parts of the solution (A_2) and 200 parts of water were added to 420 parts of RFL-1 to obtain an adhesive composition, $(A_2) + (B_1) + (C_1)$, 20% in solid content. The adhesive composition was violet black in color and showed no gel formation on long storage.

In a manner similar to that in Example 1, a polyester cord was dipped into the above adhesive composition, dried and heat-treated. The treated cord was embedded in natural rubber compound similar to that in Example 1 and vulcanization-bonded. The mean pull-out force by H-test was 13.4 kg/9 mm at room temperature and rupture took place in rubber. Further, the H-test was carried out at a high temperature of 120°C; the mean pull-out force was 11.2 kg/9 mm and rupture took place in most of the cases, in rubber. In a manner similar to that mentioned above, a cord was treated with the above adhesive composition which had been stored for five days, and then vulcanized. The H-test at room temperature gave a mean pull-out force of 13.2 kg/9 mm.

Ethylene-Propylene-Diene Terpolymers and Resorcinol-Aldehyde Resin

According to a process described by *K.F. King; U.S. Patent 3,884,862; May 20, 1975; assigned to E.I. du Pont de Nemours & Company* a liquid adhesive composition having beneficial utility as a tire cord adhesive can be prepared by

(1) dry milling an elastomer which is a copolymer of ethylene and at least one other monomer including a diene with a reinforcing carbon black so that the weight ratio of dispersed carbon black to elastomer is about 25:75 to 50:50;

(2) adding an organic solvent for the elastomer and mixing to form a liquid composition;

(3) mixing the liquid composition in a high-shear mixer with water containing an emulsifying agent to form an emulsion;

(4) removing the organic solvent to form a latex, and

(5) mixing with the latex a resorcinol-aldehyde resin component so that the elastomer:resin weight ratio is about 50:50 to 91:9.

An especially preferred elastomer is an ethylene/1,4-hexadiene copolymer in which each 100 parts by weight of the copolymer contains about 60 to 84 parts of ethylene and 16 to 40 parts of 1,4-hexadiene. This copolymer preferably has an unsaturation value of about 2 to 4. Another very useful elastomer is a terpolymer containing in each 100 parts by weight thereof about 50 to 62 parts of ethylene, 20 to 42 parts of propylene and 4 to 16 parts of 1,4-hexadiene. It is preferred in many uses of the process that the terpolymer has an unsaturation value of about 0.25 to 2.

The adhesive composition is especially useful for providing a strong bond between reinforcing components made of nylon fibers and rubber compositions whose principal elastomer component is a terpolymer of ethylene, propylene and a diene. For example, automobile tires can be made from such rubber compositions and nylon tires cord coated with the novel composition which have excellent cord-to-rubber adhesion both in a dynamic adhesion test and in a static adhesion test.

Also, when the nylon cord is replaced with rayon cord coated with the process composition, excellent adhesion is again obtained; and the harmful tendering of the rayon cord encountered with certain prior art bonding means (e.g., chlorine-liberating compositions) is eliminated. The composition also has beneficial utility in the manufacture of industrial belts, elastomer coated fabrics, and rubber hose.

Example 1: All quantities shown are on a weight basis unless otherwise indicated. A tire cord adhesive is prepared in this example. First an elastomeric terpolymer is provided; it is composed of 43.0% ethylene, 43.8% propylene and 13.2% 5-ethylidene-2-norbornene (unsaturation value about 1.0). Such a terpolymer can be made in solution by reacting the monomer mixture in tetrachloroethylene in the presence of a vanadium type coordination catalyst; and the solvent can then be removed. The terpolymer has an inherent viscosity of 2.31 and a Mooney viscosity of 54. The Mooney viscosity values given herein are at ML1 + 4/121°C.

A liquid/adhesive composition useful as a tire cord adhesive is prepared by

(a) mixing 190 g of the terpolymer with 3,000 ml of benzene and 450 ml of isopropanol to form a 6% solution of the terpolymer in the solvent mixture;

(b) forming an emulsion of the terpolymer solution in water by mixing 1,540 g of the solution (which is at a temperature of 48°C) in a high-shear, high-speed turbine-stator or homogenizer type of mixer with 770 g of water containing 1.2% of an emulsifying agent (sodium lauryl sulfate), the water being gradually added to the terpolymer solution during high-shear mixing and the mixing being continued for 5 minutes after adding all the water;

(c) removing the benzene and isopropanol in a distilling apparatus which is allowed to operate until the head temperature reaches about 99°C;

(d) cooling the resulting latex to 23°C;

(e) forming a more concentrated or creamed latex having a total solids content of about 50% by mixing with the step (d) latex 0.1 part of 1% sodium alginate for each 100 parts of water in the latex, allowing the composition to stand for 16 hours, removing the creamed latex which separates out, and further concentrating the creamed latex in a centrifuge;

(f) adjusting the total solids content of the remaining latex to 41% by adding a small amount of water;

(g) mixing 62.26 g of the latex with 59.33 g of distilled water at 24°C and cooling the diluted latex to 9°C;

(h) adding slowly to the latex with stirring 2.84 g of 28% aqueous ammonium hydroxide and stirring for another 5 minutes;

(i) mixing with the resulting latex 28.88 g of a resin component (resorcinol-formaldehyde resin solution) and stirring slowly for another 3 minutes, said resin component having a pH of about 7.2 and being the product of adding (while stirring) to 158.4 g of distilled water at 24°C a 40.28 g portion of a 1.67% aqueous sodium hydroxide solution and stirring for another minute, slowly adding 24.84 g of resorcinol flakes and stirring for another 5 minutes, adding 36.36 g of 37% aqueous formaldehyde solution and stirring for another 2 minutes, and allowing the composition to age without stirring for 6 hours;

(j) aging the resin-modified latex at 24°C for 12 hours, the composition now having a pH of 10.2;

(k) adding to the composition while stirring 38.3 g of a 33.3% dispersion of SAF carbon black (a reinforcing type) in water containing a small amount of an anionic dispersing agent composed of highly purified partially desulfonated sodium lignosulfonate; and

(l) adding to the resulting composition 25.6 g of distilled water.

The weight ratio of carbon black to terpolymer in the composition is 33.67. The resulting composition has the utility described in the paragraph just prior to Example 1. It is very useful as a nylon tire cord adhesive; and it is useful as an adhesive for tire cords and fabrics made of fibers or filaments of other materials, for example polyesters, glass, steel, and rayon.

Example 2: A composite reinforcing cord for use in the manufacture of rubber tires and the like is prepared by

(a) providing a quantity of nylon tire cord made from continuous filament 66 nylon yarn having a denier of 1,260 g per 9,000 m by twisting one of these yarns in a ply twist (in the Z direction) and then bringing two ends of the resulting yarn together and applying a cord twist (in the S direction);

(b) passing the cord at the rate of 20 yards per minute through a container of the adhesive composition prepared in Example 1 at a temperature of 25°C;

(c) passing the coated cord through a 146°C heat zone for 2.6 minutes while the cord is stretched 1%;

(d) passing the cord through a 215°C heat zone for 0.8 minute while the cord is stretched 12%;

(e) passing the cord through a 204°C heat zone for 1.25 minutes while the cord is allowed to relax 2%; and

(f) cooling the composite cord to 24°C and winding it on a storage spindle to await use in the manufacture of tires.

Example 3: The coated tire cord prepared in Example 2 is used in the manufacture of rubber tires as follows. A quantity of carcass skim stock is prepared by first mixing the ingredients shown below in Formula 3.1 in a water-cooled intensive internal batch mixer using a ram pressure of 100 psi and a speed of 40 rpm. Ingredients 1 through 7 are added in the order shown; the mixer is closed and run until the temperature in the mixer reaches 121°C; ingredient No. 8 is added in small portions over a period of 10 minutes while the mixture is kept at about 113°C; then the completed Formula 3.1 composition is removed from the mixer and allowed to cool to 25°C.

Ingredient No. 1 is an elastomer terpolymer of 53% ethylene, 40% propylene and 7% 1,4-hexadiene, it has a Mooney viscosity of 70 and an inherent viscosity of about 3.0. Such a terpolymer can be made in solution by reacting the monomer mixture in tetrachloroethylene in the presence of a premixed coordination catalyst formed by combining $VOCl_3$ with diisobutyl aluminum monochloride; then the solvent is removed.

Ingredient No. 2 is an elastomeric terpolymer of 59% ethylene, 36% propylene and 5% 1,4-hexadiene; it has a Mooney viscosity of 40; inherent viscosity is 2.2. Such a terpolymer can be made in solution by reacting the monomer mixture in tetrachloroethylene in the presence of a coordination catalyst formed in situ by combining VCl_4 and diisobutyl aluminum monochloride; hydrogen modification can be used as described in U.S. Patent 3,051,690 for control of molecular weight.

Ingredient No. 3 is a high temperature, high abrasion furnace black known as HS-HAF black. Ingredient No. 4 is a fine extrusion furnace black known as FEF black. Ingredient No. 7 is Reichhold Chemicals Company's Betaprene H-100, which is understood to be the product of (a) adjusting the olefin-containing by-product of a hydrocarbon cracking process so that it contains about 6% by weight of each of 2-methyl-2-butene and dicyclopentadiene with the remainder characterized by substantially equal weight proportions of isoprene and piperylene, and (b) heating the mixture at about 70° to 80°C in the presence of $AlCl_3$ until the reaction is substantially completed.

Ingredient No. 8 (Sunpar 2280 from Sun Oil Co.) is a paraffinic petroleum oil having a viscosity-gravity constant of about 0.796, a specific gravity at 60°F of 0.8916, a Saybolt viscosity at 210°F of 165 seconds and a flash point of 595°F.

Formula 3.1

Ingredient No.		Parts
1	Elastomer	75
2	Elastomer	25
3	Furnace black	50
4	Furnace black	40
5	Zinc oxide	5
6	Stearic acid	1
7	Isoprenoidal resin	40
8	Paraffinic oil	40

The preparation of the skim stock is completed by mixing for a period of 1.25 minutes a quantity of the resulting Formula 3.1 composition in a Banbury Mixer operating at 28 rpm with the following curing agents: 1.5 phr (parts per hundred of elastomer present) of zinc dimethyldithiocarbamate, 1.5 phr of tetramethyl thiuram disulfide, 0.8 phr of 2,2'-dithiobisbenzothiazole and 3 phr sulfur (Crystex 90, a rubber compounding grade of amorphous sulfur from Stauffer Chemical Co.); passing the resulting mixture twice through a water-cooled two-roll rubber mill adapted to form a 0.5 inch thick rubber sheet; and cooling the resulting skim stock by passing it through an aqueous soap-stone dispersion, and then drying it.

Automobile tires are manufactured by combining the coated tire cord prepared in Example 2 with the skim stock prepared in the previous paragraph on a calendar whose rollers are at about 82°C; positioning the resulting composite material in the usual manner on a tire building drum and forming a green tire carcass drum; and, by a conventional tire building procedure, adding ethylene/propylene/diene rubber tread stock and curing the composite structure in a tire-curing press. The tire cord of each tire is now bonded to the rubber of the tire with a very durable bond.

Aqueous Polyurea Adhesive

A process described by *L.L. Wood; U.S. Patent 3,897,585; July 29, 1975; assigned to W.R. Grace & Co.* provides improved polyurea adhesive components prepared by reacting polyisocyanates with polyamines in a reaction moderating solvent. The resultant reactant mixture of noncrosslinked polyureas may then be disposed onto a filament or cord as desired with final curing, i.e., by use of a crosslinking agent; by application of external energy such as heat, oxidative environments and the like; and accompanied by removal of the reaction moderating solvent. The cured crosslinked polyureas are characteristically tough, abrasion resistant resins.

The adhesive systems are particularly useful as bonding agents for nonwoven and woven natural or synthetic fibers and fabrics; cords, yarns, strands, ropes, filaments and the like; plastics, metals, and glasses; as crosslinking agents for various other resins including epoxy resins, isocyanated, carboxylated, and halogenated polymers; and as functional additives to conventional cord-elastomeric adhesives such as especially those used for tire cords presently. Tire elastomers of particular merit are those comprising styrene/butadiene copolymers, polybutadiene, polyisoprene, natural rubber, ethylene/propylene copolymers, and ethylene/propylene/nonconjugated diene multipolymers.

In bonding polyester tire cord to rubber, the aqueous polyureas adhesives can be used either in a one-step dip, i.e., the polyureas adhesive is added to the RFL formulation prior to treatment of the cords, or by a two-step dip in which the polyureas adhesive is used in a first dip which acts as a tie coat for a second conventional RFL dip.

Example 1: SBR (styrene-butadiene-rubber) was compounded according to ASTM D15-59T for sample preparation for physical testing of rubber products using the following recipe:

Ingredients	Parts by Weight
SBR 1502 (General Tire & Rubber Co.)	100
Zinc oxide	5
Sulfur	1.5
MBTS (Atlax by R.T. Banderbilt Co.)	1.75
EPC Black (Spheron 9 by Cabot Corp.)	40

The prepared rubber was sheeted on a two-roll mill and stored at 0°C to prevent premature vulcanization.

Example 2: EPDM (ethylene-propylene-diene monomer) rubber was compounded according to the procedure outlined in DuPont booklet, *Nordel,* April 1964, p. 8, using the following recipe:

Ingredients	Parts by Weight
EPDM (Nordel 1070 by DuPont)	100
HAF Black (Vulcan by Cabot Corp.)	80

(continued)

Ingredients	Parts by Weight
Zinc oxide	5
Process Oil (Shellflex 411 by Shell)	40
Tetramethylthiuram disulfide	0.75
Mercaptobenzothiazole (MBT)	1.5
Sulfur	2
Dipentamethylenethiuram tetrasulfide	0.75

Again, the prepared rubber was sheeted on a two-roll mill and stored at 0°C to prevent premature vulcanization.

Example 3: RFL adhesives were prepared generally according to Moult, "Bonding Textiles to Rubber", Chap. 42, *Handbook of Adhesives,* Reinhold Publishing Co., 1962, using the following:

(1) Resorcinol-formaldehyde solution (F:R mol ratio = 2.5):

	% Active	Grams Active	Grams Total
Resorcinol	100	21.5	21.5
Formaldehyde	36.9	14.6	39.6
NaOH, 10% w/w	10.0	1.3	13.0
Water, distilled	-	-	299.9
	10.0	37.4	374.0

The NaOH, resorcinol, and formaldehyde were added in that order to the water with stirring. The solution was stirred for 4 hours with cooling to keep the temperatures at 15° to 25°C.

(2) RFL adhesive:

	% Active	Grams Active	Grams Total
Latex	t	143.0	(143) (100/t)
Water, distilled	-	-	626—(143) (100/t)
RF solution (1)	10.0	37.4	374
	18.0	180.4	1,000

The latex t containing 8% solids was added to the water with stirring. This diluted latex was added to the RF solution (1) with stirring, and the mixture allowed to age overnight before use. Generally, the RFL adhesives were then used within 3 days. Additives to the RFL were generally added after aging, and water content was adjusted to keep the overall concentration at 18%.

Example 4: 1,000 grams polyoxyethylene glycol having a weight average molecular weight of 1,000 was outgassed by stirring at 110°C for about 3 to 5 hours. The remaining glycol having a weight of about 980 grams was cooled to 50°C and then added as a liquid over a period of 15 to 20 minutes under an inert nitrogen atmosphere, to 325 grams of tolylene diisocyanate. The tolylene diisocyanate had 80% by weight of the 2,4 isomer, and 20% by weight of the 2,6 isomer. The reaction temperature was maintained between about 50° and 60°C with stirring continuing for 3 to 5 hours. Isocyanate capped polyoxyethylene glycol was recovered from this reaction.

Example 5: A solution of 103 grams of diethylene triamine in 500 ml dry acetone was stirred at 60° to 65°C under a dry nitrogen atmosphere for 2 hours. The resultant solution was cooled to 5° to 10°C. Next, 980 grams of the isocyanate capped polyoxyethylene glycol prepared in Example 4 in 1,000 ml dry acetone at 50° to 10°C was added, with good stirring. The resultant clear, pale yellow solution was stirred at 10° to 15°C for 2 hours.

The product solution was next warmed to 60° to 70°C, and approximately three-fourths of the acetone representing about 1,100 ml was removed by distilling the solution. To keep the product fluid, 1,000 ml water was added and most of the remaining acetone representing about 300 to 400 ml was distilled at less than 80°C. The remaining yellow product syrup was diluted to 20% solids content by adding 4,400 ml water. A noncrosslinked amino-containing polyurea polymer in solution having a pH 9.5 was recovered.

Example 6: 5,400 grams of the aqueous polyurea polyamine polymer solution recovered in Example 5 was warmed to 60°C. 555 grams of epichlorohydrin were added to the aqueous prepolymer solution over a period of about 5 minutes with good stirring. The resultant solution was stirred at 60°C for about 1 to 4 hours until the Brookfield viscosity at 25°C reached 500 to 600 cp. A turbid tan colored product resulted which was then diluted with 12,400 ml cold water. The diluted product solution was cooled to 25°C after which the pH was adjusted to 5.5 using dilute aqueous hydrochloric acid.

Although the polyureas composition generally is applied to the cord by dipping, it may also be applied during spin finishing, weaving or twisting operations. In processing metal cords, it is useful to first shot blast the cord, clean the blasted cord with a suitable solvent especially for grease or oil removal, and thereafter apply the adhesive. A useful metal cord is brassed steel (1 x 3 + 5 x 7 + 1) x 0.0059 kn, with a brass coating of 70/30 copper/zinc ratio.

Example 7: Polyester tire cord (1260/2) was stretched across an aluminum frame and dipped in 1,000 ml of the product solution from Example 6. The sample was then dried at 110°C for 20 minutes and dipped in the RFL preparation of Example 3. After drying again at 110°C for 20 minutes, the sample was imbedded with SBR rubber compound of Example 1 on a one-fourth inch H-test frame (ASTMD 2138-67). The sample was then cured for 30 minutes at 150°C. The cord pull-out strength at 23°C was 19 pounds. By contrast, when the RFL was used alone, the pullout value was 11 pounds.

Example 8: Example 7 was repeated except that the product solution (500 ml) from Example 6 was admixed with RFL adhesive (1,000 ml) from Example 3 prior to dipping of the polyester cord. After dipping, drying, imbedding and curing, the cord pull-out strength was measured and found to be greater than 15 pounds.

Example 9: The procedures of Examples 7 and 8 were repeated except using as the tire cord, both polyamide (nylon 6.6) and Fiber B, a trademarked product by du Pont de Nemours for a modified polyamide tire cord. Comparable results were realized. Also it was noted that placements of the tire cord in bias, radial, or bias belted configurations resulted in good adhesion indicating that the particular configurations were not critical for adhesion using the present compositions.

Polyamide-Rubber Bonding

G.E. Van Gils and E.F. Kalafus; U.S. Patent 3,888,805; June 10, 1975; assigned to The General Tire & Rubber Company have found that an aqueous alkaline dispersion of a rubbery vinyl pyridine copolymer, trimethylol phenol, and a resorcinol compound, in certain amounts is useful in forming an adhesive for bonding polyamide or nylon cords to rubber compounds or stocks. After dipping the cord in the adhesive dip, the coated cord is heated to dry it and heat cure or heat set the adhesive on the cord. Thereafter, the adhesive containing cord is combined or laminated (calendered) with a curable rubber compound and the resulting assembly is cured.

High H-adhesions are obtained with this one-step or single treatment coating dip. It is particularly useful for bonding crystalline polyamides having a large number of aromatic groups in the polymeric structure, e.g., Fiber B. The following examples illustrate the process. In these examples the parts are parts by weight unless otherwise indicated.

Example 1: Cord Dip Compositions –

Material	A	B
Vinyl pyridine rubber latex*	100 g	100 g
Trimethylol phenol**	5 g	10 g
Resorcinol-formaldehyde resin***	5 g	10 g
Concentrated ammonium hydroxide (28% NH_3 in H_2O)	10 drops	10 drops
Water	50 g	50 g

*Aqueous alkaline (about pH 10.5 to 10.7) dispersion of a terpolymer of 70 parts by weight of butadiene-1,3, 15 parts of styrene and 15 parts of 2-vinyl pyridine, 41% by weight total solids (about 38% by weight rubber solids, balance surfactant, stabilizer, etc.).
**60% total solids in H_2O, Bakelite BRL-1031.
***50% total solids in H_2O, preformed water-soluble resorcinol-formaldehyde novolak resin, Penacolite R-2170.

Part of the water was used to dilute the viscous trimethylol phenol and another part after the addition of the ammonium hydroxide was used to dilute the resorcinol-formaldehyde resin. Thereafter the diluted solutions were mixed with the vinyl pyridine latex.

Polyamide tire cards were then dipped in the foregoing dips and were heat-treated under stretch (about 2%) at 450°F for 90 seconds in air, laminated with the standard rubber stock described in the table below, and molded; and the resulting assembly was cured for 30 minutes at 315°F. The polyamide tire cords used were of duPont Fiber B (1500/3), a long chain synthetic aromatic polyamide (poly-p-phenylene terephthalamide) in which at least 85% of the amide linkages are attached directly to aromatic rings. Fiber B is highly crystalline, has a denier of 4500, a breaking strength of 180 lb, a tenacity of 18 grams per denier, an elongation at break of 4%, a modulus of 350 grams per denier, a loop tenacity of 8.0 grams per denier, a growth and creep of 0.49 (lgpd, 30 minutes, 75°F), and a 160°C shrinkage of 0%.

After curing, the assembly was removed from the mold, cooled and tested according to the H-adhesion test. The following H-adhesions were obtained for the cords treated as described above: Dip A, 63 lb; Dip B, 67 lb.

Standard Type Rubber Composition

Material	Parts by Weight
Natural rubber	46.64
Butadiene-styrene rubber copolymer, average 23.5% bound styrene (SBR-1500)	38.5
Polybutadiene (solution polymerized BD, about 93% cis-1,4, Raw Mooney ML-4 at 212°F, about 40-50)	15.0
Carbon black, fast extrusion furnace	45.0
Hydrated silica (Hi-Sil 233, Pittsburgh Plate Glass)	15.0
BLE-25 antioxidant (a high temperature reaction product of diphenylamine and acetone, Naugatuck Chemical Division of Uniroyal)	2.0
Processing oil (a blend of highly aromatic petroleum fractions)	5.0
Zinc oxide	3.0
Stearic acid	1.5
Cohedur RL (a 1:1 mixture of resorcinol and Cohedur A (the pentamethyl ether of hexamethylol melamine which is a colorless, viscous liquid which liberates formaldehyde on heating, Naftone, Inc.)	4.7
N-tert-butyl-2-benzothiazole-sulfenamide (Santocure NS, Monsanto Chemical Co.)	1.2
Sulfur	3.0

Example 2: The method of this example was the same as that of Example 1, above, except that the cords were dipped only in Dip B. The dipped and dried cords were then cured and tested in the same type rubber stock and also in the same type rubber stock which omitted the Cohedur RL. H-adhesion tests were run on the cured samples and the results are shown below.

Stock	H at RT*	H at 250°F
Same as Example 1	70 lb	54 lb
Same as Example 1 without Cohedur RL	66 lb	50 lb

*Room temperature, about 25°C.

Example 3: The method of this example was the same as that of Example 1, above, except that Dip B was diluted with more water to lower the total solids (TS) content of the dip. After curing in the same type rubber stock, H-adhesion tests were run and the results of the tests are shown below.

TS in % (by weight of aqueous dispersion)	RT Adhesion	250°F Adhesion
17½	68 lb	52 lb
15	68 lb	58 lb
13	70 lb	60 lb

Example 4: A concentrated tire cord dip was prepared by adding to 100 parts of a 41% by weight (total solids) vinyl pyridine latex 15 parts of BRL 1031 and then 5 parts of Penacolite R 2170 which was first made alkaline with a few drops of concentrated ammonia. The viscosity of the resulting aqueous dip (40% TS), which was initially 12 cp, increased to 20 cp after 20 days and to 24.5 cp after 31 days. The adhesive properties of the dip were tested by taking a sample of the dip, diluting it with 200 parts of water to 15% by weight total solids and using it as a cord dip for Fiber B. The freshly prepared dip showed an H-adhesion of 71 lb at RT. After 10 days, the H-adhesion was 69 lb at RT, and after 34 days, it was 68 lb at RT. In this example, except where noted, the chemical compounds, the cord used, and the processing and testing procedures were the same as in Example 1, above.

TIRE CORD—STEEL WIRE

Styrene-Butadiene-Carboxylic Acid Latex and Epoxy

There are many products that require rubber bonded to metal such as electrical components like rubber isolators, pry bars, switch gear; industrial products like bridge bearings, dock bumpers, and hangers; and automobile parts like bushings, engine mounts, shock absorbers, and foot pedals. In this latter group is the growing importance of steel cord reinforced pneumatic tires.

Bonding rubber to different substrates involves a variety of factors such as the polarity between the respective substrates, the composition of the metal, the presence of lubricants, the difficulties encountered with foreign matter at the interface, to name just a few. Specifically with respect to steel and steel alloys and further with respect to steel cord used in reinforced rubber tires, bonding the rubber to the steel has posed extremely difficult problems.

Not only is the surface of steel somewhat incompatible with many adhesives and many rubber compounds, but steel has a tendency to corrode in the presence of water-containing adhesives and rubbers so that both initial and long-term bonding has to be conducted under rather anhydrous conditions.

There are two useful methods of bonding rubber to steel; one is to plate the steel surface with brass for direct contact to the rubber, and the other is by the use of a nonaqueous, i.e., organic solvent-based adhesive on the steel prior to pressing the rubber into contact therewith. These practices develop acceptable bonding but involve a variety of disadvantages. Brass plating is extremely expensive, tends to fail in areas of curvature in the steel, and is very sensitive to the composition of the rubber.

The organic solvents in the adhesives produce a toxic environment and the risk of fire. Further, the bond between rubber and brass-plated steel and between rubber and conventional adhesive coated steel deteriorate with time; this occurs when the adhesive coated steel and the rubber coated steel is exposed to air. These adhesion problems have increased to the point that the compounder must first compound the rubber to obtain usable adherence to the metal and then alter the compound to enhance rubber properties. Thus, compounders are severely restricted in their use of rubbers. There appears in the art, therefore, a real need for a better adhesive for bonding rubber to metal.

G.E. Van Gils; U.S. Patent 3,859,239; January 7, 1975; assigned to The General Tire & Rubber Company describes a composition for bonding bright steel and bright alloy steel to natural rubber and synthetic replacements therefor comprising the following mixture where the ingredients are in parts by weight per 100 parts by weight of latex solids:

> A styrene-butadiene-carboxylic acid polymeric latex having a styrene to butadiene ratio of from 30:70 to 70:30 and containing the polymerization product of from 0.5 to 2.5 parts of a polymerizable carboxylic acid;
>
> An epoxy compound, that is water-soluble to the extent of at least 97% by weight, in an amount from about 10 to 50 parts; and
>
> At least a curable amount of an epoxy curing agent.

A number of steels and alloy steels are in a variety of rubber-containing products such as tire belts, tire beads, drive belts, and bushings. By bright is meant that the steel is not brass plated but is treated to prevent or at least reduce corrosion such as by phosphatizing, chromic acid soaking, or nitrate treatment, etc. where required. That is, some steels are highly corrosion resistant without extra treatment; in these cases no treatment is needed.

In addition, these bright steels and alloy steels are generally sandblasted and pickled in acids to remove traces of foreign matter. Specific among the alloy steels usable with the curable adhesive composition of this process is the type known as E4130 which is an electric furnace steel containing carbon, manganese, chrome, and molybdenum as alloying agents. It is one of the types that does not require a corrosion resistant treatment.

The steel or alloy steel may be used in a variety of sizes and shapes such as plates, billets, bar stock, formed products, cords and braided steel wire, and other configurations. Particularly important, however, are steel wires and cords useful as reinforcement in the carcass of pneumatic tires. Among these tire cord steels is the type known as Bekaert which has the following composition.

Carbon	0.701%
Manganese	0.490%
Sulfur	0.020%
Phosphorus	0.018%
Silicon	0.280%
Copper	0.021%
Chromium	0.010%
Nitrogen	0.005%
Iron	Balance

The following examples illustrate the process. Unless otherwise noted, all parts are parts by weight and all percentages are percentages by weight.

Example 1: An aqueous composition was prepared according to the process by first making a latex of the following recipe.

Ingredients	Parts by Weight
Styrene	42.50
Butadiene	42.50
Water	93.00
Surfactant	
Monawet MB45	1.25
Dowfax 2A1	0.25
Catalyst, ammonium persulfate	0.25
Chain terminator, Sulfol 120	1.00
Chelator, Sequestrene N3T	0.05

The emulsion was agitated at 60°C for 6 hours after which 1.00 part of methacrylic acid, dissolved in 10.00 parts water and 0.5 part (surfactant) Siponate DBS 10, was added and the agitation and heat continued thereafter for a total of 12½ hours. Conversion was 99+% and total solids were 45%. To 222 parts of the 45% solids latex (100 parts total solids) was then added 200 parts of a 10% aqueous solution of a water-soluble epoxy compound, NER-010A, which is a diglycidyl ether of glycerol in a clear liquid form having an epoxy content of

153, a viscosity of 110 cp at 25°C, a specific gravity of 1.23. The epoxy solution was stirred until fully blended into the latex. Thereafter, 160 parts of a 10% aqueous solution of tetraethylenepentamine (epoxy curing agent) was stirred into the latex until a homogeneous blend was formed.

Pieces of E4130 electric furnace steel of a size of 2⅜" x 1" x 1/16" were sandblasted, pickled for 5 minutes at 80°C in a 5% aqueous solution of itaconic acid, and then rinsed in water and dried. A curable natural rubber tire stock compound was prepared according to the following formulation.

Ingredients	Parts by Weight
Natural rubber	100.00
Carbon black	50.00
Sulfur	3.80
Accelerator	1.00
Activator	12.00
Plasticizer	3.00
Antioxidant	2.00

The rubber was milled until workable and the rest of the ingredients added and dispersed therein per general rubber compounding procedures. The aqueous composition was thereafter brushed onto one surface of the pieces of steel and the excess was allowed to run off. The coating was dried at room temperature and then the pieces were heated to 450°F for 3 minutes to cure the adhesive.

Thereafter, strips of the curable natural rubber stock compound were pressed tightly against the adhesive coated surfaces to form a rubber-adhesive-steel laminate. The laminate was heated at 287°F for 1 hour to cure the rubber. Testing was done by stripping the rubber from the metal. Values of 120 to 150 pounds of stripping force were obtained for the different pieces; any failure was always in the rubber phase. In a similar test wherein no adhesive was used, there was no adhesion at all between the rubber and the metal.

Example 2: An aqueous composition was prepared by first making a latex of the following recipe.

Ingredients	Parts by Weight
Styrene	48.00
Butadiene	50.00
Water	93.00
Surfactant	
Monawet MB45	1.25
Dowfax 2A1	0.25
Catalyst, ammonium persulfate	0.25
Chain terminator, Sulfol 120	1.00
Chelator, Sequestrone N3T	0.05

The emulsion was agitated at 60°C for 6 hours after which 1.50 parts of itaconic acid and 0.5 parts of methacrylic acid, dissolved in 10.00 parts water and 0.5 parts (surfactant) Siponate DBS10, was added and the agitation and heat continued thereafter for a total of 12½ hours. Conversion was 99+% and total solids were 48%. To 208 parts of the 48% solids latex (100 parts total solids) was then added 160 parts of a 10% aqueous solution of a water-soluble epoxy compound Araldite RD-2 and blended therein. Then 64 parts of a 10% aqueous solution of tetraethylenepentamine was blended therein. A sample of Bekaert steel wire, a bright steel alloy wire rubber tire reinforcement, was sandblasted and cleaned of adhering dust and other foreign matter and then rendered corrosion resistant by dipping it in a heated (90°C) 1% solution of sodium dihydrogen phosphate, rinsing with water, and drying.

The construction of the wire was 1 x 3 + 5 x 7 x 0.0059 inches. This is a tire cord terminology and means that the cord is made of 1 center bundle (strand) of 3 wires (filaments) and is surrounded by 5 bundles of 7 wires wherein all of the wires are 0.0059 inch nominal diameter.

The tire cord was cut into 16, 12" segments and each segment dipped into the adhesive composition, shaken to remove the excess, air dried at room temperature, and then cured at 450°F for three minutes. One portion of each of the 16 pieces of adhesive coated wire was then sandwiched (perpendicularly) in between strips of curable polyisoprene rubber stock to form a pad 9½" long, ½" wide and ⅝" high.

Thereafter the pad with the wires embedded therein was cured in an oven at 287°F for 35 minutes. The pad containing the wires was trimmed and placed in the top vise grip of a testing machine and the free ends of the wires clamped into the bottom grip of the machine. All 16 wires were individually pulled from the rubber pad and the average of the 16 pulls calculated to be the pull-out adhesion strength.

The experiment was repeated three more times using (1) brass-plated Bekaert wire cord (without adhesive), (2) brass-plated Glanzstoff wire cord (without adhesive), and (3) bright Bekaert steel wire cord with Thixon adhesive (a commercial adhesive) in lieu of the adhesive of this process. Pull-out tests were conducted on each of these four different wires and the averages are shown in the table below. Separate samples of each were made and tested later, after exposure to air at room temperature for various periods; these values also appear in the table.

	- - - Average Pull-Out Adhesion in Pounds - - -			
	0	1 Day	2 Days	8 Days
(1) Bright tire cord with adhesive of this process	165	156	151	160
(2) Brass-plated Bekaert wire	158	–	147	98
(3) Brass-plated Glanzstoff wire	189	–	148	83
(4) Bright tire cord with commercial adhesive	87.4	–	–	–

This example shows that the composition of this process not only provides a high degree of bonding but the bond strength is maintained at or near its original level after prolonged exposure to a normal atmospheric environment whereas the other bonding methods show rapid deterioration in bond strength under the same conditions.

Cobalt Naphthenate Promoter

According to a process described by *C. Bellamy; U.S. Patent 3,897,583; July 29, 1975; assigned to Uniroyal, SA, Mexico* adhesion of rubber to metal (e.g., brass-plated wire tire cord) is improved by incorporating a cobalt salt (e.g., cobalt naphthenate) into a rubber stock containing an adhesive resin forming system, particularly a methylolated nitroalkane (methylene donor) and a resorcinol type methylene acceptor, prior to vulcanizing the rubber in contact with the metal.

In the process the rubber is mixed with the usual vulcanizing and other compounding ingredients appropriate to the particular article being manufactured (e.g., pneumatic tire, conveyor belt, hose) along with the cobalt salt and adhesive-resin forming ingredients, that is, the methylolated nitroalkane (A) and resorcinol-type methylene acceptor (B).

The stock is then laminated or coated (e.g., by calendering, extrusion, etc.) onto the metal reinforcement, which is frequently in the form of wire, as in wire tire fabric (e.g., for use as breaker, belt or carcass plies, including radial plies) or tire bead wire. The resulting assembly or laminate is fabricated or shaped and cured as in conventional practice appropriate to the specific article being manufactured. The adhesive bond is developed at elevated temperature during the curing or vulcanizing step.

Example: An example of suitable metal reinforcement for use in the process is wire tire cord of a 5 x 7 x 0.0058 inch over twisted 3 x 1 x 1 construction commonly employed in making the plies of a pneumatic tire. The surface of each strand is coated with brass [an alloy of 70% copper and 30% zinc (± 5%)] at the rate of 5.5 ± 1.5 grams of brass per kilogram of steel.

A suitable rubber stock may be prepared by mixing 30 parts of natural rubber, 25 parts of cis-polybutadiene rubber (95% cis content), 67.5 parts of oil extended styrene-butadiene rubber (made up of 45 parts of copolymer containing 23% styrene and 22.5 parts of naphthenic petroleum hydrocarbon extender oil), 60 parts of carbon black, 5 parts of zinc oxide, 1.5 parts of stearic acid, 10 parts of naphthenic petroleum hydrocarbon process oil and 1 part of an antioxidant, namely, a reaction product of diphenylamine and diisobutylene. To this masterbatch may be added 0.7 part of N-tert-butyl-2-benzothiazole sulfenamide, 0.2 part of diphenylguanidine and 2 parts of sulfur.

Adhesion test samples may be prepared and tested according to a procedure described in U.S. Patent 3,707,178. Using a basic recipe such as described above, a series of stocks may be prepared, containing various adhesion promoting additives as noted in the following table.

	Stock	
Additive	A	B
Cobalt*	—	0.07
2-Methylol-2-nitropropane	1.5	1.5
Resorcinol-acetaldehyde resin	1.0	1.0
Test Results		
Adhesion pull (pounds)	33	53
Appearance rating	1	5

*As cobalt naphthenate composition containing 7% cobalt.

In the table the amount of each additive is expressed as parts per 100 parts by weight of rubber in the stock. The stocks are formed into adhesion test specimens using wire as described above, and cured for about 8 minutes at 350°F in a mold. Adhesion tests conducted as noted above at 250°F produce results as set forth in the table.

Stock A, which is outside the process, represents a control having no adhesive additive other than 2-methylol-2-nitropropane and resorcinol-acetaldehyde resin (prepared as described in Example 1 of U.S. Patent 3,266,970) as a resorcinol-type methylene acceptor. Stock B represents the process and contains, in addition to the resorcinol-acetaldehyde resin, 0.07 phr (parts per hundred of rubber) of cobalt (as cobalt naphthenate preparation) and 1.5 phr of 2-methylol-2-nitropropane.

To facilitate dispersion of the cobalt naphthenate in the rubber stock, it may be dissolved in pine tar (for example, by mixing 100 parts of the 10% cobalt commercial preparation with 43 parts of pine tar at 80° to 90°C to provide a solution containing 7% cobalt).

The table shows the improvement in adhesion made possible by the process (Stock B), as evidenced by increased pull necessary to separate the wire from the rubber in the cured laminate, and especially by the increased appearance rating (a rating of 5 indicating that the wire as pulled from the test laminate was covered with rubber, signifying failure in the stock rather than failure of the adhesive bond at the metal-rubber interface.

RUBBER-TO-METAL BONDING

Polymethylene Polyphenylisocyanate and Chlorosulfonated Polyethylene

A process described by *L.G. Manino and F.H. Sexsmith; U.S. Patent 3,879,337; April 22, 1975; assigned to Lord Corporation* relates to a solvent-based composition containing at least one polyisocyanate normally reactive at room temperature, at least one acidic halogen-containing polymer, at least one inert filler, which composition has an extended shelf life and which is capable of bonding a variety of metallic and organic substrates such as leather, fabrics, cords, and the like, to a wide variety of vulcanizable elastomers during vulcanization. The preferred adhesive composition is a solution of polymethylene polyphenylisocyanate and chlorosulfonated polyethylene having a dinitrosobenzene suspended therein and carbon black.

Example 1: An adhesive was prepared by dissolving 10 parts of chlorosulfonated polyethylene (Hypalon, Grade No. 40) and 10 parts of polymethylene polyphenylisocyanate (PAPI) in a combination of 43 parts of xylene and 104 parts of trichloroethylene. One part of p-dinitrosobenzene and 5 parts of carbon black were dispersed in the polymer solution. The solids content was 15%.

The composition had an apparent viscosity of 225 cp (Brookfield Viscometer, No. 2 spindle, 30 rpm, 77°F). A sample thereof stored for 6 months in closed container at 70°F showed no evidence of gel formation. The stability of the adhesive composition was further evaluated by determing the viscosity of the adhesive after accelerated storage in a closed container for 30 days at 130°F.

The viscosity was found to be 265 cp, indicating good stability. The bonding strength of the adhesive was not adversely affected by the accelerated storage conditions. It was not necessary to scrupulously free the total system of water. Nor did open exposure to moisture-laden air seem to hurt. When an open container of the adhesive was exposed to the ambient atmosphere for 21 days, it did not gel.

Example 2: This example illustrates the versatility of the adhesive composition of Example 1 in bonding fabrics coated therewith to a variety of vulcanizable elastomers. The various rubber stocks employed are designated A, B, C, D and E. Rubber stock A is a natural rubber stock composed of 100 parts, by weight, of smoked sheet, 3 parts of zinc oxide, 2 parts of stearic acid, 1 part of N-phenyl-β-naphthylamine, 40 parts of carbon black (high abrasive furnace), 0.6 part of N-oxydiethylene-2-benzothiazolesulfenamide and 2.15 parts of sulfur.

Rubber stock B is a polychloroprene stock composed of 100 parts, by weight, of polychloroprene (Neoprene GNA), 75 parts of carbon black (fine thermal), 10 parts of naphthenic process oil (specific gravity of 0.9230, SUS viscosity at 100°F of 156), 5 parts of zinc oxide, 4 parts of magnesia, 2 parts of N-phenyl-β-naphthylamine, 1 part of petrolatum, 1 part of stearic acid and 1 part of sulfur.

Rubber stock C is buna N stock composed of 100 parts of buna N (butadiene-acrylonitrile copolymer), 1 part of stearic acid, 10 parts of terpene polymer, MP 10°C, predominantly a polymer of β-pinene (Piccolyte S-10), 10 parts of dibutyl phthalate, 5 parts of zinc oxide, 65 parts of carbon black (furnace), 0.3 part of tetramethylthiuram monosulfide and 1 part of sulfur.

Rubber stock D is an SBR stock composed of 100 parts SBR (butadiene-styrene copolymer), 50 parts of carbon black (high abrasive, furnace), 5 parts of zinc oxide, 1 part of stearic acid, 8 parts of saturated polymerized petroleum hydrocarbons (specific gravity 0.95 to 1.02, viscosity, Saybolt Furol at 140°F, 250 to 350 seconds, Para-Flux), 1.75 parts of sulfur and 1.25 parts of N-cyclohexyl-2-benzothiazolesulfenamide.

Rubber Stock E is a butyl rubber stock composed of 100 parts of butyl rubber, 1 part of stearic acid, 5 parts of zinc oxide, 50 parts of carbon black (high abrasive furnace), 1 part of benzothiazyl disulfide, 1.5 parts tellurium diethyldithiocarbamate and 1 part of sulfur.

The adhesive composition was applied to several 4.5" x 6.5" pieces of polyester fabric and dried for 30 minutes until tack-free. The coated fabric was sandwiched between two 5" x 7" x ⅛" pads of the particular elastomer and the resulting assembly was wrapped with masking tape. The assembly was cured for 30 minutes at 307°F and allowed to stand overnight. One inch strips of the cured assembly were cut and pulled at an angle of 180° on a tensile tester at a speed of 2 inches/minute. The pull values in pounds per inch are set forth in the table below.

Elastomer	Pull (lb/in)
Natural rubber (stock A)	98
Styrene-butadiene rubber (stock D)	96
Neoprene (stock B)	103

(continued)

Elastomer	Pull (lb/in)
Butyl rubber (stock E)	77
Butadiene-acrylonitrile rubber (stock C)	72

Polydiene-Aminosilane Compositions

E.L. Polaski; U.S. Patent 3,894,982; July 15, 1975; assigned to Lord Corporation has found that compositions comprising at least one polydiene characterized by a high vinyl content and at least one polyfunctional aminoorganosilane are effective as adhesive materials for bonding vulcanizable elastomers, including sulfur-vulcanizable and free radical-vulcanizable elastomers, to themselves or other solid structural substrates. If desired, conventional additives such as fillers, dyes, pigments, extenders, and the like can be incorporated into the compositions.

The compositions are characterized by the unexpected ability to provide strong rubber-to-metal bonds without the necessity of first priming the metal surface; however, the use of conventional substrate primer compositions does enhance the strength of the adhesive bond. Thus the compositions have been found to be effective as one-coat adhesive materials, particularly for peroxide-curing elastomers such as EPDM, EPR, and silicone elastomers, and as two-coat adhesive materials when used with conventional substrate primers.

Representative aminoorganosilane compounds which can be employed in the process include:

γ-aminopropyltriethoxysilane;
γ-aminopropyltripropoxysilane;
γ-aminoisobutyltriethoxysilane;
γ-aminopropylmethyldiethoxysilane;
γ-aminopropylethyldiethoxysilane;
γ-aminopropylphenyldiethoxysilane; and
Δ-aminobutyltriethoxysilane.

The following examples illustrate the process. Parts are parts by weight unless otherwise expressed. In the examples, in bonding the elastomer to a metal substrate, unless otherwise noted the substrate surface is primed with a conventional chlorinated rubber-base primer coat. The assembly is then cured and the adhesive bond tested according to ASTM standard D-429, Method B, modified to 45° angle of pull.

The bonded structures are subjected to various tests, including room temperature (RT) pull, the boiling water test, and the salt spray test. In the RT pull test, the rubber body is peeled from the metal at a 45° angle using a Scott tensile tester and the force required in pounds per inch is recorded.

In the boiling water test, bonded samples after having been scored at the bond line and prestressed by bending the rubber body back from the metal, are immersed in boiling water for 2 hours; and in the salt spray test, the samples, after scoring and prestressing, are exposed to a spray of salt solution (5% sodium chloride) for 48 hours at 100°F. The samples so treated are tested for relative bond strength by pulling the rubber body from the metal. In the data given in the examples, reference is made to failure in the rubber body (R), between the adhesive composition and the rubber (RC), between the adhesive composition and the primer (CP), between the primer and the metal (PM), or between the adhesive composition and the metal (CM) when no primer is listed. Failure is expressed in terms of percent, e.g., with no metal primer, 95R means that 95% of the failure occurred in the rubber body itself and not between the adhesive composition and the rubber (RC) or between the adhesive composition and the metal (CM).

Example 1: An adhesive is prepared from 100 parts by weight methylenebis-(4-phenyl isocyanate) chain-extended 1,2-polybutadienediol (MW about 45,000), 0 to 20 parts by weight γ-aminopropyltriethoxysilane, 0 to 0.2 parts by weight black azo dye and sufficient toluene to provide a total solids content (TS) of 20%. The proportions of materials, in parts by weight is set forth as follows.

	A	B	C	D	E	F
Methylene bis(4-phenyl isocyanate) chain-extended 1,2-polybutadienediol	100	100	100	100	100	100
γ-aminopropyltriethoxysilane	0	2	6	10	20	20
Black azo dye	0	0	0	0	0	0.2
Toluene (to 20% TSC)						

These adhesives are used to bond to nonprimed grit-blasted degreased cold-rolled steel a sulfur-vulcanizable propylene oxide/allyl glycidyl ether elastomer stock having the following composition.

	Parts by Weight
Propylene oxide/allyl glycidyl ether elastomer	100.00
Carbon black	30.00
Zinc oxide	5.00
Stearic acid	1.00
Nickel dibutyldithiocarbamate	1.0
Tetramethylthiuram monosulfide	1.5
2-mercaptobenzothiazole	1.5
Aromatic process oil (ASTM D-2222 type 102)	5.0
Sulfur	1.25

The elastomer stock is cured at 320°F for 25 minutes. Peel adhesion tests are performed in accordance with ASTM D-429, Method B modified to 45° with the following results.

Run	Adhesive	Peel Adhesion Lb./In.	Failure	
1	A	8	90 CM,	10 RC
2	A	12	85 CM,	15 RC
3	B	30	10 R,	90 CM
4	B	31	10 R,	5 RC, 85 CM
5	C	34	10 R,	90 CM
6	C	33	5 R,	95 CM
7	D	36	99 R,	1 CM
8	D	37	75 R,	25 CM
9	E	43	30 R,	70 CM
10	E	38	80 R,	20 CM
11	F	61	100 R	
12	F	70	99 R,	1 CM

Example 2: The adhesive compositions of the previous example are applied over grit-blasted degreased cold-rolled steel primed with a conventional chlorinated elastomer primer. The metal is then bonded to a propylene oxide/allyl glycidyl ether stock of Example 1. Peel adhesion tests are performed according to the procedure of Example 1 with the following results.

Run	Adhesive	Peel Adhesion Lb./In.	Failure	
1	A	50	95 R,	5 RC
2	A	48	100 R	
3	B	54	99 R,	1 CP
4	B	56	100 R,	
5	C	51	100 R	
6	C	45	70 R,	30 CP
7	D	59	98 R,	2 CP
8	D	60	100 R	
9	E	60	100 R	
10	E	56	100 R	
11	F	72	100 R	
12	F	65	100 R	

Example 3: An adhesive formulation comprising 100 parts by weight methylene diphenylisocyanate chain-extended 1,2-polybutadienediol (MW about 45,000), 20 parts by weight γ-aminopropyltriethoxysilane, 0.2 part by weight azo black dye and sufficient toluene to provide a total solids content of 15% was applied over grit-blasted degreased cold-rolled steel primed with a conventional chlorinated rubber base primer.

The metal was bonded to a sulfur-vulcanizable propylene oxide/allyl glycidyl ether elastomer stock comprising 100 parts by weight elastomer, 50 parts by weight carbon black, 5 parts by weight zinc oxide, 1 part by weight stearic acid, 1 part by weight nickel dibutyldithiocarbamate, 1.5 parts by weight tetramethylthiurammonosulfide, 1.5 parts by weight 2-mercaptobenzothiazole and 1.25 parts by weight elemental sulfur. The elastomer stock is cured at 320° for 25 minutes. Peel adhesion, boiling water and salt spray tests are made with the following results.

Peel Lb./In.	Adhesion Failure	Boiling Water,2 hrs.	Salt Spray (48 hrs., 5% salt)
117	100 R	98 R, 2 CM	100 R

Example 4: An adhesive is prepared from 100 parts by weight methylene diphenylisocyanate chain-extended 1,2-polybutadienediol (MW 40,000), 20 parts by weight γ-aminopropyltriethoxysilane, 0.2 parts by weight black azo dye and sufficient toluene to provide a total solids content of 15%.

The adhesive is used to bond grit-blasted degreased cold-rolled steel to various elastomers in the presence of various vulcanization agents. Except where noted, the steel substrate is primed with chlorinated rubber adhesive.

The results obtained by using the adhesive of Example 4 are shown in the table on the following page.

Run	Elastomer	Vulcanization Agent	Peel Strength Lbs./In. ASTM D 429-B Mod. to 45°	Failure
1[a]	EPDM[b]	Peroxide	80	100 R
2[a]	EPDM	Sulfur & Peroxide	16	50 RC 50 PM
3	EPDM	Sulfur	29	100 RC
4	EPM[c]	Peroxide	52	100 R
5[a]	Silicone	Peroxide	22	100 R
6	Natural rubber	Sulfur	26	100 RC
7	Epichlorhydrin	Sulfur	0	100 RC
8	Poly(alkylene oxide)-type d	Sulfur	104	100 R
9	EVA[e]	Peroxide	150	100 R

[a]Primer not used on metal surface.
[b]Ethylene/propylene/nonconjugated diene rubber.
[c]Ethylene/propylene rubber.
[d]Propylene oxide/allyl glycidyl ether rubber.
[e]Ethylene/vinyl acetate rubber.

Chlorosulfonated Polyethylene, Nitroso Compound and Polyisocyanate

R.-N. Ozelli, G. Klement and E. Lieske; U.S. Patent 3,878,134; April 15, 1975; assigned to Henkel & Cie GmbH, Germany describe an adhesion assistant or binder composition for use in the production of composite bodies by vulcanizing a vulcanizable rubber mixture on a substrate which is stable under vulcanization conditions consisting essentially of the following binder ingredients dispersible in inert organic solvents:

(a) from 10 to 30 parts by weight of chlorosulfonated polyethylene;
(b) from 5 to 25 parts by weight of a di-C-nitroso aromatic compound;
(c) from 20 to 40 parts by weight of an organic polyisocyanate;
(d) from 10 to 30 parts by weight of chlorinated rubber;
(e) from 5 to 20 parts by weight of a component selected from the group consisting of a phenol resin of the resol type having reactive groups, a compound having more than one epoxide group in the molecule and the mixtures thereof;
(f) from 0 to 25 parts by weight of a pigment component; and
(g) from 0 to 20 parts by weight of additional tack producing compounds.

The process can advantageously be employed using the adhesion assistants and binders for vulcanizing vulcanizable rubber mixtures based on styrene-butadiene rubber, nitrile rubber, polychloroprene, natural rubber and butyl rubber on metal substrates. Moreover, it can also be used with so called stable substrates, i.e., those which are stable under vulcanization conditions.

The bonds obtained are marked by a high adhesive strength and high stability against corrosive media and against commercial oils and fats. The composite bodies produced by means of the process may be used in a large variety of technical fields, especially in the automobile industry as flexible bearings, clutches, shock absorbers and sealing parts, in the manufacture of apparatus for the construction of chemical plants or in the building sector.

In the following examples the vulcanizable rubber mixtures designated by A, B, C, D and E were vulcanized on substrates. The compositions of the mixtures and the vulcanization conditions are indicated below. All parts are by weight unless otherwise indicated.

Mixture A*

	Parts by Weight
Natural rubber (smoked sheets)	100
Zinc oxide	10
Stearic acid	2
Phenyl-β-naphthylamine	1
Pine tar	2
EPC carbon black	25
Zinc dimethyldithiocarbamate	0.33
Dibenzothiazyldisulfide	0.58
Sulfur	2.75

*Vulcanization conditions: 10 minutes at 153°C

Mixture B*

Styrene-butadiene rubber	100
Zinc oxide	5
Stearic acid	1
HAF carbon black	50
Saturated polymeric peanut oil hydrocarbons	8
N-cyclohexyl-2-benzothiazylsulfenamide	1.25
Sulfur	1.75

*Vulcanization conditions: 30 minutes at 153°C

Mixture C*

Polychloroprene rubber	100
Magnesium oxide	4
Zinc oxide	5
Phenyl-β-naphthylamine	2
MT carbon black (Sterling)	80
Mixture of high-boiling hydrocarbons (BP over 300°C)	1
Naphthenic oil	5
Tetramethylthiuram monosulfide	0.5
Di-o-tolylguanidine	0.5
Ethylenethiourea	0.5
Sulfur	1

*Vulcanization conditions: 30 minutes at 153°C

Mixture D*

Nitrile rubber (42% acrylonitrile content)	100
Zinc oxide	5
Stearic acid	1
SRF carbon black	65
Terpene resin	10
Dibutyl phthalate	10
Tetramethylthiuram monosulfide	0.31
Sulfur	1

*Vulcanization conditions: 25 minutes at 153°C

(continued)

Mixture E*

	Parts by Weight
Butyl rubber	100
Zinc oxide	5
Stearic acid	1
HAF carbon black	50
Dibenzothiazyldisulfide	1
Tellurium diethyldithiocarbamate	1.5
Sulfur	1

*Vulcanization conditions: 30 minutes at 160°C

The isocyanate components are as follows:

(a) Dianisidine diisocyanate;
(b) Dimerized tolylene diisocyanate;
(c) Reaction product from the diglycidyl ether of bisphenylolpropane with diphenylmethane diisocyanate in the molar ratio 1:2;
(d) Reaction product from the diglycidyl ether of diphenylolpropane with tolylene diisocyanate in the molar ratio 1:2;
(e) Reaction product of epoxidized bis(hydroxymethylcyclohexenyl)-tetrahydrobenzaldehyde acetal with tolylene diisocyanate in a molar ratio such that one molecule of tolylene diisocyanate is present per epoxide oxygen group;
(f) Reaction product from the diglycidyl ether of diphenylolpropane with naphthylene diisocyanate-(1,5) in the molar ratio 1:2;
(g) Reaction product from N,N-diglycidylaniline with tolylene diisocyanate in the molar ratio 1:2;
(h) Reaction product from N,N-diglycidylaniline with diphenylmethane diisocyanate in the molar ratio 1:2; and
(i) Reaction product from vinylcyclohexene dioxide with diphenylmethane diisocyanate in the molar ratio 1:2.

Example 1: Adhesion binders were prepared by dispersing the following components in a mixture of 7.36 kg of xylene and 0.400 kg of trichloroethylene.

	Grams
Chlorosulfonated polyethylene, having a chlorine content of 34.5% and a sulfur content of 0.9%	400
Chlorinated rubber having a chlorine content of 65 to 68%	500
p-Dinitrosobenzene	100
Isocyanate component according to (a), (b), (c), (d) and (e) above	560
Phenol formaldehyde resol resin, dilutable with water, viscosity measured in 80% ethanol solution at 20°C was about 4,500 cp	400
Inert carbon black	280
Zinc oxide	140

Metal sheet samples of cold-rolled steel were degreased with trichloroethylene vapor, blasted with chill casting grit and treated again with trichloroethylene vapor. Then the adhesion binder was applied to the sheets by dipping. After drying, the sheets were bonded to the rubber mixtures A, B, C, D and E according to ASTM/D 429 Method B within the conditions of a vulcanization in the press under a pressure of 60 kp/cm^2. The amount of rubber used was sufficient to give a layer of approximately 5 mm thickness. After storing the test body for 24 hours at room temperature, the rubber coating was peeled off at an angle of 45°. The peeling resistance and the tearing behavior were determined according to the normal method.

In the tearing behavior 100 R indicates that 100% tearing was observed in the rubber; at 50 RC, 50% tearing was found in the rubber and 50% in the layer of binder. The number before M gives the percentage rate of exposure of the metal together with the binder after tearing.

In Table 1, proceeding from left to right, the rubber mixture used is given in the first column followed by the adhesion binder means used and then the peeling resistance in kilograms/inch. Then the tearing behavior of the sample not exposed to the hot water follows. In the last column the tearing behavior is given which was found after 2 hours storage with application of load in water at 95°C (KWT).

TABLE 1

Rubber Mixture	Adhesion Assistant	ASTM-D 429, B Peeling Resistance kg/in.	Tearing Behavior	Tearing Behavior K.W.T.
A	a)	38	100 R	95 R - 5M
	b)	20	100 R	100 R
	c)	33	100 R	100 R
	d)	25	100 R	100 R
	e)	22	100 R	100 R
B	a)	65	100 R	100 R
	b)	74	100 R	100 R
	c)	71	100 R	100 R
	d)	67	100 R	100 R
	c)	71	100 R	100 R
C	a)	49	100 R	95 R - 5M
	b)	37	100 R	100 R
	c)	43	100 R	100 R
	d)	43	100 R	97 R - 5M
	e)	47	100 R	100 R
D	a)	48	100 R	100 R
	b)	55	100 R	95 R - 5M
	c)	47	100 R	100 R
	d)	54	100 R	100 R
	e)	54	100 R	100 R
E	a)	47	100 R	100 R
	b)	50	100 R	95 R - 5M
	c)	43	100 R	100 R
	d)	44	100 R	97 R - 5M
	e)	43	100 R	100 R

Example 2: Utilizing a procedure analogous to that described in Example 1, the rubber mixtures were bonded to cold-rolled steel with the binder which contained dianisidine diisocyanate [isocyanate component (a) above]. The test bodies were stored at room temperature for seven days in brake fluid liquids customarily used in cars (ATE brake fluid liquid) and a lubricating oil, which consisted of a mixture of hydrocarbons. Further samples were stored in the same oil for seven days at 200°C. Then all the samples were torn and the tearing behavior was determined.

With these conditions of storage practically no alteration of the bond had occurred over that of Example 1.

Example 3: Binders were prepared by dispersing the following components in a mixture of 7.2 kg of xylene and 0.4 kg of trichloroethylene.

	Grams
Chlorosulfonated polyethylene as in Example 1	400
Chlorinated rubber as in Example 1	500
Zinc oxide	140
p-Dinitrosobenzene	300
Inert carbon black	100
Isocyanate component according to (f), (g), (i) and (h) above	560
Phenol formaldehyde resol resin, dilutable with water, viscosity measured in 80% ethanol solution at 20°C was about 5,000 cp	400

Using this adhesion binder means, various rubber mixtures were vulcanized on steel sheets and the bonds were examined according to Example 1. The results are compiled in Table 2.

TABLE 2

Rubber Mixture	Adhesion Assistant	Peeling Resistance kg/in.	Tearing Behavior	Tearing Behavior K.W.T.
A	f)	24	100 R	100 R
	g)	27	95R-5RC	100 R
	i)	21	100 R	100 R
	h)	26	100 R	100 R
B	f)	70	100 R	100 R
	g)	73	100 R	100 R
	i)	72	100 R	100 R
	h)	81	100 R	90 R - 10M
C	f)	59	100 R	100 R
	g)	45	100 R	98 R - 2M
	i)	47	100 R	100 R
	h)	50	100 R	60 R - 40M
E	f)	49	100 R	100 R
	g)	49	100 R	100 R
	i)	46	100 R	100 R
	h)	46	100 R	100 R

POLYURETHANES FOR RUBBER-TO-RUBBER BONDING

Aromatic Nitro Compounds as Additives

A process described by *W.R. Hausch, J.W. Fieldhouse and E.L. Kay; U.S. Patent 3,880,810; April 29, 1975; assigned to The Firestone Tire & Rubber Company* relates to adhesive compositions containing an organic isocyanate, natural and synthetic elastomers and resins and as an adhesion promoter an aromatic nitro compound. Simple aromatic nitro compounds such as nitrobenzene, nitrotoluene and nitroxylenes are not particularly good adhesion promoters. However, for some purposes, even these simple aromatic nitro compounds impart sufficient adhesion to satisfy a specific end use which does not require an extraordinarily strong adhesive bond. For example, the simple aromatic nitro compounds would be adequate adhesion promoters for adhering decorative rubber articles to various substrates providing the articles are not exposed to high shear stress. The preferred aromatic nitro compounds of this process are compounds which contain an activated nitro function. These preferred aromatic nitro compounds are illustrated by the following generalized formula:

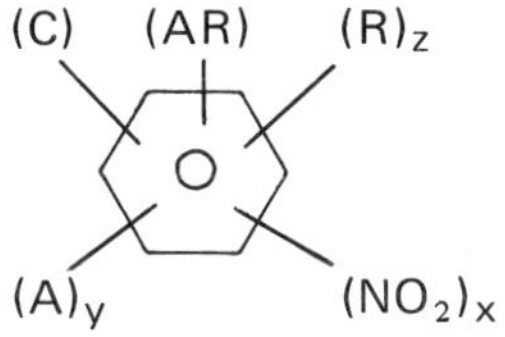

wherein (NO_2) represents a nitro function directly attached to the aromatic ring; (A) is a function attached to the aromatic ring which activates the nitro function such as hydroxyl, mercaptyl, carboxyl, alkoxy, alkylthio, amino, alkylamino, halogen and sulfonic acid substituents; (R) is a hydrocarbyl substituent of 1 to 16 carbon atoms; (AR) in conjunction with the depicted ring contemplates condensed aromatic structures such as naphthalene, anthracene, phenanthracene and the like; (C) contemplates bis and tris compounds; x is an integer of 1 to 5; y is an integer of 1 to 5; and z is an integer of 0 to 4.

The adhesion test results reported hereafter were obtained on laminates of olefinic rubber and polyurethane rubber, a cement in accordance with the process having been interposed between the two rubbers. The procedure unless otherwise stated was as follows. First, a sheet 6" square and 0.100" thick was prepared from vulcanizable olefinic rubber compound. One side of the olefinic rubber sheet was backed with reinforcing nylon tire fabric treated for adhesion. This sheet was cured (usually of the order of 20 to 30 minutes at about 300°F).

The cured olefinic rubber sheet was buffed to a satiny finish on the side opposite the tire fabric, the buffed surface being washed with acetone or white gasoline. A 2" x 6" portion of the buffed surface in a direction perpendicular to the direction of the tire cord of the fabric was covered with holland cloth, the holland cloth being securely taped down on all four edges.

The exposed buffed surface of the cured olefinic rubber sheet was then coated with the adhesive cement of the process being tested and allowed to dry. The olefinic rubber sheet was then placed in the bottom of a 6" x 6" x 0.250" mold with the adhesive cement coated surface up. A curable liquid polyurethane rubber compound was poured into the mold in a sufficient quantity to fill the mold. After the polyurethane material had gelled a 6" x 6" piece of square woven nylon was placed on the surface of the polyurethane. A cover plate was placed on the top of the mold and the mold was heated 3 hours at 212°F to complete the curing of the polyurethane rubber.

Flashing was trimmed from the resulting test laminate and the edge containing the holland cloth was cut back to expose the cloth. One inch wide test strips were cut from the laminate in a direction parallel to the direction of the reinforcing tire cord. After 24 hours room temperature test results were obtained by clamping the free ends of a given test strip to the jaws of a testing machine and the jaws of the test machine were pulled apart at a rate of 2 inches per minute, the force in pounds per inch at tear as well as the nature of the tear being observed. With regard to the nature of the tear, the term "pad tear" means that the tear was along the fabric; "rubber tear" means that the tear was in the olefinic rubber; "urethane tear" means that the tear was in the polyurethane rubber. Adhesive failure means that the tear was along the interface between the polyurethane rubber and the olefinic rubber. Where the percentage of tear reported hereafter is less than 100%, the balance was tear along the interface.

For elevated temperature testing the test samples were placed in an oven maintained at the desired elevated temperature at 5 minute intervals. Each test sample was then removed after being in the oven a total of 15 minutes and immediately subjected to the pulling test.

The following examples illustrate the process. In the examples, the vulcanizable olefinic rubber sheets employed were prepared of the following formulation.

	Parts by Weight
SBR	60.0
Polybutadiene	40.0
Carbon black	70.0
Oil	44.0
Zinc oxide	2.0
Stearic acid	2.0
Wax	3.0
Sulfur	1.7
Accelerator	1.2
Antioxidant	1.0

Following the procedure outlined above, such sheets were backed, cured, buffed and one surface coated by brushing with a cement. The adhesive coated sheets were in each instance formed into a laminate by the above described procedure with a polyurethane rubber composition of the following formulation.

	Parts by Weight
Adiprene L-167*	100.00
Silicone DC-200**	0.10
Methylenebis(o-chloro-p-aniline)	20.98
Dioctyl phthalate	20.00

*A liquid polyurethane rubber containing 6.15 to 6.55% available isocyanate and being curable to a solid rubber with polyamine or polyol compounds.

**A silicone oil.

Example 1: A cement containing an adhesion promoter of the process was prepared of the following formulation.

	Grams
2,4-Dinitrophenol	0.75
Adiprene L-167	3.00
Toluene diisocyanate	1.50
Toluene	3.00

Following the procedure outlined above, adhesion test strips were produced employing this cement to adhere the described olefinic rubber to the described polyurethane rubber. The test results obtained were as follows.

Test Temperature	Adhesion (lb/in)	Nature of Adhesion
Room	110	100% rubber tear
212°F	52	95% rubber tear
250°F	38	10% rubber tear

Example 2: A cement containing an adhesion promoter of the process was prepared of the following formulation.

	Grams
2-Chloro-4-nitrobenzoic acid	0.75
Adiprene L-167	1.50
Toluene diisocyanate	3.00
Toluene	3.00

Following the procedure outlined above, adhesion test strips were produced employing this cement to adhere the described olefinic rubber to the described polyurethane rubber. The test results obtained were as follows.

Test Temperature	Adhesion (lb/in)	Nature of Adhesion
Room	78	100% rubber tear
212°F	43	90% pad tear
250°F	35	100% pad tear

Example 3: A cement containing an adhesion promoter of the process was prepared of the following formulation.

	Grams
o-Nitrophenol	0.75
Adiprene L-167	1.50
Toluene diisocyanate	3.00
Toluene	3.00

Following the above outlined procedure, except that the test pads were postcured 90 minutes at 250°F, adhesion test strips were produced employing this cement to adhere the described olefinic rubber to the described polyurethane rubber. The test results obtained were as follows.

Test Temperature	Adhesion (lb/in)	Nature of Adhesion
Room	55.0	50% rubber and pad tear
212°F	13.5	100% pad tear
250°F	18.5	100% pad tear
300°F	8.5	100% pad tear

Nitro and Chloronitroalkanes as Additives

A process described by *W.R. Hausch; J.W. Fieldhouse and E.L. Kay; U.S. Patent 3,880,808; April 29, 1975; assigned to The Firestone Tire & Rubber Company* relates to adhesive compositions comprising an organic isocyanate and often for best results an organic film-forming material, the adhesives compositions contain also an aliphatic nitro compound as an adhesion promoter. Aliphatic nitro compounds employed in this process are illustrated by the following generalized formula:

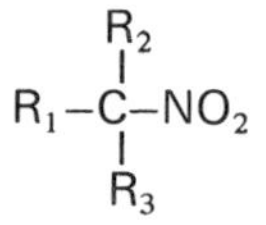

where R_1, R_2 and R_3 are selected from the group consisting of hydrogen atoms, hydrocarbyl groups and halogen. R_1, R_2 and R_3 may be the same or different; that is any combination of hydrogen atom, hydrocarbyl group, and halogen atoms are suitable.

The following examples illustrate the process. The adhesion test is described in U.S. Patent 3,880,810 noted above. In the examples the vulcanizable olefinic rubber sheets employed were prepared of the following formulation.

	Parts by Weight
SBR	60.0
Polybutadiene	40.0
Carbon black	70.0
Oil	44.0
Zinc oxide	2.0
Stearic acid	2.0
Wax	3.0
Sulfur	1.7
Accelerator	1.2
Antioxidant	1.0

Following the procedure outlined above, such sheets were backed, cured, buffed and one surface coated by brushing with a cement containing the adhesion promoter of the process. The adhesive coated sheets were in each instance formed into a laminate by the above described procedure with a polyurethane rubber composition of the following formulation.

	Parts by Weight
Adiprene L-167*	100.00
Silicone DC-200**	0.10
Methylenebis(o-chloro-p-aniline)	20.98
Dioctyl phthalate	20.00

*A liquid polyurethane rubber containing 6.15 to 6.55% available isocyanate and being curable to a solid rubber with polyamine or polyol compounds.

**A silicone oil.

Example 1: A cement containing an adhesion promoter of the process was prepared of the following formulation.

	Grams
Nitroethane	0.75
Adiprene L-167	1.50
Toluene diisocyanate	3.00

Following the procedure outlined above, adhesion test strips were produced employing this cement to adhere the described olefinic rubber to the described polyurethane rubber. The test results obtained were as follows.

Test Temperature	Adhesion (lb/in)	Nature of Adhesion
Room	120	100% rubber tear
212°F	60	100% rubber and pad tear
250°F	60	100% rubber tear

Example 2: A cement containing an adhesion promoter of the process was prepared of the following formulation.

	Grams
2-Nitropropane	0.75
Adiprene L-167	1.50
Toluene diisocyanate	3.00

Following the procedure outlined above, adhesion test strips were produced employing this cement to adhere the described olefinic rubber to the described polyurethane rubber. The test results obtained were as follows.

Test Temperature	Adhesion (lb/in)	Nature of Adhesion
Room	78	100% rubber tear
212°F	20	100% rubber tear
250°F	35	50% rubber tear

Example 3: A cement containing an adhesion promoter of the process was prepared of the following formulation.

	Grams
1-Chloro-1-nitropropane	0.75
Adiprene L-167	1.50
Toluene diisocyanate	3.00
Toluene	3.00

Following the procedure outlined above, adhesion test strips were produced employing this cement to adhere the described olefinic rubber to the described polyurethane rubber. The test results obtained are shown in the table on the following page.

Test Temperature	Adhesion (lb/in)	Nature of Adhesion
Room	120	100% urethane tear
212°F	50	75% rubber tear and 25% pad tear
250°F	83	100% urethane tear

METAL-TO-METAL ADHESION

EPOXYSILANE ADDITIVE FOR EPOXY RESINS

A process described by *E. Leumann and H. Lehmann; U.S. Patent 3,867,322; February 18, 1975; assigned to Ciba-Geigy Corporation* relates to epoxysilanes of the formula (1)

(1)

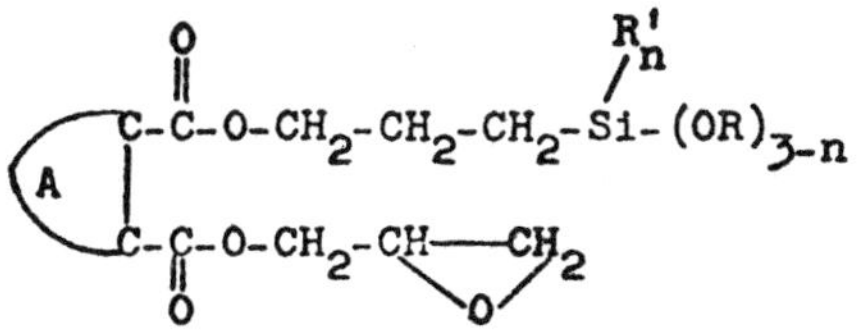

in which A denotes a cycloaliphatic five-membered or six-membered ring or an aromatic ring, R and R' each denote an alkyl group with 1 to 4 carbon atoms and n denotes 0, 1 or 2, and their use in adhesives, to which they impart increased resistance to tropical conditions and to water.

The compounds of the formula (1) can be manufactured by reacting compounds of the formula (2) with silanes of the formula (3).

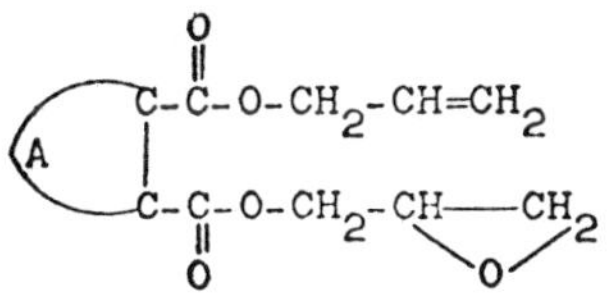

(2)

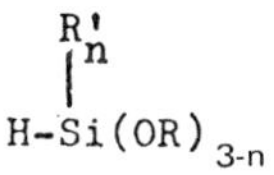

(3)

The reaction preferably takes place at elevated temperature, for example between 100° and 150°C and in the presence of a catalyst, for example based on metallic platinum or platinum compounds, such as platinum hexachloride. Advantageously, the more easily volatile component is used in a certain excess in this addition reaction and can, after completion of the reaction, be recovered by distillation and reemployed for a subsequent operation. Preferably, allylglycidyl-hexahydrophthalic acid ester or allylglycidyl-tetrahydrophthalic acid ester is reacted with triethoxysilane or trimethoxysilane.

It has been found that these epoxysilanes, if they are mixed into adhesives, especially those based on epoxide resin, which can be filled or unfilled, in amounts of 0.1 to 5%, preferably 1%, relative to the resin component, or are spread on the surfaces to be glued before application of the adhesive, lead to improved strength of the adhesive bonds as compared to epoxysilanes previously commercially available, especially after storage in water and storage in a tropical climate.

DIVINYL COMPOUNDS AND RUBBER MATERIAL AS TWO-COMPONENT ADHESIVES

S. Kusayama, K. Ohashi and S. Takada; U.S. Patent 3,870,675; March 11, 1975; assigned to Sumitomo Chemical Company, Limited, Japan describe a two-component adhesive consisting of (1) a main composition comprising about 40 to 80 parts by weight of at least one of divinyl compounds of the formula:

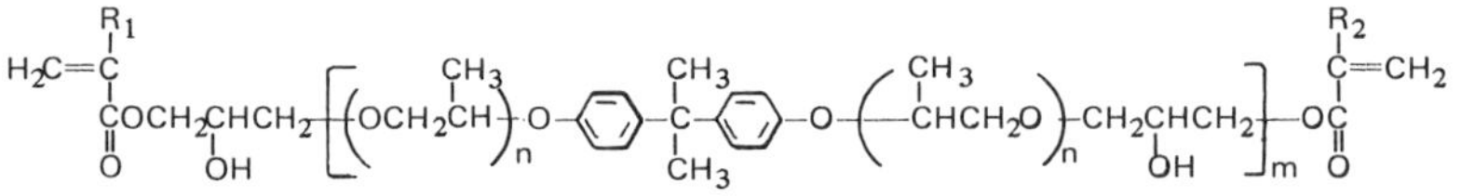

where R_1 and R_2 are each hydrogen, methyl or ethyl, n is an integer of 1 to 5 and m is an integer of 1 to 10; 0.05 to about 10 parts by weight of an organic peroxide, about 2 to about 120 parts by weight of at least one mono-vinylic monomer selected from the group consisting of vinyl acetate, methacrylic acid, acrylic acid, glycidyl methacrylate and β-hydroxyethyl methacrylate (the amount of each vinylic monomer not exceeding 40 parts by weight); and about 0.5 to about 15 parts by weight of at least one rubber material selected from the group consisting of acrylonitrile-butadiene-styrene polymer (ABS), acrylonitrile-butadiene rubber (NBR) and carboxylated NBR; and (2) an accelerator composition comprising about 0.01 to about 10 parts by weight of a thiourea compound; about 0.01 to 10 parts by weight of α,α'-dipyridyl; 0.001 to 5 parts by weight of at least one of copper salts, cobalt salts, manganese salts, chromium salts and vanadium salts of organic and inorganic acids; and 75 to 100 parts by weight of an organic volatile solvent. The two components of the adhesive are each applied to separate articles and then the articles are combined together.

In the adhesive composition of the process, the combined use of the main composition and the accelerator composition contributes to the assurance of the instantaneous curing property of the adhesive composition and to the attainment of satisfactory results in the primary adhesion strength, the impact strength, the peel strength, the heat-resistance and the like. Thus the main composition and the accelerator composition are usually applied separately. In use, these components are each applied on separate articles and then the articles are combined together. About 0.5 to about 1.0 parts by weight or volume of accelerator composition is usually used to about 1.0 part by weight or volume of main composition. These represent workable ratios. The instantaneous curing property used means that a considerable adhesion strength (e.g., 10 kg/cm^2 or more in tensile strength in case of the adhesion of steel-made substances) can be produced within several 10 seconds after contacting the surfaces of the materials to be bound which have been applied an adhesive composition.

The divinyl compound to be used as the main component in the main composition is a compound of a transparent, light yellow, viscous liquid and may be prepared, for instance, by reacting 1 mol of a glycol diglycidyl ether (Adeka Resin EP-4000) with about 2 mols of acrylic compounds such as methacrylic acid, acrylic acid and α-ethylacrylic acid in the presence of a catalyst such as triethylamine at a temperature of 40° to 80°C for 10 to 35 hours.

Example 1: In a three-necked flask equipped with an agitator there are charged Adeka Resin EP-4000 (207 parts) and methacrylic acid (48 parts), and triethylamine (3.5 parts) is added at 45°C while stirring. After maintaining at 50°C for 10 hours, acrylic acid (12 parts) is added to the resultant mixture, and the reaction is carried out at 55°C for 9 hours to give a light yellow viscous liquid of 3,000 cp in viscosity (referred to as liquid A). The liquid A (100 parts) is mixed with a vinylic monomer as shown in Table 1, and NBR (6 parts) and cumene hydroperoxide (3 parts) are added to make a main composition. Separately, tetramethylthiourea (0.8 part), cobalt acetate (0.01 part), cobalt naphthenate (0.05 part) and α,α'-dipyridyl (1.0 part) are dissolved in an acetone ethanol mixture (8:2) (100 parts) to make an accelerator composition.

On the top surface of a steel-made cylindrical piece of 1 cm^2 in section area, the accelerator composition is applied. After about 30 seconds, the applied surface is closely contacted with the bottom surface of another steel-made cylindrical piece of 1 cm^2 in section area, the said bottom surface having been applied the main composition prior to the contacting. The tensile strength is measured 5 minutes, 15 minutes and 24 hours after the contacting.

The results are shown in Table 1, from which it is seen that the incorporation of the mono-vinylic monomer into the main composition results in the increase of the tensile strength and the primary adhesion strength.

TABLE 1

Monovinylic Monomer		Tensile Strength After Contact*		
Kind	Amount Added**	5 min	15 min	24 hr
None	0	20	85	96
Vinyl acetate	40	90	110	132
Methacrylic acid	35	78	96	115
Acrylic acid	35	84	105	120
β-Hydroxyethyl methacrylate	40	93	103	126

*kg/cm^2
**part

Example 2: The liquid A obtained in Example 1 (100 parts) is mixed with vinyl acetate (35 parts), acrylic acid (35 parts) NBR (6 parts) and a peroxide as shown in Table 2 to make a main composition. Using the main composition and the accelerator composition obtained in Example 1, two steel-made cylindrical pieces are adhered to each other in the same manner as in Example 1. The tensile strength is measured 5 minutes, 15 minutes and 24 hours after the adhering. The results are shown in Table 2, from which it is seen that a considerable adhesion strength is attained in an extremely short time.

TABLE 2

Peroxide		Tensile Strength After Contact*		
Kind	Amount Added**	5 min	15 min	24 hr
Cumene hydroperoxide	3	102	118	128
p-Methane hydroperoxide	2	78	105	125
t-Butyl hydroperoxide	4	90	103	114

*kg/cm²
**part

Example 3: The liquid A obtained in Example 1 (100 parts) is mixed with cumene hydroperoxide (4 parts), acrylic acid (40 parts) and vinyl acetate (30 parts), and a rubber material as shown in Table 3 is added to make a main composition. Separately, tetramethylthiourea (0.5 part), cobalt acetate (0.05 part), cobalt naphthenate (0.03 part) and α,α'-dipyridyl (0.2 part) are dissolved in acetone (100 parts) to make an accelerator composition.

Using the thus obtained main composition and the accelerator composition, two steel-made cylindrical pieces are contacted with each other, and the tensile strength is measured as in Example 1.

Also, the accelerator composition is applied on the surface of a steel-made rectangular strip of 25 mm in width, 100 mm in length and 1.6 mm in thickness, the application being made 12.5 mm width from the one edge. The main composition is applied on the one surface of another steel-made rectangular strip as above in the similar manner. The strips are contacted with each other at the applied portions. After 24 hours, the shear strength is measured by the aid of a tensile tester (Omnipotent Tester RS-2) at a velocity of 45 mm/min. The results are shown in Table 3, from which it is seen that the shear strength is markedly increased by the incorporation of NBR, carboxylated NBR (Hycar 1072) or ABS into the main composition.

TABLE 3

Rubber Material		Tensile Strength After Contact*		Shear Strength (kg/cm^2)
Kind	Amount Added**	5 min	24 hr	24 hr at Room Temperature
None	0	104	133	85
NBR	8	95	127	273
Carboxylated NBR	8	111	141	256
Carboxylated NBR	15	100	121	232
Carboxylated NBR	30	77	108	196
NBR +	4	------	------	------
Carboxylated NBR	4	98	122	269
ABS	8	108	146	260

*kg/cm²
**part

ETHYLENE-ACRYLIC ACID COPOLYMER AND PHENOLIC RESIN

R.E. Parkinson; U.S. Patent 3,843,576; October 22, 1947; assigned to United States Steel Corporation describes aqueous dispersions in which ethylene-carboxylic acid copolymers and a low molecular weight phenolic resin are combined to produce a material useful as an adhesive, insulator, primer or coating.

Example 1: The simplest composition of the process was prepared by mixing the phenolic resin solution into the ammonium solution of the ethylene-acrylic acid copolymer. As an example, 27 parts of Arofene 352 (phenolic resin) was slowly added to 100 parts of a 20% solids ammonium solution of EAA 9300 with vigorous stirring, to give approximately equal parts of phenolic resin and copolymer resin solids. The addition of 10 to 30 parts of ethanol or 2-propanol reduces foaming and improves wetting and stability of the solution somewhat but is not necessary for the performance of the compound. The resulting mixture is a very fluid, light brown, milky colloidal solution. It has low surface tension and a tendency to foam.

The shelf life of this mixture at 24°C is limited to about 2 to 15 days after which the solution becomes green in color and a brown precipitate of phenolic resin settles. When refrigerated at about 8°C, the composition is stable for longer than 6 months.

Coatings of these compositions, dried and heated at about 300°F for 20 minutes or 450°F for 1 minute the thermoset as indicated by loss of tack and increased hardness at elevated temperatures. When applied to metals, such as carbon steel, galvanized steel or stainless steel, the solution wets the surface well, and the cured coating is very adherent, tough, and heat and moisture resistant. When used as a primer (usually as a more dilute solution) applied to metals, wood, or paper, these compounds improve the strength and durability of bonds of certain other thermoplastic adhesives, especially those containing ethylene copolymers, such as USS Nexus P-1001 or P-1003, and Surlyn A.

Example 2: Coal tar pitch is added to the ethylene-acrylic acid-phenolic dispersions as extenders to improve durability, moisture resistance, and adhesive properties of the compounds. Blends with coal tar pitch are more stable at room temperatures than the ethylene copolymer-phenolic dispersions without pitch. A typical formulation (SX 9001-6870-50A) is as follows:

100 parts by weight of ethylene-acrylic acid copolymer (EAA 9300) dissolved in aqueous ammonia (20% solids);
135 parts phenolic resin (about 70 to 75% solids) (Arofene 352);
10 parts coal tar pitch (70°C fiber grade);
20 parts ethanol or 2-propanol (to assist wetting of substrate and drying of coating);
10 parts xylene (to dissolve pitch);
5 parts pyridine (to aid in dissolving and dispersing pitch);
and 1 part 1-octanol (antifoam).

This compound may be blended by milling in a ball mill or by dissolving the pitch in the xylene and pyridine and mixing it into the other components with a high-shear mixer until the pitch is colloidally dispersed. When the pitch is adequately dispersed, this compound has a shelf life of at least several months at a room temperature of about 25°C.

Example 3: The composition of Example 2 is coated on steel as a primer for an adhesive prepared according to Example 1 of U.S. Patent 3,361,692. Without the primer the adhesive has a peel strength of 19 to 37 lb per inch; with the primer, the peel strength is 69 to 77 lb per inch (ASTM Test D 903).

Example 4: The composition of Example 2 is used as a primer for steel-to-plywood bonds using the adhesive of Example 3. Using 0.018 inch carbon steel, the peel strength increases from 32 to 50 lb per inch to 90 to 130 lb per inch.

Example 5: The heat resistance of the bond produced in Example 3 is tested. A lap adhesive joint is made with and without the primer and heated in an oven. Without the primer the bond fails at 77°C. With the primer the bond fails at 90°C. Thus, heat resistance of the bond when primer is used is found to be improved.

Example 6: When used as an adhesive or coating by itself (rather than as a primer for another adhesive or coating), it is often desirable to have a higher solids content in the dispersion in order to obtain thicker coatings. The formula of such an adhesive is as follows in parts by weight:

100 parts ethylene-acrylic acid copolymer (30% ammoniacal dispersion of EAA 9300);
7 parts phenolic resin (BRL-1031);
5 parts 2-propanol;
7 parts high-boiling tar oil (95% boiling above 355°C);
3.5 parts xylene; and
1 part morpholine (optional, to adjust viscosity).

The high-boiling tar oil is combined with the xylene and morpholine, mixed with the phenolic and added to the copolymer dispersion to which the 2-propanol has previously been added.

This compound is used for bonding steel wire to fiber pads. The wire is dipped in the dispersion, dried, and then heated to about 200°C while being pressed against the fiber. This compound gives a sufficient thickness of coating on the wire and during the heating it first becomes tacky and soft so that it penetrates and adheres to the fiber and then thermosets, becoming rigid and tack-free at 200°C.

POLYURETHANE CURED WITH POLYAMINE-ACID ANHYDRIDE ADDUCT

J.C. Bolger; U.S. Patent 3,886,228; May 27, 1975; assigned to Amicon Corporation has found that a special class of amine condensation products is effective as a curing agent for polyisocyanates at elevated temperature, particularly when used in conjunction with epoxy resins, but is virtually inactive at temperatures as high as 40°C, making practicable single component compositions which can be shipped and stored in a single container for extended periods of time before use.

Such unitary compositions are sensitive to water, alcohols, and other compounds containing active hydrogen atoms, so that in order to achieve maximum stability and shelf life, it is necessary that they be made and stored under anhydrous conditions and that they not contain or be mixed with fillers which have adsorbed water or with other ingredients, such as various resins, containing active hydrogen atoms. It has further been found that partial blocking of the isocyanate groups of aromatic polyisocyanates (including isocyanate terminated prepolymers) with phenols, alkyl phenols, polyphenols or phenolic novolacs, eliminates sensitivity of the compositions of the process to compounds containing active hydrogen atoms without significantly decreasing the shelf life or the curing rate at elevated temperature and without deterioration of the physical properties of the products after cure.

The amine condensation products which can be used are those obtained by condensing, with removal of water, approximately equimolecular proportions of an amine selected from the group consisting of diethylenetriamine and triethylenetetramine with a carboxylic anhydride selected from the group consisting of phthalic anhydride, hexahydrophthalic anhydride, maleic anhydride and succinic anhydride, the condensation product having a melting point from 75° to 130°C. Phthalic anhydride and maleic anhydride are preferred.

The amine condensation products can be prepared as described in U.S. Patents 3,488,742 and 3,636,657 by heating equimolecular proportions of the amines and the anhydrides at a temperature from 50° to 160°C while permitting the water by-product to evaporate, or by using an excess of the amine and evaporating the excess along with by-product water after an initial reaction period. If desired, a solvent inert to the reactants, such as benzene, may be used and removed along with the excess amine and water.

Example 1: There are provided a polyisocyanate prepolymer, identified as urethane A (Adiprene L-167) made by reacting 1 mol of polytetramethylene glycol (molecular weight 1,000) with at least 2 mols of toluene diisocyanate. The prepolymer, which has a viscosity of 12,000 cp at 25°C contains 6.7% by weight of isocyanate groups and has an isocyanate equivalent weight of approximately 670.

There was also provided an amine condensation product, identified as amine A, made by heating 9,064 grams diethylene triamine with 3,256 grams phthalic anhydride (in a molar ratio of 4:1) at about 100°C for 2 hours, after which the unreacted diethylene triamine (6,456 grams) together with about 1 mol of water (394 grams) per mol of phthalic anhydride were removed by vacuum distillation at a pressure of 20 mm mercury, as described in Example 1 of U.S. Patent 3,488,742. The condensation product, which has a softening point of about 104°C and contains nitrogen in the form of imide as well as amide and primary and secondary amine groups. It has an equivalent weight (based on total amino groups) of 100, an equivalent weight (based on primary amino groups) of 187, and an equivalent weight (based on total amino hydrogen atoms) of 66.

One hundred grams of urethane A were mixed with 18 grams of amine A, which had been pulverized to pass through a 325 mesh screen. A portion of this paste was tested for cure rate by stroking several grams on the surface of a 320°F hot plate. The stroke cure time so measured was approximately 20 minutes, although a sharp end point was hard to measure due to volatilization of amine A during the test and also because a rapid thickening, or false gel occurs within the first 10 to 15 seconds of the test. A 10 gram mass of the mixture was placed in an aluminum dish in a 250°F oven.

The mixture was not hard after 3 hours but after 6 hours had cured to a very soft urethane elastomer, Shore A hardness approximately 40, which exhibited excellent adhesion to the aluminum dish. The remainder of the paste was divided into two sealed jars; one stored at 40°C (104°F) and the other at room temperature. At the end of 28 days the stored mixtures were inspected and found to be still liquid, curable and usable, indicating their long shelf life.

Example 2: There was provided as an epoxy resin component, identified as epoxy A, a material sold under the trade name Epon 828 which was a diglycidyl ether of 2,2-bis(4-hydroxyphenyl) propane having a viscosity of 14,000 centipoises at 25°C and an epoxy equivalent weight of 190.

Fifty grams of epoxy A were mixed with 100 grams of urethane A and 35 grams of powdered amine A. Separately, 17 grams of amine A were mixed with 50 grams of epoxy A. Stroke cures and shelf lives of the two mixtures, determined as in Example 1, were as follows.

	Stroke Cure 320°F.	Shelf Life 40°C.	Hardness, Shore D
Amine A + urethane A + epoxy A	15 sec.	>14 days	60
Amine A + epoxy A	35 sec.	>14 days	84

The results clearly show not only that the addition of the epoxy dramatically accelerates the cure rate without loss of shelf life, but also that the three component mixture is faster than the epoxy-amine mixture alone, indicating some synergistic coreaction is taking place.

CHLOROSULFONATED POLYETHYLENE AND VINYL MONOMERS

P.C. Briggs, Jr. and L.C. Muschiatti; U.S. Patent 3,890,407; June 17, 1975; assigned to E.I. du Pont de Nemours & Company have found that solutions of chlorosulfonated polyethylene or mixtures of sulfonyl chlorides and chlorinated polymers in polymerizable vinyl monomers having a Brookfield viscosity of up to 1 million cp form excellent adhesive compositions which cure to high bond strengths. The preferred monomers are acrylic monomers, especially lower alkyl acrylates and methacrylates and ethylene glycol diacrylate and dimethacrylate.

The chlorosulfonated polyethylene suitable in the process can be prepared in a manner known to those skilled in the art by reaction of linear or branched polyethylene and sulfuryl chloride, or sulfur dioxide and chlorine. Chlorosulfonated polyethylene is also available commercially, for example, under the trade name Hypalon. In practice, the chlorosulfonated polyethylene may also be a chlorosulfonated copolymer of ethylene with small proportions of propylene or other olefins. Various chlorosulfonated polyethylenes and methods of their preparation are discussed, for example, in U.S. Patent 2,982,759.

Alternatively, sulfonyl chloride and chlorinated polyethylenes of suitable molecular weight can be used. The sulfonyl chlorides can be mono- or poly-functional and can be C_1—C_{12} alkyl sulfonyl chlorides, such as methane or butane sulfonyl chloride, C_6—C_{24} aromatic sulfonyl chlorides such as benzene or toluene sulfonyl chloride. Some sulfonyl chlorides containing hetero atoms have also been found to work, such as diphenylether-4,4'-disulfonyl chloride.

The proportions of chlorosulfonated polyethylene and polymerizable vinyl monomer can vary in a broad range. In the case of acrylic polymers, the practical range is about from 25 to 2,000 parts by weight of the monomer per 100 parts of chlorosulfonated or chlorinated polyethylene. The preferred range is 50 to 500 parts by weight of the monomer per 100 parts of polymer, whether chlorosulfonated polyethylene or a mixture of sulfonyl chloride and chlorinated polyethylene is used.

Example 1: To a mixture of acrylic monomers comprising 85 grams of methyl methacrylate (containing 50 to 90 ppm of hydroquinone inhibitor), 15 grams of glacial methacrylic acid (containing 250 ppm of 4-methoxyphenol), and 2 grams of ethylene glycol dimethacrylate, there was added 100 grams of chlorosulfonated polyethylene, made from branched polyethylene having a melt index of 100 and containing 43% chlorine and 34 mmols sulfonyl chloride:100 grams of polymer. The mixture was rolled in a jar at room temperature until solution of the polymer was complete (24 to 48 hours).

An adhesive composition was prepared by stirring into 50 grams of the above solution, in turn, 1.5 grams of cumene hydroperoxide and 0.5 grams of N,N-dimethylaniline. Lap shear bonds were prepared by pressing a small amount of the adhesive composition between 1 inch x 3 inch x 0.064 inch grit blasted and perchloroethylene vapor degreased hot rolled steel coupons in a mold so that a glue line of 0.010 inch x 1 inch x 1 inch was obtained. The specimens were stored for 18 hours in a nitrogen atmosphere (optional) and 14 days in air at room temperature. The specimens were tested in shear at a separation rate of 0.5 inch/minute and failed adhesively at 2,850 psi (ASTM-D-1002-64).

Aluminum T peel specimens were prepared by pressing the adhesive composition between 1 inch x 10 inches x 0.018 inch etched aluminum strips so that a glue line thickness of 0.005 to 0.010 inch was obtained. The samples were stored as described above and tested in 180° peel at a separation speed of 10 inches per minute (ASTM D-1876-61T). The average peel strength obtained was 35 lb per linear inch with mixed adhesive and cohesive failure.

Example 2: A second, shelf-stable adhesive composition was prepared by stirring 0.130 gram of cumene hydroperoxide into 65.3 grams of the chlorosulfonated polyethylene/acrylic monomers solution described in Example 1. Lap shear bonds were prepared by applying to steel coupons (treated as described in Example 1) an accelerator which was a mixture of butyraldehyde and aniline condensation products (sold as Du Pont Accelerator 808). The accelerator was applied with a cotton swab and then wiped to a thin film with a piece of cloth. The adhesive was pressed between the treated steel coupons to give a layer 0.005 to 0.010 inch thick. The following rate of shear strength development was observed.

Time, min	Lap Shear Strength, psi
3	1,450
10	2,030
30	2,050
60	2,100

Examples 3 through 8: Polymer solutions containing the following were prepared as described in Example 1: 100 grams chlorosulfonated polyethylene (the polymer described in Example 1) and the acrylic monomer listed for each example in the table. Adhesive compositions were prepared as follows: to 25 grams of each solution were added 0.125 gram cumene hydroperoxide, 0.06 gram N,N-dimethylaniline, and 0.04 gram cobalt naphthenate. Lap shear and T-peel test assemblies were then prepared as described in Example 1. The test results are shown in the table below.

Example	Monomer Compositions	Lap Shear (psi)	T-Peel (pli)
3	Methyl methacrylate 100	1,100	*
4	Ethyl methacrylate 100	1,725	*
5	Methacrylic acid 10		
	Methyl methacrylate 90	2,650	19
6	N-butyl methacrylate 90		
	Methacrylic acid 10	2,500	15
7	Methacrylic acid 10		
	Methyl methacrylate 45		
	N-butyl methacrylate 45	2,850	5.0
8	Methacrylic acid 10		
	Methyl methacrylate 45		
	2-Ethylhexylmethacrylate 45	2,780	22

*Not tested.

CYANOACRYLATE ADHESIVE CONTAINING MALEIC ANHYDRIDE

Cyanoacrylate adhesive compositions are extremely sensitive, and great care must be exercized in their formulation. Cure (polymerization) is generally considered to be initiated by an anionic mechanism, with water being a sufficiently strong base to initiate cure under most circumstances. The adhesives remain shelf stable items of commerce as long as they are suitably packaged, but when placed on a substrate to be bonded and exposed to atmospheric and surface moisture in so doing, cure generally is instituted in a relatively short period of time, generally in less than one minute, and on many surfaces within a matter of a few seconds. Typical examples of patents relating to cyanoacrylate adhesives are U.S. Patents 2,784,215 and 2,794,788 and British Patent 1,196,069.

This exceptional cure speed offers numerous advantages, particularly to those whose use adhesive bonding in production line applications. However, a major shortcoming which has limited the areas of applicability of cyanoacrylate adhesives has been the relatively low thermal resistance of the cured bonds. Bonded assemblies frequently are exposed to continuous operating temperatures substantially above normal room temperature and the adhesive assembly must retain reasonable strength for substantial periods of time at these elevated temperatures to retain its usefulness.

A process described by *D.J. O'Sullivan and D.P. Melody; U.S. Patent 3,832,334; August 27, 1974; assigned to Loctite Limited, Ireland* concerns adhesive compositions comprising one or more monomeric esters of 2-cyanoacrylic acid, and at least about 0.1% of a maleic anhydride by weight of the adhesive composition. These compositions additionally contain an inhibitor of anionic polymerization. The maleic anhydride additive does not produce undue stability problems in the cyanoacrylate adhesive composition, and hence can be added at the time of manufacture.

Example 1: Two separate batches of methyl-2-cyanoacrylate monomer were used to prepare cyanoacrylate adhesive compositions. Each batch contained, as stabilizers, 0.5% by weight propane sultone and 0.01% by weight hydroquinone. The first batch of the methyl-2-cyanoacrylate was divided into three parts, and designated Adhesives A, B and C, respectively. Adhesive A was allowed to remain unaltered as a Control. Approximately 0.6% by weight maleic anhydride and approximately 1.2% by weight maleic anhydride were added to Adhesives B and C, respectively.

The second batch of methyl-2-cyanoacrylate was divided into two portions; the first was designated Adhesive D and allowed to remain unaltered as a second Control sample. Approximately 0.8% by weight maleic anhydride was added to the second portion, which was then designated Adhesive E. Each of these five cyanoacrylate adhesives then was used to bond a series of 1 inch by 5 inch by 1/16 inch thick steel lap strips.

The bonded assemblies then were tested according to ASTM Test D-1002-64 to determine the thermal resistance characteristics of the bonds prepared with the various adhesives.

To form the assemblies, a drop of the adhesive in question was placed near the end of one of the lap strips, and the mating surface of the second lap strip was placed thereover in an aligned relationship with a one-inch overlap. It was found that within substantially less than a minute a firm bond was obtained. All the bonds were allowed to age overnight at room temperature before testing to guarantee that essentially full cure had taken place.

Thereafter, the assemblies were aged for 24 hours at 100°C and then were allowed to return to room temperature, after which the tensile-shear force required to separate the bonded assemblies was determined. This test is designated to be a measurement of the resistance of the adhesive to heat degradation. The results are presented below.

Adhesive and % Maleic Anhydride	Tensile Shear Strength of Bond, lb/sq in*
A 0%	460
B 0.6%	1,650
C 1.2%	3,000
D 0%	620
E 0.8%	2,450

Example 2: Three cyanoacrylate adhesive compositions were prepared, each using a different monomeric ester of 2-cyanoacrylic acid. The monomeric esters were, respectively, methyl-2-cyanoacrylate, ethyl-2-cyanoacrylate and allyl-2-cyanoacrylate. In each case the monomer contained 0.5% by weight propane sultone and 0.01% by weight hydroquinone. One half of each adhesive was set aside to serve as Control samples, and 1% by weight maleic anhydride was added to each of the other three portions. Each of these adhesives then were used to prepare a series of bonded assemblies, substantially as described in Example 1, above. Assemblies prepared by use of the various adhesives then were used in

one or both of the following tests, which are designed to measure the strength of the adhesives at elevated temperatures. In the first test, the assemblies in question were aged for thirty minutes at 125°C, and the assemblies then were immediately transferred to a tensile tester. The tensile-shear force required to separate the assemblies was determined without allowing substantial cooling of the assembly to take place. In the second test, essentially the same procedure was followed as for the first test, except that the aging was conducted at 105°C. The results of the tests are presented below.

Adhesive Monomer	Weight Percent Maleic Anhydride	Tensile Shear Strength of Bond, psi: Aged at 125°C Tested Hot	Tensile Shear Strength of Bond, psi: Aged at 105°C Tested Hot
Methyl-2-cyanoacrylate	0	100	1,400
	1	2,650	2,950
Ethyl-2-cyanoacrylate	0	0	850
	1	1,450	1,950
Allyl-2-cyanoacrylate	0	---------	0
	1	---------	950

ANAEROBIC ADHESIVE CONTAINING DIAZONIUM SALT

M.M. Skoultchi; U.S. Patent 3,880,956; April 29, 1975; assigned to National Starch and Chemical Corporation has found that anaerobic curing compositions characterized in possessing an extended shelf life of a half-year or more in the presence of air or oxygen together with the ability to polymerize or set up rapidly when excluded from air or oxygen and in contact or presence of certain metals can be formed comprising the mixture of a polymerizable acrylic or substituted acrylic monomer and diazonium salts. These properties render the composition particularly useful in the anaerobic bonding of adjacent or closely facing surfaces at least one of which is metal, for example, the mating threads of a nut and bolt.

Typical diazonium salts useful in the compositions include, for example, bis-(2-methyl-4-chlorobenzenediazonium) naphthalenedisulfonate, bis(2-methoxy-4-nitrobenzenediazonium) naphthalenedisulfonate, bis[-4-(N,N-diethylamino)-2-methoxybenzenediazonium] tetrafluoroborate and bis(3-nitro-4-chlorobenzenediazonium) tetrachlorozincate.

In these examples the compositions were prepared by mixing the indicated ingredients in amounts given as parts by weight unless otherwise specified. The ingredients are identified in the following tables.

Catalyst	Ingredient
Catalyst A	bis[4-(N,N-diethylamino)benzenediazonium] tetrachlorozincate
Catalyst B	bis[4-(N,N-diethylamino)-2-methoxybenzenediazonium] tetrafluoroborate
Catalyst C	bis[4-(N,N-dimethylamino)benzenediazonium] tetrachlorozincate
Catalyst D	bis[4-(N-ethyl-N-hydroxyethylamino)benzenediazonium] tetrachlorozincate

(continued)

Catalyst	Ingredient
Catalyst E	bis(2-methyl-4-chlorobenzenediazonium) naphthalene-disulfonate
Catalyst F	bis(2-chlorobenzenediazonium) tetrachlorozinacate
Catalyst G	bis(2,5-dichlorobenzenediazonium) tetrachlorozincate
Catalyst H	2-methoxy-4-(3'-methoxy-4'-diazobenzene)benzene-diazonium tetrachlorozincate
Catalyst I	bis(3-nitro-4-chlorobenzenediazonium) tetrachloro-zincate
Catalyst J	bis(2-methoxy-4-nitrobenzenediazonium) naphthalene-disulfonate
Catalyst K	bis(2-methoxy-4-nitrobenzenediazonium) tetrachloro-zincate

Monomer	Ingredient
Monomer AA	Ethyleneglycol dimethacrylate
Monomer BB	Hydroxyethyl methacrylate
Monomer CC	Hydroxypropyl acrylate
Monomer DD	Hydroxypropyl methacrylate
Monomer EE	Hydroxyethyl acrylate
Monomer FF	Pentacrythritol triacrylate
Monomer GG	Trimethylolpropane triacrylate
Monomer HH	Triethyleneglycol diacrylate
Monomer II	Triethyleneglycol dimethacrylate
Monomer JJ	Neopentylglycol diacrylate
Monomer KK	Ethyleneglycol diacrylate
Monomer LL	Methyl acrylate
Monomer MM	Ethyl methacrylate
Monomer NN	Butyl acrylate
Monomer OO	Octyl acrylate
Monomer PP	Poly(butylene maleate) dimethacrylate (MW 2,000)
Monomer QQ	Poly(propylene glycol) dimethacrylate (MW 1,025)
Monomer RR	Reaction product of 2 mols hydroxyethyl acrylate and 1 mol toluene diisocyanate
Monomer SS	Reaction product of 2 mols toluene diisocyanate, 1 mol poly(butylene adipate-phthalate) and 2 mols hydroxyethyl methacrylate
Monomer TT	Reaction product of 2 mols toluene diisocyanate, 1 mol poly(propylene glycol) and 2 mols hydroxyethyl acrylate

Examples 1 through 11: These examples illustrate representative compositions of this process prepared with a variety of diazonium salts.

	Ingredients (parts by weight)	
Example	Monomer	Catalyst
1	100 AA	0.5 A
2	100 AA	0.5 B*
3	100 AA	0.5 C
4	100 AA	0.5 D*
5	100 AA	0.5 E*
6	100 AA	0.5 F
7	100 AA	0.5 G
8	100 AA	0.5 H*
9	100 AA	0.5 I
10	100 AA	0.5 J
11	100 AA	0.5 K

*Required methanol to dissolve the diazonium salt

The compositions of the above examples were each evaluated in a finger-tight lock test known in the art. About 2 or 3 drops of each composition were placed on the exposed threads of separate ⅜-24 iron bolts (degreased) and immediately thereafter a nut (degreased) with mating threads was run onto the bolt so that the nut was directly in the thread area of the applied composition. Measurements were made of the time-periods necessary with each composition at room temperature to achieve a finger-tight lock such that the nut could not be moved on the threads with fingers. The results obtained with the various compositions are given in the following table.

Example	Lock Test Time (minutes)
1	8
2	13
3	20
4	10
5	20
6	7
7	7
8	35
9	5
10	5
11	30

Examples 12 through 21: These examples illustrate the use of various concentrations of catalyst in the compositions of this process.

Example	Ingredients (parts by weight)		
	Monomer	Catalyst	
12	100 BB	0.125	A
13	100 BB	0.25	A
14	100 BB	0.50	A
15	100 BB	1.00	A
16	100 BB	1.50	A
17	100 BB	0.125	B
18	100 BB	0.25	B
19	100 BB	0.50	B
20	100 BB	1.00	B
21	100 BB	1.50	B

The compositions of these examples were each evaluated in the finger-tight lock test. The bond strength of the bond between the nut and bolt formed with the composition after curing at room temperature for 2 hours was measured by means of a torque wrench. The torque required to move the nut on the threads initially, one-half turn and a full turn is noted in the table below, together with results obtained with the finger-tight lock test.

Example	Time (minutes)	Torque (inch-pounds)		
		Initial	½ Turn	Full Turn
12	10	120	300	250
13	5	125	125	105
14	10	100	210	125
15	5	120	170	100
16	4	100	190	200
17	17	30	100	120
18	13	110	230	200
19	13	150	225	230
20	7	120	220	150
21	5	110	200	190

Examples 22 through 44: The following examples illustrate the use of various monomers. Results obtained with the respective compositions in the finger-tight lock test are given in the last column.

Example	Ingredients (parts by weight) Monomer	Catalyst	Lock Test Time (minutes)
22	100 CC	0.5 A	10
23	100 CC	0.5 E	10
24	100 DD	0.5 A	7
25	100 DD	0.5 E	50
26	100 EE	0.5 A	8
27	100 EE	0.5 E	3
28	100 BB	0.5 A	15
29	100 BB	0.5 E	15
30	100 FF	0.5 A	60
31	100 GG	0.5 A	90
32	100 HH	0.5 E	3
33	100 II	0.5 E	120
34	100 JJ	0.5 E	40
35	100 KK	0.5 A	15
36	100 LL	0.5 A*	overnight
37	100 MM	0.5 A*	overnight
38	100 NN	0.5 A*	overnight
39	100 OO	0.5 A*	overnight
40	100 PP	1.0 E	60
41	100 QQ	1.0 E	30
42	100 RR	1.0 E	35
43	100 SS	1.0 E	30
44	100 TT	2.0 E	25

*Required methanol to dissolve the diazonium salt.

CARBOXY-MODIFIED OLEFIN COPOLYMER FOR CAN SEAMS

According to a process described by *K. Ishikawa and T. Shibata; U.S. Patent 3,900,126; August 19, 1975; assigned to Nippon Paint Co., Ltd., Japan*, metallic can is glued with a synthetic resin composition comprising an olefin copolymer having a carboxyl group in the side-chain, preferably together with a polyurethane resin. The product not only has an excellent adhesive strength and a superior sealing property but also such other advantages that the glued part is hardly affected with its contents and the whole surface is available for printing.

Example 1: In a 0.5 liter volume autoclave equipped with an agitator, there were charged ethylene-acrylic acid copolymer (acrylic acid content, 18% by weight; melt index, 300 dg/min) (100 g), 28% ammonia water (10 g) and water (20 g) and the resulting mixture was heated at 150°C for 10 minutes while agitating. After the portionwise addition of water (130 g), the resultant mixture was agitated at 90°C for 30 minutes and cooled to room temperature. Then, bisphenol A type epoxy resin (average molecular weight, 380; epoxy equivalent, 190) (20 g) was added thereto while agitating to make an aqueous dispersion to be used as an adhesive agent.

A chromium plated steel sheet (0.32 mm thick) was cut to make the plates of predetermined sizes for the trunk, ceiling and bottom parts of a 18-liter volume can, and the plates were subjected to press processing. The said adhesive agent was applied uniformly on the seam portions of the plates in 3 mm wide and 0.25 mm thick and by the aid of a can lining machine. After keeping in a dryer at 130°C for about 3 minutes in order to evaporate the water, the plates were assembled by the use of a can seamer according to the procedure as described in JIS (Japanese Industrial Standard) Z-1602, and the assembled product was kept in a dryer at 200°C for 1 minute whereby a 18-liter can [temper: 4 (for trunk), 2.5 (for ceiling and bottom)] was obtained. No abnormality was found on the can in the compression test and the leakage test carried out according to the procedure as described in JIS Z-1602 No. 6. Comparison was made between the above manufactured can and a can manufactured in a conventional soldering procedure on some physical and mechanical properties. The results are shown in the table below.

Can	Process	Soldered can
Adhesive strength between ceiling and trunk (kg/in)	70	80
Adhesive strength between trunk and trunk (kg/in)	above 100	60
Leakage test (JIS-Z-1602 No. 6–1)	accepted	accepted
Compression test (JIS Z-1602 No. 6–2)	accepted	accepted
Drop test (2 m)	deformed without leakage	broken at glued part with leakage
Content test		
Toluene	no change	no change
Methylethylketone	no change	no change
Ethyl acetate	no change	no change
N-Butanol	no change	no change
Salad oil	no change	no change

Example 2: Powdery ethylene-acrylic acid copolymer (acrylic acid content, 22% by weight; melt index, 120 dg/min; maximum particle size, 100 μ) (about 200 g) and liquid nitrogen (about 10,000 g) were put in a 20-liter volume turbo mill, pulverized for about 30 minutes and sieved to remove coarse particles of more than 100 μ in diameter. The resulting powder was used as an adhesive agent.

A tin plate (0.32 mm thick) was cut to make the plates of predetermined sizes for the trunk, ceiling and bottom parts of a 18-liter volume can, and the plates were subjected to press processing. The adhesive agent was sprayed on the seam portions of the plates preheated at 120°C by the aid of an infrared heater to make a uniform coating of about 0.1 mm in thickness. After cooling, the plates were assembled by the use of a can seamer as in Example 1, and the assembled product was kept in a dryer at 200°C for 1 minute whereby a 18-liter can [temper: 4 (for trunk), 2.5 (for ceiling and bottom)] was obtained.

No abnormality was found on the can in the compression test and the leakage test carried out according to the procedure as described in JIS Z-1602 No. 6.

Example 3: In a one-liter volume autoclave equipped with an agitator, there were charged polyvinyl alcohol (degree of polymerization, 1,000; saponification degree, 88%) (4 grams) and water (400 grams), and the resulting mixture was heated at 90°C while agitation to dissolve the polyvinyl alcohol.

Ethylene-acrylic acid copolymer (acrylic acid content, 10% by weight; melt index, 55 dg/min) (100 grams) and 28% ammonia water (5 grams) were added thereto, and the resultant mixture (5 grams) were added thereto, and the resultant mixture was kept at 100°C for 30 minutes while agitation occurred. Then bisphenol A type epoxy resin (average molecular weight, 380; epoxy equivalent, 190) (5 grams) was added thereto while agitating to make an aqueous dispersion to be used as an adhesive agent.

A tin plate having an epoxy resin coating on the surface (0.32 mm thick) was cut to make the plates of predetermined sizes for the trunk, ceiling and bottom parts of a 18-liter volume can, and the plate was subjected to press processing. The adhesive agent was applied uniformly on the seam portions of the plates in 3 mm wide and 0.2 mm thick by the aid of a can lining machine.

After keeping in a dryer at 130°C for about 3 minutes in order to evaporate the water, the plates were assembled by the use of a can seamer according to the procedure as described in JIS Z-1602, and the assembled product was kept in a dryer at 200°C for 1 minute whereby a 18-liter can (temper: 4 (for trunk), 2.5 (for ceiling and bottom) was obtained. No abnormality was found on the can in the compression test and the leakage test carried out according to the procedure as described in JIS Z-1602 No. 6.

Example 4: In a 0.5 liter volume autoclave equipped with an agitator, there were charged ethylene-acrylic acid copolymer as in Example 1 (100 grams), 28% ammonia water (10 grams) and water (250 grams), and the resulting mixture was heated at 95° to 98°C for 60 minutes while agitation occurred.

After cooling to 80°C, trimethylolpropane triglycidyl ether (15 grams) was added thereto, and agitation was carried out for 10 minutes. The resultant mixture was cooled to room temperature, a 40% aqueous dispersion of linear polyurethane resin (200 grams) was added thereto, and agitation was further continued for 20 minutes to make an aqueous dispersion to be used as an adhesive agent.

A tin plate (0.32 mm thick) was cut to make the plates of predetermined sizes for the trunk, ceiling and bottom parts of a 18-liter volume can, and the plates were subjected to press processing. The adhesive agent was applied uniformly on the seam portions of the plates in 3 mm wide and 0.25 mm thick by the aid of a can lining machine. After keeping in a dryer at 100°C for about 3 minutes in order to evaporate the water, the plates were assembled by the use of a can seamer according to the procedure as described in JIS Z-1602, and the assembled product was kept in a dryer at 200°C for 1 minute whereby a 18-liter can; temper: 4 (for trunk), 2.5 (for ceiling and bottom) was obtained. No abnormality was found on the can in the compression test and the leakage test carried out according to the procedure as described in JIS Z-1602 No. 6.

Comparison was made between the above manufactured can and a can manufactured in the similar manner but using an adhesive agent excluding the linear polyurethane resin on some physical and mechanical properties. The results are shown below.

Can	Linear Polyurethane Resin Used	Linear Polyurethane Resin Not Used
Elongation in film, %	390	300
100% modulus (kg/cm^2)	58	120

(continued)

Can	Linear Polyurethane Resin Used	Linear Polyurethane Resin Not Used
T-type peeling strength (kg/in)	22	20
Defective rate in seaming	0/200	3/200
Critical compressive force (kg/cm^2)	3.0	1.5

From the above results, it is understood that the softening in the adhesive layer results in the improvement of the defective rate in seaming and the critical compressing force.

SPECIALTY APPLICATIONS

In recent years much of the research effort in the adhesives area has been devoted to the development and optimization of bonding systems for use in high performance structural adhesives and the very demanding service areas such as dental, underwater and circuit board applications. This chapter provides formulations for many different modern adhesives processes in a variety of end uses.

DENTAL FORMULATIONS

Cationic Polymer and Anionic Compounds

Traditionally adherent powders used to secure dentures within the mouth were prepared from such materials as finely powdered natural gums, i.e., karaya, acacia or tragacanth gum. These materials have the particular property of swelling to many times their original volume upon the addition of water to form a gelatinous or mucilaginous mass.

Denture adhesive powders may be a combination of natural gums, flavored with pleasant tasting volatile oils. Many additives may be included, such as antiseptics, stabilizers, bactericides, special deodorants, plasticizing agents, fillers and coloring agents. Cream forms of the denture adherent, prepared from finely ground particles of the natural gums dispersed in a cream base may be used. When wet with water, the natural gum in either the cream or powder formulation, expands to become a viscous gel which acts as a cushion and an adherent between the denture plate and the gum tissue. While these relatively simple formulations are effective in securing dentures within the oral cavity for a short period of time, more than one application per day is necessary. This is most undesirable.

There have been numerous improvements in the above described simple denture adhesive formulations. For example U.S. Patent 3,003,988, a water-insoluble, water-sensitized copolymer is disclosed as the adhesive or stabilizing component of a denture composition. Actually, this patent describes mixed partial salts of copolymers of maleic anhydride with lower alkyl-vinyl-ethers, and partial esters of these salts, as the adhesive material. For example, the calcium-sodium partial salts of a partial isopropyl ester of methyl vinyl ether/maleic anhydride copolymer is said to provide a superior denture adhesive.

In U.S. Patent 2,997,399, the principal ingredient of the denture adhesive is a hydroxyethylcellulose having a certain degree of substitution of ethylene oxide groups per anhydro-glucose unit of the cellulose molecule; preferably, there is also present a lower alkyl ether of cellulose, such as methylether of cellulose, exhibiting temperature-dependent viscosity properties, i.e., showing an increase in viscosity with an increase in temperature. As an optional ingredient in the denture adhesive of U.S. Patent 2,997,399, there may be used a polyacrylamide. This optional ingredient is to produce an improved feel, texture or body in the denture adhesive.

U.S. Patent 3,440,065 describes still another denture adhesive where a powdered insoluble absorptive cellulose material, such as alpha-cellulose wood flour or microcrystallinecellulose, is incorporated into a gum-petrolatum base adhesive. As the gum in the adhesive sodium carboxymethylcellulose, hydroxyethylcellulose, karaya gum, guar gum, tragacanth and mixtures thereof are suggested.

Another improvement over conventional denture adhesives is shown in U.S. Patent 3,511,791, which teaches that copolymers of acrylamide and acrylic acid, or homopolymers of acrylamide can form the principle adhesive component in a denture adherent to provide sustained adherence in use without swelling. Whenever a homopolymer of acrylamide, which is nonionic in character, is used, gum karaya is added to the formulation in order to provide a synergistic improvement in gel strength and adhesiveness.

Further, U.S. Patent 3,575,915 provides a superior denture adhesive containing polyvinyl acetate, water, ethanol, a mineral hydrocolloid, such as colloidal silicate, and a plasticizer. U.S. Patent 2,978,812 describes the addition of ethylene oxide homopolymers to denture adhesive gums, including cellulose derivatives, to improve adhesive qualities. U.S. Patent 3,736,274 describes a denture adhesive containing three essential ingredients: a maleic anhydride and/or acid copolymer (with a lower alkyl vinyl ether), a polymeric N-vinyl lactam, and sodium carboxymethylcellulose.

While all of the above denture adhesives provide some improvement over simple formulations containing only finely powdered natural gums, it is generally recognized that no one product has yet been developed which can accommodate over a long period of time, the many variations in temperature, pH and mechanical agitation which are quite normal in the oral cavity.

J.J. Keegan, H. Rubin and R.N. Gidwani; U.S. Patents 3,868,432, February 25, 1975; 3,833,518, September 3, 1974; 3,868,339, February 25, 1975; all assigned to Warner-Lambert Co., describe an improved anhydrous denture adhesive which is prepared to contain a mixture of a cationic polymeric material and a synthetic anionic gum material as the adhesive ingredient. Suitable cationic materials include at least one of the following: copolymers of substituted or unsubstituted acrylamide with vinyl quaternary ammonium salts, and copolymers of substituted or unsubstituted acrylamide with vinyl and alkyl substituted vinyl pyridinium salts. As the anionic component there may be used at least one synthetic gum which is a copolymer of maleic acid with vinyl-lower alkyl ethers.

Example 1: Preparation of Denture Adhesive Cream — (A) Weigh 217.6 lb of heavy mineral oil into a steamjacketed Hobart mixer and heat to 90°C. At 90°C add 32.4 lb of a polyethylene wax of average molecular weight 2,000, and disperse the wax thoroughly in the mineral oil while force cooling the batch to 70°C during mixing. At 70°C, add 0.50 lb of methyl paraben and 0.50 lb of propyl paraben, and mix well, while force cooling the batch to 45°C.

(B) In a Hobart mixer, add 62.5 lb of a copolymer of 85% by weight acrylamide with 15% by weight β-methacrylolyloxyethyltrimethyl ammonium methyl sulfate; 62.5 lb of a methyl vinyl ether/maleic acid copolymer; and 0.50 lb of FD and C Red No. 3 Lackolene (19% pure dye); mix these dry ingredients until complete blending is achieved. At 45°C, add B to A and mix for about 15 min with the Hobart mixer.

(C) Add 0.50 lb of peppermint oil, U.S.P. to the mixture of A and B and mix at a high speed until a homogeneous cream is obtained.

Example 2: Preparation of Denture Adhesive Cream — A denture adhesive cream is prepared from the following ingredients:

	Pounds
Polyvinyl alcohol	125.0
Copolymer of acrylamide and methacrylolyloxymethyltrimethyl ammonium methyl sulfate	62.5
Copolymer of methyl vinyl ether and maleic acid	62.5
Petrolatum	149.0
Mineral oil	100
Propyl paraben	0.50
Methyl paraben	0.50
FD & C Red No. 3 Aluminum Lake	0.50
Imitation peppermint oil	0.50

The petrolatum and mineral oil are added to a steamjacketed mixture and heated to 70°C. At 70°C, the methyl paraben and propyl paraben are added and the mixture is force cooled at 45°C. The procedure of Example 1, parts B and C, is followed for all of the remaining ingredients.

Example 3: Preparation of Denture Adhesive Powder — A denture adhesive powder is prepared to contain the following ingredients:

	Pounds
Copolymer of acrylamide and γ-methacryloyloxypropyltrimethyl ammonium methyl sulfate	150
Copolymer of methyl vinyl ether and maleic acid	848
Micronized silica	1
Peppermint oil	1

The peppermint oil is well dispersed in about 25 lb of the copolymer of methyl-vinyl ether and maleic anhydride. The micronized silica is similarly dispersed in about 25 lb of the copolymer of methyl vinyl ether and maleic anhydride. The two premixes are then added to the balance of the copolymer of methyl vinyl ether and maleic anhydride which has previously been placed in a ribbon blender mixing apparatus. The whole is then mixed for about 15 min, after which the 150 lb of the acrylamide copolymer is added and the batch mixed for an additional 15 min.

In related work *J.J. Keegan, H. Rubin and R.N. Gidwani; U.S. Patent 3,868,340; February 25, 1975; assigned to Warner-Lambert Company* describe an improved anhydrous denture adhesive which is prepared to contain a mixture of a cationic polymeric material and a natural anionic gum material as the adhesive ingredient.

The mixture may be formulated as a powder or anhydrous paste, which, when applied to dentures and exposed to moisture, develops superior adhesive properties. Suitable cationic materials include one of the following: copolymers of substituted or unsubstituted acrylamide with vinyl quaternary ammonium salts and copolymers of substituted or unsubstituted acrylamide with vinyl and alkyl substituted vinyl pyridinium salts. As the anionic component there may be used at least one natural gum, such as gum karaya, gum arabic, gum Shiraz, gum tragacanth, pectin, pectinates, algin, and alginates.

J.J. Keegan, G. Patel and H. Rubin; U.S. Patent 3,878,135; April 15, 1975; assigned to Warner-Lambert Company also describe an anhydrous denture adhesive which is prepared to contain a mixture of a cationic polymeric material and an anionic protein material as the adhesive ingredient. The mixture may be formulated as a powder of anhydrous paste, which, when applied to dentures and exposed to moisture, develops superior adhesive properties.

Suitable cationic materials include at least one of the following: copolymers of substituted or unsubstituted acrylamide with vinyl quaternary ammonium salts; copolymers of substituted or unsubstituted acrylamide with vinyl and alkyl substituted vinyl pyridinium salts; and an O-lower alkyltrimethylammonium chloride-substituted-anhydroglucose polymer. As the anionic component there may be used at least one protein material, such as soybean meal, sunflower seed meal or peanut meal.

J.J. Keegan, H. Rubin and R.N. Gidwani; U.S. Patent 3,868,259; February 25, 1975; assigned to Warner-Lambert Company describe an anhydrous denture adhesive which is prepared to contain a mixture of a cationic polymeric material and a synthetic anionic gum material as the adhesive ingredient.

Suitable cationic materials include at least one O-lower alkyltrimethylammonium chloride-substituted-anhydroglucose polymer. As the anionic component there may be used at least one synthetic gum including carboxymethylcellulose and cellulose sulfates.

Polyacrylic Acid, Zinc Oxide and Stannous Fluoride

J. Foster and E. Henry; U.S. Patent 3,856,737; December 24, 1974; assigned to The Amalgamated Dental Company, Ltd., England describe a pack for the preparation of a surgical cement which comprises a solid component (a) comprising zinc oxide together with at least 2% by weight, based on the weight of zinc oxide, of stannous fluoride, and a liquid component (b) comprising an aqueous solution of polyacrylic acid. The surgical cement is prepared by mixing component (a) with component (b) suitably in a weight ratio of 0.05 to 4:1.

Example 1: A polyacrylic acid solution having a viscosity average molecular weight of 70,000 and a concentration of 42% w/w was prepared by solution polymerization of acrylic acid monomer using the method described in British Patent 1,139,430.

A powder consisting of 100 parts zinc oxide, 15 parts magnesium oxide, 4 parts silica and 2 parts aluminum oxide was then fired to obtain a suitable reactivity. This was then blended with 8% w/w of stannous fluoride. When mixed at a powder/liquid ratio of 0.95 g/0.50 g the components reacted to give the cement having a compressive strength of 970 kg/cm^2 and a diametral compression of 115 kg/cm^2. Without the addition of stannous fluoride to the powder the set cement had a compressive strength of 600 kg/cm^2.

Example 2: A liquid was prepared having the same characteristics as that detailed in Example 1. A powder was prepared as in Example 1 but was then blended with 10% w/w stannous fluoride. The compressive strength of the set cement was found to be 1,260 kg/cm^2 and its diametral compression 145 kg/cm^2.

Refractory Protective Composition

C.P. Mabie, Jr.; U.S. Patent 3,892,579; July 1, 1975; assigned to American Dental Association describes an adhesive refractory composition useful in investment casting to secure wax or plastic preforms to a pattern, and as a protective coat on the inner surface of a mold. The composition consists of a tacky organic solvent solution of a refractory metal salt (zirconium octoate in methylene chloride) having dispersed therein fine granules of a refractory ceramic material (zircon).

Example 1: A coating slip was initially prepared by precipitation of zirconium octoate from a 350 ml sample of a 22% mineral spirits solution (22% ZrO_2) by repeated absolute ethanol rinses. The salt obtained by precipitation, which resembles a white taffy, was dissolved in 350 ml of methylene chloride. This solution contained about 60% of zirconium octoate (30% ZrO_2). To the solution was added 600 g of precision casting grade zircon having a particle size of from 2 to 5 microns.

The slurry was mixed by repeated pouring back and forth between two polyethylene containers. A Ticonium refractory cast, consisting of a gypsum-quartz investment material (Investic) of a partial denture was dipped into the slurry to coat one half the surface and then permitted to dry for about 30 min before application of plastic or wax preforms. Plastic preforms may be lifted off the refractory coated models with care. Engineered plastic and wax patterns may be dipped again in the coating mixture. Excess coat must be knifed around the edges of the pattern in order to eliminate fin formation in the final cast part. Outer investment was then applied and normal burn out and alloy casting was executed.

A low fusing chromium base alloy which fuses below 1,400°C was melted and injected centrifugally into the casting investment. Subsequently, the resulting casting and its surrounding investment was let to cool on standing to room temperature. The resulting metal casting was knocked free of surrounding investment with a hammer and then sandblasted to eliminate adhering investment and scale.

Comparative profilometer measurements were made over the sandblasted partial denture casting surface with and without the coating of the process. The average center of line values (CTA) were measured and their significance computed. On the outer investment side, the CTA value was 1.351 ± 0.123 microns for the surface which was cast against the coating of the process.

The CTA value for a surface cast against the investment was 2.002 ± 0.127. Profilometer values on the refractory cast casting side was the same for both the coated and uncoated areas, 1.880 microns. Prior to sandblasting the surface cast against the coating was shiny and showed no sign of scaling. This was in contrast to the surface cast directly against the investment which was a dull gray and exhibited abundant scaling.

The absence of scaling, when the coat of this process is used, is due to the fact that the zircon-zirconium octoate composition after burnout does not liquefy during the casting and thus react with oxides and suboxides of vaporized casted alloy constituents. Thermal and corrosive effects which have been observed by petrographic examination of spent gypsum, phosphate and silica bonded investments and currently marketed protective coats were not present when the composition of this process is employed as a protective coating.

Example 2: The zirconium octoate solution of Example 1 was replaced in the amount of 50% by a methylene chloride solution of cobalt octoate. The cobalt octoate improved the flux capability of the coating although the refractoriness of the coating was reduced. The improved flux capability of the cobalt octoate addition aides in the adherence of the invented protective coat to the investment both during and after burnout, thus reducing the chance of spalling into the mold.

Example 3: Partial denture plates were made from Ticonium plastic triforms using a single dip of the composition of Example 1 as an adhesive coat on the refractory cast. Two castings were made from the low fusing chromium-nickel-base alloy which fuses below 1,400°C and investing technique and three were made from the high fusing chromium-cobalt-base alloy which fuses above 1,440°C and phosphate bonded investment. In all castings high detail resolution and nearly complete scale removal were effected.

1,3-Bis[2,3-Di(Methacryloxy)Propoxy] Benzene

According to a process described by *R.J. Gander; U.S. Patent 3,845,009; October 29, 1974; assigned to Johnson & Johnson* improved dental restorative cements are prepared from a new class of methacrylate monomer 1,3-bis[2,3-di-(methacryloxy)-propoxyl] benzene (RGTMA). When the monomer is utilized as a binder for inorganic filler materials such as silane-treated crystal quartz filler, composite dental restoratives having substantially improved compressive strength and related physical strength properties are produced.

For convenience in use, the composite dental filling compositions can be formulated in the form of pastes adapted for ready mixing by the dentist or other user. Thus, a paste (A) can be formulated containing the resin-forming monomer, inorganic filler and activator while a second paste (B) can contain the monomer, filler and peroxide, approximately the same proportions of monomer and filler being present in each paste for convenience.

Upon mixing of the two pastes, polymerization of the monomer or monomers is initiated with the working or hardening time being variable and controllable by use of more or less of the activator. A typical formulation for pastes A and B in this embodiment would be as follows:

Paste	----Proportion, Weight----	
	A	B
Component:		
Silane-treated quartz	80.0	80.0
1,3-bis[2,3-di(methacryloxy)-propoxy] benzene (RGTMA)	20.0	20.0
N,N-di(2-hydroxyethyl)-p-toluidine	0.2	–
Benzoyl peroxide	–	0.2

Orthodontic Bonding Agent

Present orthodontic practices generally include the securement of wire-receiving brackets to the patient's teeth, and the engagement of wire in such brackets, as by the light-wire, edgewise and other techniques, to effect the desired tooth movement. Securement of a bracket to its corresponding tooth is generally accomplished by means of a metal band which is attached to the bracket and circumposed about the tooth.

This securement system, while providing an excellent transfer of forces from the bracket to the tooth, has the very important disadvantage that it creates a high risk of tooth decay since food particles often lodge in the difficult to clean regions between the metal bands and their corresponding teeth.

The profession has recognized the decay problem and suppliers of orthodontic materials have sought to obviate the necessity of the metal bands by providing a chemical bonding composition for securing the brackets to the teeth. This composition, although of substantial utility has, because of the lack of permanence of the bond formed with the tooth, been restricted in use to those instances where the desired tooth movement can be achieved without the application of severe forces or can be achieved over a period of several months.

This product is the mixture of a 1:1 powder blend of polymethyl methacrylate and polycyclohexyl methacrylate in liquid hexyl methacrylate. The ratio of powder to liquid is about 2:1. This type of product is referred to as a thermoplastic bonding composition.

Another development in the field of dentistry has been the formulation of enamel adhesive sealers. An adhesive composition is bonded to the pits and fissures of teeth to prevent dental caries in these highly susceptible areas. These adhesives are usually based on a material which is the reaction product of bisphenol A and an epoxy ester of methacrylic acid, which may optionally be compounded with a filler, see U.S. Patent 3,066,112. This type of product is referred to as thermosetting sealer composition.

In some recent effort *E. Silverman and M. Cohen; U.S. Patent 3,895,445; July 22, 1975* describe an adhesive or bonding system which comprises a thermosetting sealer composition and a thermoplastic bonding composition. The combination of these two compositions may be adopted to provide a chemical bracket-to-tooth bond which has a high bond strength and a long life in orthodontic applications.

With regard to the bonding of a bracket to a tooth utilizing the bonding system of this process, the first step involves the coating of a section of the tooth with the thermosetting portion of the adhesive since these thermosetting sealers are well-known to the dentist, their selection in accordance with this process will present no problem to the practitioner, as he will be able to select that sealer with which he is most familiar.

Furthermore, the practitioner will be familiar with the procedure for applying the selected sealer to the tooth, since the procedure is essentially the same as that employed where the sealer is applied for the purpose of sealing fissures and the like. Thorough cleaning of the tooth and application of a thin coating of sealer comprises the preferred procedure.

After the resin has set, the bracket of a type having a plastic or metal surface is bonded to the coated tooth by means of the thermoplastic bonding composition. Like the thermosetting sealer, the thermoplastic bonding composition also is a material with which the orthodontist is familiar; the previously mentioned Bracket Bond is a preferred material.

The thermoplastic material is applied to either the surface of the bracket or the coated portion of the tooth, and if desired may be applied to both; the surface of the bracket is then positioned adjacent the coated portion of the tooth and held in position until the bonding composition has hardened.

Example 1: A tooth was cleaned with polishing brushes and pumice, and air dried. The tooth was further conditioned by rubbing a 50% phosphoric acid solution containing 7.0% dissolved zinc oxide on the tooth with a small cotton pellet, washing with water and drying with compressed air.

The sealer selected contained the following major ingredients: 3 parts by weight of the reaction product of bisphenol A and glycidyl methacrylate, and 1 part by weight of methylmethacrylate monomer. This sealer is a relatively stable, thick liquid. As per the instructions of the manufacturer, immediately before using, about 2.0% benzoin methyl ether was dissolved in the liquid to form an ultraviolet light-sensitive composition.

This composition was then painted on the surface of the tooth and exposed to long-wave ultraviolet light to cause setting of the resin. After setting, which required only a few seconds, the surface of the sealer was wiped with a cotton roll to remove any traces of unpolymerized surface sealer.

Bracket Bond was then applied to the coated tooth with a fine brush and a plastic bracket positioned adjacent the tooth. The bracket was held in place until the Bracket Bond had air-hardened. The bond formed between the bracket and tooth was strong, meeting all the requirements of orthodontic therapy.

Example 2: The process was carried out in accordance with Example 1 except that the sealer also contained 5% of a finely powdered suspension of synthetic calcium hydroxyapatite and 2.5% calcium fluoride. Results were similar to those obtained in Example 1. The thermosetting resin useful in the process comprises the reaction product of bisphenol A with glycidyl methacrylate, which is depicted as:

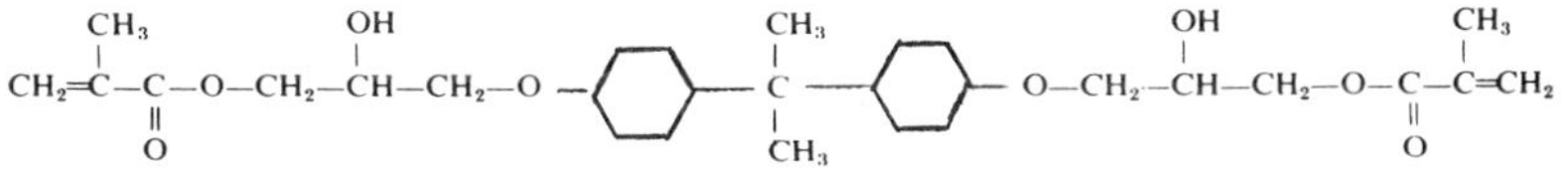

Thermosetting sealer compositions comprise this product with an ester of methacrylic acid, such as methyl methacrylate, glycidyl dimethacrylate or triethyleneglycol bismethacrylate. An organic or inorganic filler, such as quartz, silica, alumina, barium oxide, barium fluoride, boric oxide, may be added into the thermosetting resin if desired. A wide variety of such compositions has already been extensively tested as sealers and reported in the literature.

The thermoplastic bonding composition comprises a mixture of a polymer and monomer of a methacrylic acid ester which hardens rapidly with a minimum of shrinkage. It has been found that a mixture of such methacrylic acid esters gives the best combination of bonding properties. The preferred group of methacrylic acid esters contain from one to eight carbon atoms in the ester portion. They are depicted as follows:

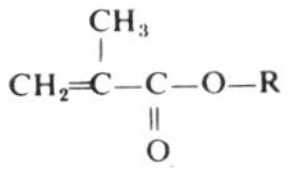

where R is a hydrocarbon group having from one to eight carbon atoms, such as methyl, hexyl, and cyclohexyl. The composite bond of the process is a layer of the thermosetting sealer composition adjacent the thermoplastic bonding composition. A sectional study under an electron microscope (12000X magnification) of the boundry layer between the two compositions, prepared as set forth in Example 1, showed that the thermoplastic bonding composition penetrated the thermosetting sealer composition.

CIRCUIT BOARDS

Flexible Printed Circuit

In recent years, with the development of electronic and electric instrument industry, reduction in size, reduction in weight, high reliance of instruments for communications industry, appliances and the like and simplification of the packaging system have been required and a printed circuit board having a light and flexible plastic film as an insulating base board has been desired.

The plastic films include a polyester film, a polyethylene film, a polyvinyl chloride film, a polypropylene film, a polyimide film and the like and each of them has excellent mechanical, electrical and chemical properties. Furthermore, they are excellent in flexibility, and hence, they are preferably adhered to a metal foil and then used as a printed circuit base board.

However, in general, these plastic films are poor in surface activity, and accordingly, it is very difficult to obtain an adhesion to a metal foil strong enough to satisfy requirements for a printed circuit. Furthermore, it is very difficult to afford to the base board excellent circuit qualities and resistance to the severe treating chemicals encountered in the production and assembling of a printed circuit and there have been very few satisfactory adhesives for a flexible printed circuit.

T. Watanabe, S. Yamaoka and K. Tanaka; U.S. Patent 3,896,076; July 22, 1975; assigned to Sumitomo Bakelite Company, Ltd., Japan describe an adhesive used for a laminate for a flexible printed circuit. The adhesive composition comprises as the essential components a polyester resin, an isocyanate compound, and a B-stage epoxy resin.

This composition is excellent in flow property and rapid curability and is suitable for use in a continuous adhesion between a plastic insulating film and a conductive foil by means of a roll-laminator and it affords an excellent characteristic to the laminate for a printed circuit.

There is also provided an adhesive composition which contains, in addition to the above ingredients, a styrene copolymer comprising maleic anhydride, or an alkyl maleate as a structural unit, or a metallic salt of an aliphatic carboxylic acid in order to prevent blocking and promote curing of the adhesive in a roll laminating process.

The addition of a small amount of a pigment results in an adhesive composition which is effective in the economical production of a laminate for a colored flexible printed circuit, and which has the effect of preventing the blocking and improving the weather resistance of the adhesive.

Example 1: In methyl ethyl ketone were dissolved 100 parts by weight of polyester resin having a number average molecular weight of about 20,000 obtained by condensation of 0.6 mol of terephthalic acid and 0.4 mol of adipic acid with 1 mol of ethylene glycol, 8 parts by weight of tolylene diisocyanate, 10 parts by weight of a B-stage epoxy resin having a number average molecular weight of about 4,500 obtained by reacting 0.8 mol of diaminodiphenylmethane with 1 mol of diglycidyl ether of bisphenol A in a mixed solvent of methyl ethyl ketone and methyl Cellosolve at a temperature ranging from 80° to 100°C for 2 hr, and 2 parts by weight of a half-esterified product with n-heptyl alcohol of an equimolar copolymer of maleic anhydride and styrene to prepare an adhesive varnish having a concentration of 20% by weight.

The resulting adhesive was applied in a thickness of about 30μ to a polyethylene terephthalate film having a thickness of 50μ by means of a roll-laminator in which an applying roll portion, a drying zone and a pressing roll portion are connected.

The film coated with the adhesive was dried at 120°C for 5 min and then adhered to a copper foil having a thickness of 35μ by pressing them at 150°C at a pressure of 15 kg/cm² between the pressing rolls consisting of a metal roll and a rubber roll while passing the film and the copper foil through between the pressing rolls in 2 sec so that the film coated with the adhesive was brought into contact with the metallic roll and the copper foil was in contact with the rubber roll, and that the copper foil was allowed to wrap around the rubber roll from the position on the circumference of the rubber roll for $\pi/4$ or more radians toward the contact line of both rolls, to form a flexible, copper-clad laminate.

The properties of the resulting flexible copper-clad laminate are shown in the table, p. 246. Though the adhesion was effected at a low pressure in a short period of time, the laminate was excellent in bonding strength, resistance to chemicals and electrical properties. In contrast thereto, when a conventional adhesive was used (in Comparative Examples 1 and 2), the resulting laminate was inferior in resistance to chemicals, heat resistance, bonding strength and the like and no sufficient performance was obtained.

Example 2: The following four ingredients were dissolved in tetrahydrofuran to prepare an adhesive varnish having a concentration of 15% by weight in the same manner as in Example 1:

(1) Polyester resin (co-condensation product of ethylene glycol, terephthalic acid and sebacic acid in a molar ratio of 1:0.5:0.5) 100 parts by weight

(2) Triphenylmethane triisocyanate—4 parts by weight

(3) Epoxy resin (B-stage reaction product of bisphenol A diglycidyl ether and metaphenylene diamine in a molar ratio of 1:0.8) 10 parts by weight

(4) Copolymer of maleic anhydride and dimethylstyrene in a molar ratio of 1:1—1 part by weight

The resulting adhesive varnish was applied in a thickness of about 30μ to a polyethylene-2,6-naphthalate film having a thickness of 50μ (Q film manufactured by Teijin Co.) and the film coated with the adhesive was dried at 130°C for 3 min and then adhered to a copper foil having a thickness of 35μ by passing the film and the copper foil through between the pressing rolls at 160°C at a pressure of 20 kg/cm² in 1 sec in the same manner as in Example 1.

The properties of the resulting laminate are shown in the following table. Although the pressing was effected in a short period of time, the resulting laminate was excellent in each property, and in particular, excellent in soldering heat-resistance as compared with a laminate prepared by use of a polyethylene terephthalate film as the base.

Example 3: Equimolar amounts of polyglycidyl ether of novolak resin and dicyandiamide were reacted in a mixed solvent of methyl ethyl ketone and dimethylformamide at a temperature ranging from 100° to 120°C for 3 hr to obtain a B-stage epoxy resin having a number average molecular weight of about 6,000. The following components shown including the epoxy resin were dissolved in methyl ethyl ketone to prepare an adhesive varnish having a concentration of 20% by weight.

(1) Polyester resin (the same as in Example 1)—100 parts by weight addition reaction product of 1 mol of trimethylol propane and 3 mol of toluene diisocyanate—5 parts by weight

(2) Epoxy resin (the aforesaid B-stage reaction product)—15 parts by weight

(3) Dibutyltin dilaurate—0.1 part by weight

The resulting varnish was applied in a thickness of about 25μ to a polyimide film having a thickness of 50μ in the same manner as in Example 1, and dried at 110°C for 5 min. The film coated with the adhesive was then adhered to a copper foil having a thickness of 35μ by passing the film and copper foil through between the same pressing rolls as in Example 1 at 170°C at a pressure of 25kg/cm^2 for 3 sec in the same manner as in Example 1 to obtain a flexible copper-clad laminate.

The properties of this laminate are shown in the following table. Although the pressing was effected in a short period of time, the resulting laminate was excellent in each property, and in particular, markedly excellent in soldering heat-resistance.

Example 4: Equimolar amounts of cycloaliphatic epoxy compound and methanediamine were reacted in methyl ethyl ketone at a temperature of from 60° to 80°C for 1 hr to prepare a B-stage epoxy resin having a number average molecular weight of about 2,500. The following four components including the epoxy resin were dissolved in a mixed solvent of methyl ethyl ketone and toluene to prepare an adhesive varnish having a concentration of 30% by weight.

(1) Polyester resin (the same as in Example 2)—100 parts by weight

(2) Diphenylmethane, diisocyanate—6 parts by weight

(3) Epoxy resin (the aforesaid B-stage reaction product)—5 parts by weight

(4) Zinc octylate—0.05 part by weight

The resulting adhesive varnish was applied in a thickness of about 15μ to a hard polyvinyl chloride film having a thickness of 200μ in the same manner as in Example 1, and dried at 80°C for 3 min. Then, the film coated with the adhesive was adhered to a copper foil having a thickness of 35μ by passing the film and copper foil through the same pressing rolls as in Example 1 at a temperature of 120°C at a pressure of 10 kg/cm^2 for 1 sec in the same manner as in Example 1 to prepare a flexible copper-clad laminate.

The properties of the resulting laminate are shown in the following table. The laminate was excellent in bonding strength, electrical properties and the like though the pressing was effected in a short period of time.

Example 5: For the adhesive components in Example 1, the following three components were substituted to prepare an adhesive varnish, and the resulting varnish was applied to a polyethylene terephthalate film. The film coated with the adhesive was adhered to a copper foil by means of the same pressing rolls as in Example 1 to prepare a flexible copper-clad laminate.

(1) Polyester resin (co-condensation product of ethyleneglycol, propyleneglycol and terephthalic acid in a molar ratio of 0.5:0.5:1) 100 parts by weight

(2) Polymethylene polyphenyl isocyanate millionate MR—5 parts by weight

(3) Epoxy resin (B-stage reaction product of a bisphenol A type diepoxy compound Epikote 1001 and menthane diamine in a molar ratio of 1:1)—30 parts by weight.

The resulting adhesive composition was excellent in flow property at the time of roll-laminating, and, even when the film and the copper foil were adhered to each other under a low pressure in a short period of time, a laminate having an excellent appearance was obtained. The properties of the laminate are shown in the following table, and the laminate was excellent in bonding strength, resistance to chemicals, and electrical properties.

Example 6: A flexible printed circuit having a circular land was prepared by an etching method by use of the laminate obtained in Example 5. Separately, the same adhesive composition was applied to a polyethylene terephthalate film having a thickness of 25μ, and dried at 120°C for 5 min, after which areas corresponding to the land were punched off from the film. The thus obtained film was placed on the flexible printed circuit in a registered position and heat-pressed at a temperature of 150°C at a pressure of 30 kg/cm^2 for 40 min to adhere the film to the circuit.

The resulting covered printed circuit board had the copper foil circuit portion completely embedded in a coverlay coated with the adhesive of the process and the adhesive did not exude to the exposed portion of the circular land, and the printed circuit was covered very good.

The boundary portion of the coverlay of the printed circuit was not invaded by chemicals such as flux at the time of soldering. When soldering was effected under the controlled conditions at 240°C, the adhesive was not peeled off and the thus obtained printed circuit had very excellent quality.

Comparative Example 1: The same adhesive composition as in Example 5, except that the B-stage epoxy resin component was not used, was applied to a polyethylene terephthalate film in the same manner as in Example 1. Then, the film thus coated with the adhesive was adhered to a copper foil by means of hot rolls. The adhesive was low in flow property at the time of roll-laminating and was unable to adhere the whole surface of the laminate uniformly. As shown in the following table, the laminate obtained was inferior in resistance to chemicals and heat.

Comparative Example 2: A polyurethane solution obtained by reacting 100 parts by weight of the polyester resin used in Example 5 with 10 parts by weight of tolylene diisocyanate in tetrahydrofuran at 80°C for an hour was applied to a polyethylene terephthalate film. Separately, a mixed solution of an epoxy resin [an equimolar mixture (A-stage product which had not been subjected to previous reaction) of a diepoxy compound of bisphenol A type (Epilote 1001), and menthanediamine] and a curing agent was applied to a copper foil. The copper foil was dried at 130°C for 5 min and then placed over the polyester film coated with the polyurethane solution, and adhered to the polyester by heat-pressing the polyester film and copper foil at a temperature of 160°C at a pressure of 60 kg/cm^2 for 120 min by means of a hot press to obtain a flexible, copper-clad laminate. The properties of the resulting laminate were as shown in the table below, and the laminate of this Comparative Example 2 was inferior in adhesiveness as compared with the laminate obtained according to Example 5 of the process.

Test Items (unit)	Test Standards	Treatment and Conditions	Base Film: Polyethylene Terephthalate Example 1	Polyethylene-2,6-Naphthalate Example 2	Polyimide Example 3
Surface resistance (Ω)	JIS C-6481	A	2×10^{15}	1×10^{15}	8×10^{14}
		C-96/40/90	8×10^{14}	6×10^{14}	3×10^{14}
Volume resistance (Ω-cm)	JIS C-6481	A	4×10^{16}	1×10^{16}	9×10^{15}
		C-96/40/90	5×10^{15}	3×10^{15}	2×10^{15}
Peel strength (kg/cm)	JIS C-6481	A In the direction of 180°	2.2	1.8	1.6
Resistance of chemicals	JIS C-6481	Immersed at room temperature for 1.5 min in trichlene, acetone and methylene chloride	Not changed	Not changed	Not changed
Rending resistance of the base board (number of bendings)	JIS P-8115	Load 100 g/mm	1,000 or more	1,000 or more	1,000 or more
Soldering heat resistance (°C)	JIS C-6481	Floated in a soldering bath for 60 seconds	at 230°C Not changed	at 250°C Not changed	at 260°C Not changed

Test Items (unit)	Test Standards	Treatment and Conditions	Base Film: Polyvinyl Chloride Ex. 4	Polyethylene Terephthalate Ex. 5	Polyethylene Terephthalate Com. Ex. 1	Polyethylene Terephthalate Com. Ex. 2
Surface resistance (Ω)	JIS C-6481	A	6×10^{14}	4×10^{15}	1×10^{14}	3×10^{13}
		C-96/40/90	1×10^{14}	1×10^{15}	2×10^{13}	3×10^{12}
Volume resistance (Ω-cm)	JIS C-6481	A	5×10^{15}	6×10^{16}	4×10^{15}	7×10^{15}
		C-96/40/90	7×10^{14}	5×10^{15}	9×10^{14}	8×10^{14}
Peel strength (mg/cm)	JIS C-6481	A in the direction of 180°	2.0	2.1	1.4	0.6
Resistance of chemicals	JIS C-6481	Immersed at room temperature for 15 min in trichlene, acetone and methylene chloride	–	Not changed	The adhesive expanded	Not changed

(continued)

			Base Film			
Test Items (unit)	Test Standards	Treatment and Conditions	Polyvinyl Chloride Ex. 4	Polyethylene Terephthalate Ex. 5	Polyethylene Terephthalate Com. Ex. 1	Polyethylene Terephthalate Com. Ex. 2
Rending resistance of the base board (number of bendings)	JIS P-8115	Load 100 g/mm	200	1,000 or more	1,000 or more	700
Soldering heat resistance (°C)	JIS C-6481	Floated in a soldering bath for 60 seconds	-	at 230°C Not changed	at 200°C delaminated	at 230°C Not changed

Phenoxy and Epoxy Resin with Chelate Crosslinker

Y. Fujiwara, K. Naito and T. Odajima; U.S. Patent 3,873,637; March 25, 1975; assigned to Sony Corporation, Japan describe a structural adhesive composition which is particularly suited for bonding or laminating a metal foil to an epoxy resin impregnated glass fiber plate, for example, as in the production of copper clad painted circuit boards. The adhesive comprises a phenoxy resin having the formula

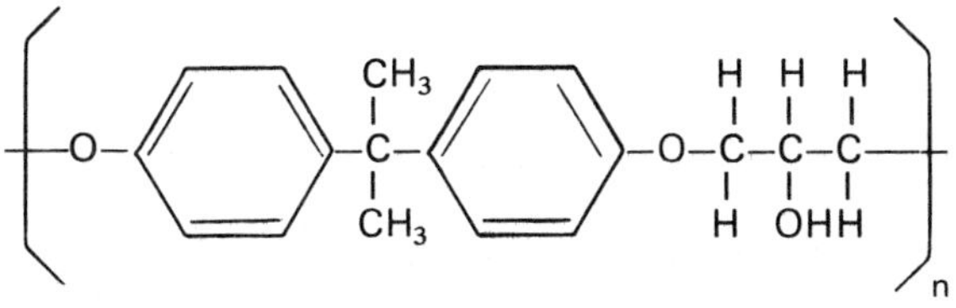

where n is 50 to 100, a low molecular epoxy resin having more than 2 epoxy groups and an expoxy value less than 4,000, and a crosslinking agent which either has at least 2 aziridine rings at the ends of the molecule to react with the hydroxyl group (OH) of the phenoxy resin and with the low molecular epoxy resin for providing a high crosslinking density, or which is a chelate of an organic metal compound having the formula

$$(R_1O)_n\text{—}M\text{—}(OR_2)_m$$

where M is a metal atom having a valence of at least one, n is an integer including zero, m is an integer, R_1 is a member selected from the group consisting of hydrogen atoms, alkyl, unsaturated aliphatic, cyclic alkyl and aryl, and R_2 is a member selected from the group consisting of alkyl and aryl which has at least one electron donor atom to form a chelate with the metal atom.

The epoxy resin is preferably present in the composition in an amount between 20 and 400 weight percent based on the amount of the phenoxy resin, and the crosslinking agent is preferably present in the composition in an amount between 2 and 30 weight percent based on the amount of the phenoxy resin.

Typical crosslinking agents are compounds having at least two aziridine rings at the ends of the molecules thereof, such as are represented on the following page.

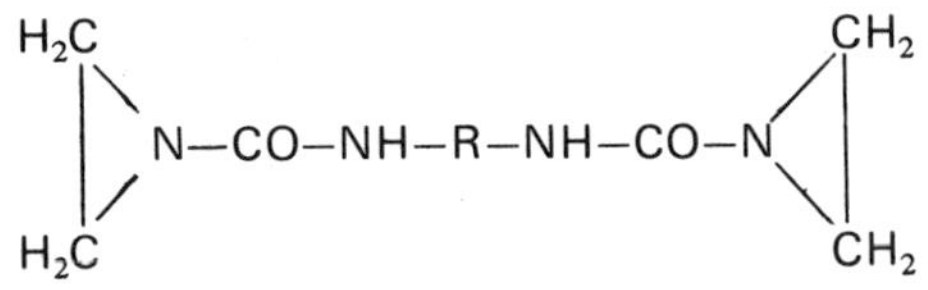

Ethylene-urea Derivative

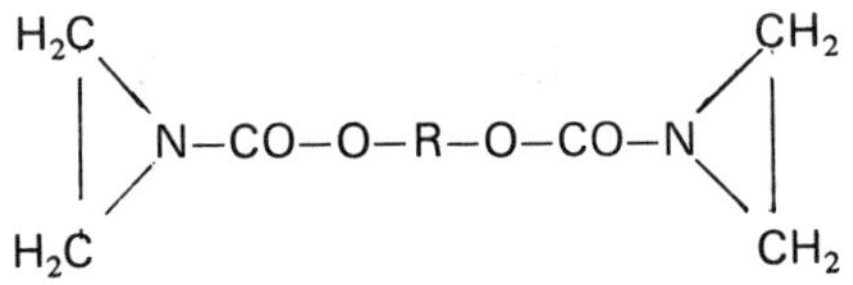

Ethylene-urethane Derivative

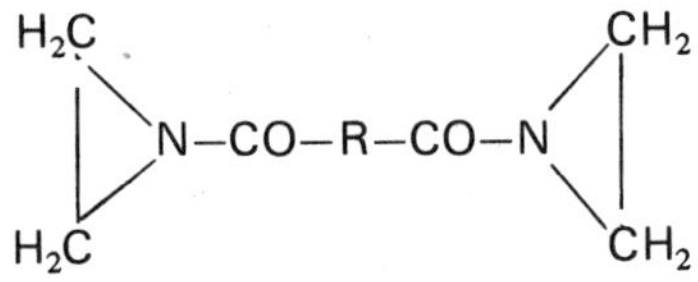

Acyl-urethane Derivative

R is shown as an aliphatic or aromatic group. Atoms such as phosphorus and sulfur may be included. Each of these compounds having aziridine rings serves as a hardener of the epoxy resin and, at the same time, reacts with active hydrogen, such as a hydroxyl group (OH) and a carboxyl group (COOH).

The chelate compounds of the organic metal compound are tetraoctyleneglycol titanate, triethanolaminetitanate diisopropylate, ethylacetoacetate aluminum diisopropylate and methylacetoacetate aluminum dibutoxylate. The adhesives including the above chelate compounds are highly stable and can be semicured by suitably selecting the conditions for coating and drying of the adhesive.

Use of the adhesive according to this process prevents the appearance of the stain which results from reaction of the copper oxide layer with the impregnated resin. In order to prove this, experiments were conducted in which two adhesive coated copper foils were laminated under the usual conditions for laminating a copper foil in an epoxy resin impregnated glass board, that is, under a pressure of 40 to 60 kg/cm^2 and at a temperature of 160° to 180°C for 1 hr.

An adhesive coated copper foil was laminated on an epoxy glass of G-10 or G-11 grade to provide a copper clad glass board. In both cases, the copper foil was etched with a solution of ferric chloride of 40° Baumé at a temperature of 25° to 30°C but no stain was found.

The metal clad printed circuit board produced with an adhesive according to this process had excellent chemical resistance, electric characteristics and adhesion between the metal foil and the glass board. Further, the metal clad glass board was free from reprint of the copper oxide, stain and discoloration.

Example 1:

Adhesive Composition	Parts by Weight
Phenoxy PAHJ (Union Carbide Corp.) a phenoxy resin that is the reaction product of bisphenol A and epichlorohydrin having a molecular weight of about 30,000 and lacking terminal highly reactive epoxy groups	100
Epon 828, a low molecular epoxy resin, a diglicydyl ether of bisphenol A based resin	70
Crosslinking agent having the formula	10

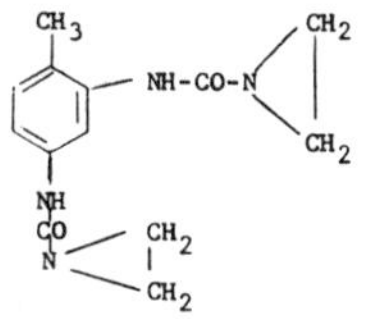

or

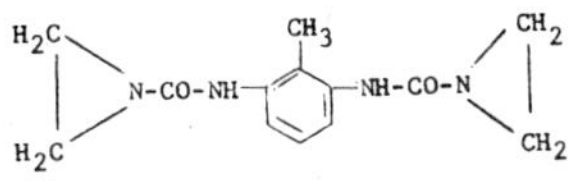

A solution of phenoxy PAHJ and Epon 828 in methyl ethyl ketone is mixed with 20% by weight of a solution of the above crosslinking agent in dimethylformamide in the above ratio and the solid content of the total solution is adjusted to be 20% by weight with respect to the total amount. The above crosslinking agent is produced by sufficiently stirring a solution of toluene diisocyanate in dehydrated acetone or toluene at 0°C while dropping ethyleneimine into the solution, in which case the reaction quantitatively proceeds.

The adhesive thus obtained is laid down on a treated electrolytic copper foil T/A (Circuit Foil Corporation) and is then treated at normal temperature for 2 min, at 90°C for 2 min and at 160°C for 4 min to produce an adhesive coated copper foil with the adhesive being 20 microns in thickness (dry thickness).

Production of Prepreg (NEMA G-10 grade)	Parts by Weight
Epikote 1001, a bisphenol A epoxy resin having epoxy values in the range of 450 to 500	100
Dicyanodiamine	4
Benzyldimethylamine	0.2

Dicyanodiamine is dissolved in a solvent containing 50% by weight, dimethylformamide and 50% by weight, ethyl cellosolve, the amount of dicyanodiamine being 20%, by weight, with respect to the solvent. Epikote 1001, the dicyanodiamine solution and benzyldimethylamine are dissolved in acetone to adjust the viscosity to about 200 cp.

The resin solution of the above composition is impregnated in a glass cloth treated with a silane coupling agent and is treated at normal temperature for 2 min, at 100°C for 2 min and at 160°C for 4 min successively to impregnate the resin in the glass cloth in order to produce a prepreg having about 40% resin.

Then, the adhesive coated copper foil is placed on several sheets of the prepregs and laminated therewith under a pressure of 10 kg/cm^2 at 160°C for 4 min and then under a pressure of 40 kg/cm^2 at 160°C for 1 hr to provide a copper clad glass board having a thickness of 1.6 mm. The characteristics of the copper clad glass board thus produced are shown in Table 1 (on p. 254).

Example 2:

Adhesive Composition	Parts by Weight
Phenoxy PAHJ	100
DEN 431, a low molecular epoxy resin, more specifically an epoxy novolac resin having an epoxy value of 172 to 179	50
Crosslinking agent having the formula	7

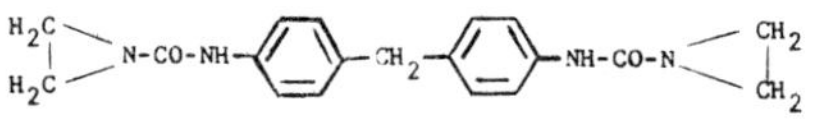

An adhesive similar to that in Example 1 is produced by using the above materials and an adhesive coated, electrolytic copper foil is produced under the same conditions as those in Example 1. The above crosslinking agent is produced by using diphenylmethanediisocyanate as a starting material under conditions similar to those employed in Example 1.

Production of Prepreg	Parts by Weight
Epikote 1001	100
Epikote 154, an epoxy novolac resin having an epoxy value of 176 to 181	100
Diaminodiphenylsulfone (hardener)	30
BF_3 400 (hardener by Shell Chemicals Corp)	1.5

The above materials are dissolved in methyl ethyl ketone and the viscosity is adjusted to about 200 cp. Epoxy resin impregnated glass cloths are produced using this solution under the same conditions as those in Example 1 to provide a prepreg of NEMA G-11 grade.

The above adhesive coated copper foil is placed on several sheets of the prepregs and laminated with the latter under a pressure of 10 kg/cm^2 at 160°C for 4 min and then under a pressure of 70 kg/cm^2 at 180°C for 1 hr, thus producing a copper clad glass board. The characteristics of this board are shown in Table 1.

Example 3:

Adhesive Composition	Parts by Weight
DER 488 N-40, a phenoxy resin that is the reaction product of bisphenol A and epichlorohydrin having a molecular weight of about 30,000 and lacking terminal highly reactive epoxy groups, with approximately 40% solids content	400

(continued)

Adhesive Composition	Parts by Weight
DER 511, a low molecular epoxy resin more specifically a brominated bisphenol A epoxy resin with an epoxy value of 445 to 520 and a bromine content of 18 to 20 weight percent	50
Crosslinked agent having the formula	10

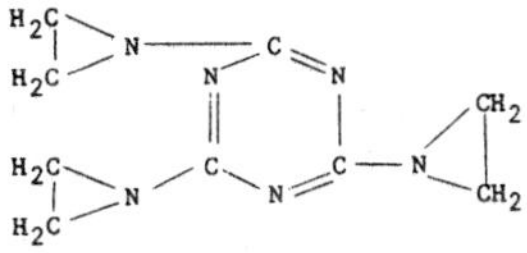

The crosslinking agent is dissolved in dimethylformamide and the resulting solution is mixed with a solution of the above resin in methyl ethyl ketone to produce an adhesive having a solids content of 20% with respect to the total amount of the mixture. An adhesive coated copper foil is produced using the adhesive under the same conditions employed in Example 1.

The preparation of the above crosslinking agent is achieved in the following manner. Cyanuryl chloride is dissolved in dioxane and suspended in ice water and the suspension is held at 0° to 2°C. An aqueous solution of ethyleneimine and calcium carbonate is dropped into the suspension for an hour while stirring, after which the solution is further stirred at 1°C for 20 min to terminate the reaction. The reactant is decolored with an active carbon powder and filtered and the filtrate is extracted with chloroform to obtain the desired substance.

Production of Prepreg	Parts by Weight
DER 511	100
DER 611, a bisphenol A epoxy resin having an epoxy value of 475 to 565	50
Dicyanodiamine	4
Benzyldimethylamine	0.3

Dicyanodiamine is dissolved in a solution containing 50% by weight dimethylformamide and 50% by weight ethyl cellosolve, and all of the above components are dissolved in acetone and the viscosity is adjusted to approximately 200 cp. Epoxy resin impregnated glass cloths are produced with the above solution under similar conditions to those in Example 1 to provide a prepreg of NEMA FR-4 grade.

Then, the adhesive coated copper foil is placed on several sheets of the prepregs and laminated therewith under a pressure of 10 kg/cm^2 at 175°C for 2 min and then under a pressure of 35 kg/cm^2 at 175°C for an hour to produce a copper clad glass board. The characteristics of the board thus produced are shown in Table 1.

Example 4:

Adhesive Composition	Parts by Weight
Araldite 684 EK-40, a phenoxy resin that is the reaction product of bisphenol A and	

(continued)

Adhesive Composition	Parts by Weight
epichlorohydrin having a molecular weight of about 30,000 and lacking terminal highly reactive epoxy groups, with approximately 40% solids content	400
DER 580, a low molecular epoxy resin, more specifically, a brominated bisphenol A epoxy resin with an epoxy value of 214 to 221 and a bromine content of 16 to 18 weight percent	40
Beckamine J820 (amine resin by Dai Nippon Ink & Chemicals Inc.)	5
APO (crosslinking agent by Sogo Yakuko) having the formula below	

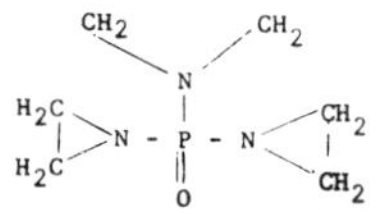

The above components are employed in the same way as in Example 1 to produce an adhesive coated copper foil.

Production of Prepreg	Parts by Weight
Epikote 1045-A-70, a brominated bisphenol A epoxy resin with an epoxy value of 450 to 500 and a bromine content of 18 to 20 weight percent	75
Epikote 154	25
Diaminodiphenylsulfone	15
BF_3 400	0.8

A prepreg of NEMA FR-5 grade is produced in the same manner as in Ex. 1, except that an epoxy resin solution of the above components in 90 parts by weight of methyl ethyl ketone is treated at normal temperature for 2 min, at 90°C for 2 min and at 130°C for 5 min.

The adhesive coated copper foil is laminated on several sheets of the prepregs under a pressure of 10 kg/cm^2 at 160°C for 4 min, and then under a pressure of 40 kg/cm^2 at 180°C for an hour to produce a coppler clad glass board. The characteristics of the board thus obtained are shown in Table 1.

Example 5:

Adhesive Composition	Parts by Weight
Phenoxy PAHJ	100
Epon 828	50
Tyzor TE (chelate compound by Du Pont)	5

A copper clad glass board is made in the same manner as in Example 1, except that the above components are dissolved in methyl ethyl ketone to produce an adhesive having a solid content of 20%. The characteristics of the glass board thus obtained are shown in Table 2 (on the following page).

Example 6:

Adhesive Composition	Parts by Weight
Phenoxy PAHJ	100
Epon 828	50
Tyzor AA (chelate compound by Du Pont)	5
$BF_3{\cdot}NH_2C_2H_5$	0.3

A copper clad glass board is produced in the same manner as in Example 1 except that the above components are dissolved in methyl ethyl ketone to provide an adhesive having a solids content of 20%. The characteristics of the glass board thus obtained are shown in Table 2.

Example 7:

Adhesive Composition	Parts by Weight
Phenoxy PAHJ	100
Epon 828	50
Methylacetate aluminum butoxylate	5
$BF_3{\cdot}NH_2C_2H_5$	0.3
Beckamine J820	5

A copper clad glass board is produced in the same manner as in Example 1 except that an adhesive is made from the above components. The characteristics of the glass board are shown in Table 2.

Example 8:

Adhesive Composition	Parts by Weight
Epikote OL-55-B-40, a phenoxy resin that is the reaction product of bisphenol A and epichlorohydrin having a molecular weight of about 30,000 and lacking terminal highly reactive epoxy groups, with a solids content of about 40%	400
Epon 828	40
Beckamine J820	5
Crosslinking agent having the formula	10

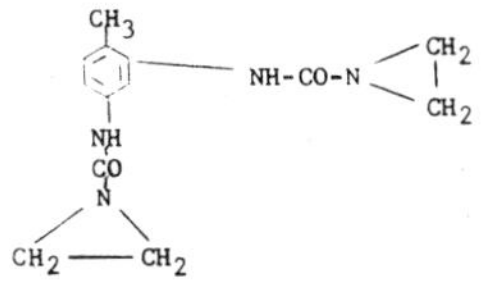

The previous components are dissolved in methyl ethyl ketone to obtain an adhesive having a solids content of 20% by weight in respect to the methyl ethyl ketone. This adhesive is coated 25 microns thick on a polypropylene laminated paper and held at normal temperature for 2 min, at 90°C for 2 min and then at 140°C for 4 min.

The coated adhesive is peeled off from the laminated paper and pressed between a copper foil and a prepreg under the same conditions employed in Example 1, thus obtaining a copper clad glass board. The characteristics of the board are shown in Table 1.

TABLE 1

Copper clad glass board	Peel Strength (kg/cm) Normal condition	After soaking in solder at 260°C for 3 minutes	After soaking in 10% KCN solution at 70°C for 30 minutes	After heating at 140°C for 7 days	After treated with boiled trichloroethylene for one hour	Surface insulation resistance (M Ω)	Treatment with sulfuric-chromic acid mixture
Example 1 Adhesive not used	1.66	1.55	1.60	1.65	1.54	Higher than 10^5	Insoluble
Conventional Sample A Adhesive not used	1.75	1.59	1.55	1.48	1.73	"	"
Conventional Sample B Adhesive not used	1.58	1.54	1.28	1.50	1.55	"	"
Butyrol phenol system adhesive used	2.15	1.90	1.76	1.80	1.89	"	Soluble
Example 1	2.38	2.17	2.20	2.25	2.17	"	Insoluble
Example 2	2.06	2.02	2.01	2.02	2.17	"	"
Example 3	2.31	2.05	2.02	2.10	2.32	"	"
Example 4	2.50	2.20	2.00	2.15	2.57	"	"
Example 8	2.42	2.20	2.30	2.30	2.42	"	"

TABLE 2

Copper clad glass board	Peel Strength (kg/cm) Normal condition	After soaking in solder at 260°C for 3 minutes	After soaking in 10% KCL solution at 70°C for 30 minutes	After heating at 140°C for 7 days	After being treated with boiled trichloroethylene	Surface insulation resistance (MΩ)	Treatment with sulfuric-chromic acid mixture
Example 5	2.17	1.90	1.90	2.05	2.10	Higher than 10^5	Insoluble
Example 6	2.36	1.95	1.84	2.34	2.30	"	"
Example 7	2.26	1.85	2.00	2.22	2.22	"	"

The surface insulation resistances were measured under the condition JIS C-96/35/90. In the treatment with a sulfuric-chromic acid mixture, the copper clad glass boards were treated with a solution of 400 g/l of CrO_3 in 70 g/l of concentrated sulfuric acid at 50°C for 15 min. The other tests were conducted with reference to the standards therefor such as MIL, NEMA and JIS.

UNDERWATER ADHESIVES

Epoxy Adhesive

There has been a need for one or more adhesive formulations suitable for bonding structures under seawater for some time. Different applications for such formulations include: attachment of explosive packages underwater, underwater salvage, and underwater construction and repair. The desired physical properties for an underwater adhesive will vary with the use.

Thus, an adhesive for attachment of an explosive device must be very rapid curing but high strength is not so important. Conversely, an adhesive used in underwater construction need not be very rapid curing, but high bonding strength is desired. It should be noted that water temperatures, and thus adhesive curing rates, may vary greatly from location to location.

Some surface waters may be at 70°F (21°C) or greater, while deep waters may get as cold as 40°F (5°C) or more. As a rule of thumb, many chemical reactions double their rates for each 10°C rise in temperature. Also, viscosity of adhesive generally varies greatly with temperatures so that an adhesive that can be manually mixed in a partitioned plastic bag underwater at 70°F might be too stiff for easy mixing at colder temperatures.

Heat packs might be advantageously used both in the mixing of adhesive components and in curing of the mixed product. In any event, practical use of an underwater adhesive requires a system that: can be readily carried, mixed, and applied underwater; will have the necessary viscosity for mixing and application; will readily displace water to coat the surfaces to be bonded together; will cure in an appropriate period of time; and will have the physical strength required for the intended purposes.

Epoxy-polyamine formulations were previously studied and shown to be readily applied underwater and to develop bonding strengths to steel of 1,000 psi or more after one week. However, such formulations were not suitable because of the slow curing rates and their limitations to use only in waters above 60°F in temperature.

R.W. Drisko and J.B. Crilly; U.S. Patent 3,891,583; June 24, 1975; assigned to U.S. Secretary of the Navy describe a quick setting adhesive to be applied underwater, cures very rapidly (i.e., in a few minutes to a few hours) at temperatures ranging from freezing to 80°F or higher to a hardened, nonflexible mass, displaces water on the substrate surfaces, and bonds tightly and permanently to the underwater surfaces. The adhesive provides a faster and simpler means for underwater joining than welding, bolting or other conventional procedures.

The adhesive consists of a two-component formulation that must be mixed dry but can then be applied underwater. Dry-mixing underwater can be achieved by use of an envelope, tube, or other container divided into two compartments. Immediately prior to use, the partition between the compartments is removed, and the two adhesive components are kneaded or otherwise mixed together to a uniform blend. The container is then opened so that the mixed adhesive can be extruded from it onto the structure or apparatus to be bonded.

The two component adhesive consists of reactive materials that will produce the desired properties after mixing. Viscous liquid epoxy polymers in the following examples are polyfunctional polymers derived by the condensation polymerization of bisphenol A and epichlorohydrin, having an epoxide equivalent of approximately 190, a viscosity of about 125 poises at 25°C, and specific gravity of about 1.18 at 25°C; and, especially for use in cold temperatures, polyfunctional polymer derived by the condensation polymerization of bisphenol A and epichlorohydrin, having an epoxide equivalent of approximately 180, a viscosity of about 5 to 7 poises, specific gravity of 1.13 at 25°C, and containing from 10.5 to 11.5% butyl glycidyl ether. Three examples of such formulations are given below:

Example 1:

Component A	Parts by Weight
Viscous liquid epoxy polymer	
Component B	100
High functionality liquid polymercaptan resin (accelerator)	100
2,4,6-Tri(dimethylaminomethyl)phenol (curing agent)	10
Blown fish oil (wetting agent)	20
Powdered asbestos	47
Carbon black	0.05
Beta-(3,4-epoxycyclohexyl)-ethyltrimethoxysilane (bonding agent)	2

Example 2:

Component A	Parts by Weight
Viscous liquid epoxy polymer	
Component B	100
High functionality liquid polymercaptan resin	100
2,4,6-Tri(dimethylaminomethyl)phenol	6
Viscous liquid polyamide resin (curing agent)	18
Blown fish oil	20
Powdered asbestos	16
Carbon black	0.05

Example 3:

Component A	Parts by Weight
Viscous liquid epoxy polymer	
Component B	100
High functionality liquid polymercaptan resin (accelerator)	100
2,4,6-Tri(dimethylaminomethyl)phenol (curing agent)	10
Blown fish oil (wetting agent)	20

(continued)

Component A	Parts by Weight
Fire-dried fumed silica having a surface area between 200 and 400 square meters per gram (thixotropic agent)	5
Carbon black	0.01

Components A and B can be mixed within a suitable container, underwater, to a uniform blend. The adhesive is then extruded from the container and spread onto cleaned surfaces of the underwater structures to be bonded. The wetting agent in the adhesive formulation helps the spreading of the adhesive onto the underwater surfaces.

Surfaces coated with the adhesive are pressed together and allowed to cure at relatively low underwater temperatures. The adhesive cures quickly, in from about 5 to 30 minutes, depending upon the temperature of the water and exact formulation to form a strong bond. Two hours of curing will produce a very strong bond.

When mixed in 100 gram masses, adhesives like Examples 1 through 3 have set up in water at 70°F in 5 to 10 minutes, and in water at 40°F in two to three times this time. When applied underwater to cleaned steel, bonding strengths of 300 lb/in^2 or more can be produced after one-half hour of curing and 700 lb/in^2 or more can be produced after 2 hours of curing. The formulation can be varied some to change both the curing time and bonding strength of the adhesive.

Cyanoacrylic Acid Esters

According to a process described by *F. Leonard and G. Brandes; U.S. Patent 3,896,077; July 22, 1975* pastes of α-cyanoacrylic acid esters and insoluble fillers can be applied to surfaces submerged in water for securing, in situ, a wide variety of metallic and nonmetallic substances.

The liquid α-cyanoacrylate monomers used for this process are of the formula $H_2C{=}C(CN){-}CO{-}OR$, where R is, preferably, alkyl having from 1 to 10 carbon atoms, e.g., methyl, ethyl, propyl, isopropyl, butyl, isobutyl, amyl, pentyl, hexyl, heptyl ethylhexyl, octyl, nonyl and decyl. From among these, it is particularly advantageous to employ those where R is alkyl having from 4 to 8 carbon atoms, either straight or branched chain. R is, alternatively, substituted alkyl having from 1 to 10 carbon atoms.

Illustrative substituents are lower alkoxy having from 1 to 4 carbon atoms (ethoxyethyl-α-cyanoacrylate) and halo, e.g., chloro and fluoro (trifluoro-isopropyl-αcyanoacrylate). (In the trifluoro-isopropyl group the three fluorine atoms are bonded to the same carbon atom.) In addition R can also be, e.g., cyclohexyl, phenyl or allyl.

The sole restrictions on R are that the monomer must be liquid at room temperature (20°C) and atmospheric pressure and a carbon atom of R must be directly bound to the $H_2C{=}C(CN){-}CO{-}O$ group. It is also essential that the α-cyanoacrylate be unsubstituted in the β-position. That means that the hydrogen atoms in the β-position cannot be replaced. The contemplated α-cyanoacrylates must bave the structure; $H_2C{=}C(CN){-}CO{-}O{-}$, without variation in this portion of the molecule. The liquid α-cyanoacrylates are monomers prepared, e.g., according to the method described by U.S. Patent 2,763,677.

The pH designated for the monomer is measured by adding one gram of monomer and 0.2 gram of sodium chloride to 50 cubic centimeters of water. The mixture is then stirred and the pH checked with a standard pH meter or other laboratory indicator. It has been found that the pH is a critical factor.

A range of from 2.45 to 4.1, preferably from 3.0 to 4.1, permits a good homogeneous mixture with inorganic fillers, such as calcium carbonate and amorphous silica. As the pH is increased above 4.1, clumping occurs (depending upon the alkalinity of the filler); at a pH below 3.7 homogeneous pastes are prepared, but delayed curing time results.

This is advantageous for those monomers which polymerize most rapidly on contact with water. These have preferred pH's, e.g., methyl-α-cyanoacrylate (pH 2.45) and propyl-α-cyanoacrylate (pH 2.7), below 3.7. It has also been found that in particular aqueous media, e.g., sea water that has contamination of a type which is not readily determined, tensile strengths may be improved by reducing the pH below the range of from 3.7 to 4.1. Particular substrates, e.g., HY-80 steel, also adhere more strongly at lower pH's. Propyl-α-cyanoacrylate has been found to yield particularly good results (strong bonds) with HY-80 steel.

The preferred fillers are either calcium carbonate alone or a mixture of calcium carbonate and amorphous silica. The preferred amorphous silica is a fine colloidal size silica of 99% purity prepared in a rapid, high temperature flame process as a white soot and sold under the name "Cab-O-Sil". For stable pastes with chemically pure barium carbonate, the pH designated for the monomer (according to the indicated testing procedure) is optimally 3.0 for higher homologs, i.e., amyl and above.

Tests were conducted with compositions of this process where (a) a second surface is immediately pressed against a first surface to which paste has been applied, (b) a second surface is pressed against a first surface one minute after a glob of paste has been placed on the first surface, and (c) a second surface is pressed against a first surface two minutes after a glob of paste has been placed on the first surface, all surfaces being submerged in water. The strength of the resulting bonds was then tested. All bond tension tests were carried out with a Baldwin Universal Testing Machine at a jaw separation speed of 0.05 inch per minute.

In the examples, which follow, a paste is made by adding the stated filler to the specified liquid α-cyanoacrylate. When a plurality of fillers are used, they can be mixed with the α-cyanoacrylate sequentially in any order or as in admixture. In the actual examples, the amorphous silica is added last. Alternatively, the monomer can be added to the filler. The α-cyanoacrylate is prepared according to the method of Jeremias (U.S. Patent 2,763,677) and contains an acidic inhibitor against anionic polymerization.

The pH designated for the α-cyanoacrylate (containing the acidic inhibitor) is the pH obtained by mixing 1 gram of the α-cyanoacrylate and 0.2 gram of sodium chloride in 50 milliliters of distilled water. The pH is 3.7 except for

Example 19, where it is 3.0; Example 24, where it is 2.7; and Example 39, where it is 4.15. The paste formed from the mixture of filler with the α-cyanoacrylate (monomer) is stirred until smooth and then placed in a squeeze tube from which it is spread, under water, on one of the two surfaces to be adhered. The other of the two surfaces is then pressed for from two to three minutes against the paste on the one surface.

Example	Monomer[j] α-Cyano- acrylate	Filler[i]	Grams	Paste Flow	Surfaces**	Average Bond Strength in Tension (psi)
1*	Octyl	$CaCO_3$	3.4			201[a]
						110[b]
2*	Heptyl	SiO_2	0.3	Fair		187[c]
		$CaCO_3$	2.7	Excellent		100[a]
						175[b]
						250[c]
						(e)
3*	Heptyl	$CaCO_3$	2.6			475[a]
		SiO_2	0.1 to 0.2			410[b]
4	Heptyl	$CaCO_3$	3.4			609[a]
						710[b]
5	Heptyl	$CaCO_3$	2.7		Rubber/steel	74[f]
					Rubber/wood	31[f]
					Glass/glass	(d)
6*	Hexyl	$CaCO_3$	2.5	Good		225[a]
						135[b]
						112[c]
7*	Amyl	$CaCO_3$	3.5	Fair		253[a]
						80[b]
						150[c]
8*	Butyl	$CaCO_3$	2.0	Very thin and fluid		420[a]
9	Nonyl	$CaCO_3$	2.7			210
		SiO_2	0.05			
10	Decyl	$CaCO_3$	2.7			200
		SiO_2	0.05			
11	Cyclohexyl	$CaCO_3$	2.7			885
		SiO_2	0.05			
12	Ethoxy- ethyl	$CaCO_3$	2.7			270
		SiO_2	0.1			
13	Isobutyl	$CaCO_3$	2.7	(g)		260
		SiO_2	0.05			
14	Isopropyl	$CaCO_3$	2.7			360
		SiO_2	0.05			
15	2-Ethyl- hexyl	$CaCO_3$	2.7			310
		SiO_2	0.05			
16	Propyl[h]	$CaCO_3$	2.6			200
		SiO_2	0.1			
17	Ethyl[h]	$CaCO_3$	2.7			75
		SiO_2	0.1			
18	Methyl[h]	$CaCO_3$	2.7			75
		SiO_2	0.1			
19	Hexyl	$BaCO_3$	2.7			750
		SiO_2	0.1			

(continued)

Example	Monomer[j] α-Cyano-acrylate	Filler[i]	Grams	Paste Flow	Surfaces**	Average Bond Strength in Tension (psi)
20	2-Ethyl-hexyl	$CaSO_4.2H_2O$	2.0			210
		SiO_2	0.1			
21	Butyl	$CaSO_4 \cdot \frac{1}{2}H_2O$	3.0			80
		SiO_2	0.1			
22	Allyl	$CaCO_3$	2.7			250
		SiO_2	0.1			
23	n-Hexyl and n-amyl	$BaCO_3$	2.7			400
		SiO_2	0.1			
24	Trifluoro-isopropyl	$CaCO_3$	2.7			155
		SiO_2	0.1			

*Salt water used in this sample.
**All examples adhere two stainless steel surfaces under fresh water, unless otherwise specified.
[a] Immediate sealing.
[b] Sealing after a 1-minute immersion.
[c] Sealing after a 2-minute immersion.
[d] Glass slides broke, precluding reading.
[e] Bond strengths as high as 580 psi obtained in individual tests.
[f] Peel strength, rubber peeled from steel and wood substrates.
[g] Held together for only 1 minute under water.
[h] Paste delivered as a glob through a polyethylene tube to one of the two steel surfaces held under water.
[i] All silica (SiO_2) used in the examples in conjunction with the filler is amorphous silica.
[j] In each example, 3.0 grams of monomer are used, except in Example 1 where 6.1 grams of monomer are used; in Example 23, 1.5 grams of each monomer are employed. Each alkyl is normal unless otherwise specified.

Heptyl-α-cyanoacrylate mixed with calcium carbonate ($CaCO_3$) and amorphous silica (as in the examples) was also found to form strong bonds (under water) of aluminum to aluminum. In the same manner wood/wood and rubber/rubber bonds were formed. Although there may be some variation in maximum tensile or shear strengths and the optimum pH may differ, strong bonds are likewise formed under water for glass/glass, plastic/plastic (phenol/formaldehyde thermoset plastic) and combinatoins of the noted substrates, e.g., steel/aluminum, plastic/glass and wood/painted steel.

Employing a paste prepared from 4.0 parts by weight of hexyl-α-cyanoacrylate (pH 3.7), 3.6 parts by weight of calcium carbonate and 0.2 part by weight of "Cab-O-Sil" additional bonds are prepared under water as set forth in Example 25 to 32. In these examples all test specimens (except those with corroded or painted surfaces) are prepared for adhesive bonding by vapor degreasing for five minutes with trichloroethylene. This is followed by a hot water rinse and then a final wash in methyl ethyl ketone. The procedure for each test (run) in Examples 25 to 32 is as follows:

(a) Water in a container is regulated at the indicated temperature;

(b) Test specimens are submerged in the water;

(c) Adhesive is applied to bonding surface of one submerged test specimen;

(d) Adhesive is spread over entire bonding area by sliding test specimens over one another;

(e) Moderate hand pressure (approximately 1 lb) is applied between bonding surfaces for a time interval A;

(f) Bonded test specimens removed from water and placed in Dillon Dynamometer for pull tests;

(g) Load applied between test specimens; time interval B is the interval between conclusion of time interval A and application of load.

The adhesive cure begins immediately upon contact with water. Up to 2 min are available following application to spread the adhesive and bring the bonding surfaces together. All ingredients are stored in closed containers. The hexyl-α-cyanoacrylate is packaged in a polyethylene bottle. Cardboard cans are used to package the amorphous silica and the calcium carbonate.

The mixed adhesive paste can be readily applied from its collapsible squeeze tube container for a period up to 4 hr. The following portions of Federal Test Standard 175 Adhesive Methods of Testing, are used for performing the physical tests:

Method 1011.1-Tensile Properties of Adhesives
Method 1033.1-T-Shear Strength Properties of Adhesives by Tensile Loading

All tests for each example are performed on the same day with the same adhesive batch. The bond area for each test is one square inch. 7.8 grams of mixed adhesive composition is adequate for approximately 20 thin film tests.

The aluminum alloy employed in each of Examples 25 through 32 is aluminum alloy AA6061-T6 anodized per Mil-A-8625 and having a surface roughness of 32. In each of the examples the steel is steel per Mil-S-16216, Grade HY-80; in Example 25, 28, 29, 31 and 32 the surface roughness is 32, in Examples 26 and 30 the surface bonded is badly oxidized to a finish of approximately 125; in Example 27 the surface bonded is painted per Navord USTD 52, System No. 48 (anti-fouling).

Example 25: Steel is adhered to anodized aluminum alloy submerged in a 3% solution of salt (NaCl) water having a temperature of 70°F. Ten repetitions produce the following results:

	A*	B**	Tensile Strengths (psi)
A	45	37	90
B	45	75	375
C	45	48	190
D	45	81	325
E	45	55	425
F	45	44	275
G	45	57	450
H	45	42	125
I	45	62	475

(continued)

	A*	B**	Tensile Strengths (psi)
J	45	47	625
		Average	335

*Time in seconds to prepare specimen bond
**Time in seconds (unless otherwise specified) from bond preparation to the application of load in testing

Example 26: Badly oxidized steel is adhered to anodized aluminum alloy submerged in tap water having a temperature of 70°F. (In this example and in Examples 27 through 32, A and B have the same meanings as indicated in the respective footnotes in Example 25). Five tests yield the following results:

	A	B	Tensile Strengths (psi)
A	45	105	390
B	45	105	290
C	45	105	440
D	45	105	450
E	45	105	475
		Average	409

Example 27: Painted steel is adhered to anodized aluminum alloy submerged in tap water having a temperature of 70°F. In each of five tests A is 45 seconds and B is 105 seconds; the respective tensile strengths are 240 psi, 225 psi, 230 psi, 265 psi and 320 psi. An average tensile strength of 256 lb/in^2 is thus obtained.

Example 28: Steel is adhered to anodized aluminum alloy submerged in tap water having a temperature of 70°F. In each of ten tests A is 45 seconds. Other data follow:

A	B	Tensile Strength (psi)
A	81	400
B	109	740
C	69	420
D	71	90
E	44	620
F	54	910
G	104	525
H	75	680
I	76	610
J	47	625
	Average	562

Example 29: Shear specimens are prepared from rectangular slides having an overlapping bond area of one square inch. Adhesive paste of the same composition as that employed for Examples 25 to 28 is applied to the slides in the same manner as it is applied to the substances therein indicated. The details of Example 29 are the same as those of Example 28 except as otherwise specified. In each of ten tests A is 45 seconds and B is 105 seconds. The respective shear strengths in pounds per square inch are: 500, 260, 345, 365, 410, 415, 230, 280, 320 and 440, the average of which is 361 psi.

HIGH-TEMPERATURE COMPOSITES

Aromatic Dicyanate Added to Epoxy

Partially cured epoxy resin compositions have been employed to provide self-supporting moldable sheets which are particularly useful in making aircraft structural components for moderately high-temperature exposure, the compositions being cured to their final hardened state by exposure to high temperature. When combined with reinforcing fibers such as glass, boron, graphite, silicon carbide, alumina, etc., prior to cure, such compositions have provided flexible sheets or films which can be molded or laminated and thereafter cured. Curing or cross-linking agents typically incorporated in such epoxy compositions, include dicyandiamide, boron trifluoride complexes, anhydrides, imidazoles, aromatic amines, etc.

While such compositions are useful for preparing molded parts and adhesives for bonding articles to be employed in moderately high-temperature environments, they have not been particularly satisfactory for use at temperatures in the approximate neighborhood of 215°C. At these temperatures, properties such as longitudinal flexural strength and overlap strength are so reduced that an article cannot withstand sufficient stress to be useful. A particularly acute commercial need exists for resin compositions suitable for molding aircraft and missile structural parts whcih are exposed to temperatures as high as 215°C. Available epoxy resin compositions have not been found suitable for such applications.

J.B. Snell and D.A. Wambach; U.S. Patent 3,876,607; April 8, 1975 describe epoxy resin compositions suitable for preparing molded articles and adhesives for bonding articles to be exposed to high-temperature environments, particularly in the neighborhood of about 215°C. Articles molded from or bonded with the resins of the process retain longitudinal flexural strength and overlap shear strength, making them particularly suited for use in aircraft and missiles exposed to high temperatures.

It has been found that aromatic dicyanate esters employed in conjunction with a conventional epoxy resin curing agent e.g., as dicyandiamide, provides a surprising synergistic effect whereby the high-temperature properties of articles and adhesives prepared from the resultant compositions are substantially and significantly improved. The amount of aromatic dicyanate ester is sufficient to provide equivalents equal to 1 to 15% (preferably 2 to 12%) of the epoxy equivalents of the composition.

The aromatic dicyanate esters useful in the process preferably have the formula NCO—R—OCN where R is a divalent aromatic moiety such as phenylene, or a plurality of such moieties (e.g., 2 to 5), which can be separated by, e.g., —O—, —C(O)—, —S—, $—SO_2—$, or divalent hydrocarbon-bridging moieties, such as lower alkylene, e.g., methylene and dimethylmethylene. R is preferably a divalent radical having at least one (e.g., 15 or more) aromatic nuclei which can be linked together or bridged with ethereal oxygen atoms. Such nuclei include phenylene, diphenylene, and radicals of diphenylether, 2,2'-diphenylpropane, benzophenone, and diphenylsulfone.

The improvement observed in the high-temperature properties of articles or adhesive made from the compositions of the process are readily ascertained by standard ASTM or military specification tests of such properties as longitudinal and transverse flexural strength, beam shear strength, tensile strength, and overlap shear strength tests. The particular tests utilized vary depending upon whether a reinforced plastic laminate, a resin coating, or adhesive bond is being prepared. The following examples, in which all parts are by weight unless otherwise indicated, illustrate the process.

Examples 1 through 5: These examples illustrate use of the compositions as adhesives and the improvement in overlap shear strength obtained at elevated temperatures. The compositions listed in the following table were prepared in sufficient methyl ethyl ketone solvent to provide a coatable viscosity, coated on a glass scrim cloth by normal knife coating techniques, and dried at 80°C for 45 minutes to remove the solvent. The scrim cloth served as a carrier web to provide a self-supporting adhesive sheet.

Ex.	Epoxy Resin Amount, Gms. Solid epoxidized cresol novolac[1], m.p. 80°C., eq. wt. 230	Triglycidyl p-amino phenol[2], eq. wt. 97–101	Primary Curing Agent Type	% by equivalents	Additive Curing Agent Type	% by equivalents
1	40	34.8	dicyandiamide	65.8	—	—
2	47.7	41.7	"	65.6	aromatic dicyanate[3]	2.3
3	46.7	40.8	"	66.2	"	4.7
4	45.7	40.0	"	66.0	"	7.2
5	44.7	39.0	"	66.2	"	9.8

Ex.	Overlap Shear Strength (psi) Room temp	120°C.	150°C.	175°C.	190°C.	215°C
1	2,410	2,710	2,530	2,210	1,910	1,210
2	2,390	2,710	2,660	2,420	2,160	1,490
3	2,310	2,820	2,710	2,540	2,170	1,550
4	2,280	2,830	2,740	2,680	2,430	1,630
5	2,380	2,840	2,840	2,590	2,330	1,540

[1]Commercially available from the Ciba Chemical Company under the trade designation "ECN 1280".
[2]Commercially available from the Union Carbide Corporation under the trade designation "ERL-0510".
[3]Bisphenol-A dicyanate "B"-staged by heating at 200°F. until a viscosity of 500 cps is obtained.

Aluminum panels (4 x 7 x 0.063 inches) were vapor-degreased with perchloroethylene, immersed in an alkaline etch bath for 10 minutes at 90°C, rinsed with tap water, etched in a sodium dichromate-sulfuric acid bath, rinsed in tap water, and dried at 65°C.

The adhesive sheet was applied to one face of a first aluminum panel and a second aluminum panel positioned thereover so that it overlapped the first panel by one-half inch. The adhesive was then cured while holding the assembly in a press at 50 psi and 175°C for 1 hour, and post-cured by heating at temperatures of 175°, 190° and 205°C each for 1 hour and at 215°C for 2 hours, after which the assembly was removed and cooled.

The assembly panels were then cut into 1-inch-wide strips extending at 90° to the overlap and the overlap shear value (psi) determined by means of an Instron tensile tester, with the tensile loading being applied at a rate of 1,300 psi/min until the bond failed. The sample specimens tested at elevated temperatures were conditioned at the test environment for not less than 15 minutes nor more than 30 minutes prior to testing. The overlap shear strength data obtained are shown in the table, significant improvement over the control being noted at the elevated temperatures when dicyanate is included in the resin formulation.

As can be seen in the table, cured compositions containing aromatic dicyanate additive in the range of 2.2 to 15.6% (by equivalent), as well as conventional curing agent, have improved flexural and overlap shear strengths. The improvement at 215°C in each of these values is consistently at least 20% more than compositions containing only conventional dicyanamide. This improvement is indicative of the improved heat resistance of compositions.

Aromatic Polyamic Acid and Aluminum Oxide

J.G. Holloway, H.W. Barch and D.M. Fahey; U.S. Patent 3,898,094; August 5, 1975; assigned to PPG Industries, Inc. describe an adhesive-sealant composition capable of withstanding surface temperatures up to 1000°F for bonding smooth surfaces.

The process involves a high-temperature, heat-hardenable, adhesive-sealant composition whose solids comprise an aluminous cement, typically 75 to 97% by weight of the composition, and containing a substantial amount as Al_2O_3, and 3 to 25% by weight of the composition of an organic compound selected from the group consisting essentially of a polyamic acid, the reaction product of a tetracarboxylic dianhydride or a tricarboxylic monoanhydride and a dihydrazine or a dihydrazide.

The aluminous cement of the adhesive-sealant composition contains a substantial amount, preferably 78 to 99% by weight, as Al_2O_3 to maintain structural, adhesive and sealing characteristics at 1000°F. In addition, the Al_2O_3 provides chemical inertness to the adhesive composition to hinder deterioration of the bonded substrates at elevated temperatures.

Example 1: To a 500-ml beaker was charged 36 g of a 60% weight solids polyamic acid solution (Skybond 700) consisting of the reaction product of benzophenone tetracarboxylic dianhydride and m-phenylene diamine in a m-pyrrole-ethanol solvent. Six grams of 28% aqueous NH_4OH was added to the Skybond 700 with manual agitation and a small amount of water was added to the solution to impart a viscosity reduction.

Two hundred twenty-five grams of an aluminous cement having 99% by weight as Al_2O_3, 0.7% by weight as SiO_2, 0.1% by weight as Fe_2O_3, 0.1% by weight as CaO, and 0.3% by weight as Na_2O was added and mixed with manual agitation until homogeneous. Three grams of active silica was added to the above adhesive-sealant composition and water sufficient to bring the total added water content of the adhesive-sealant composition to 70 grams.

A rectangular sheet of convoluted glass fiber paper was disposed into a cylindrical form and the two edges defining the length of the rectangle were joined by the adhesive-sealant composition. The convoluted glass fiber cylindrical article was placed in a forced-air oven at 400°F for 3 hours to dry the adhesive-sealant, and then the article was cooled to room temperature. The glass fiber paper article was placed in an oven at 800°F for 3 hours, and after cooling to room temperature the joint of the sealed glass fiber paper showed no deterioration of the glass and excellent cohesive and adhesive strength was observed.

Example 2: The adhesive-sealant composition of Example 1 was repeated and 340 g of this composition was poured into a stainless steel end cap having the configuration of a circular channel defining a central aperture of 4 inches in diameter. The channel was 2 inches in width and ¼ inch in depth, the width was sufficient to accommodate the convolutions of the glass fiber paper cylinder of Example 1 and the depth was sufficient to support the glass fiber paper cylinder of Example 1.

The end cap was filled to the top edge of the channel with the adhesive-sealant composition and the circular edge of the glass fiber paper cylinder of Example 1 was placed therein. The above article was placed in a forced-air oven at 450°F for 1 hour to dry the adhesive-sealant. After the article had cooled to room temperature the above procedure was repeated using another end cap on the opposite end of the convoluted glass fiber paper cylinder.

This filtration element was placed in a suitable housing and was mounted in a diesel exhaust system to filter carbonaceous exhaust particulates. The filtration unit ran for 10 hours at about 700°F without any sign of deterioration of the glass fiber to glass fiber seals or the glass fiber to stainless steel seals. No deterioration of the glass or the steel was observed at the bonding sites.

Polybenzothiazole-Based Adhesive

T.J. Aponyi and E.A. Arvay; U.S. Patent 3,832,320; August 27, 1974; assigned to the U.S. Secretary of the Air Force describe an adhesive that consists essentially of: a polybenzothiazole modified by reaction with 4-aminophthalimide, aluminum and zinc oxide. The adhesive is particularly useful in bonding structural members, e.g., those fabricated from stainless steel or titanium, that are subjected to high temperatures as in the operation of high-performance aircraft.

The modified polybenzothiazole is prepared by reacting mixed toluidines with elemental sulfur at a temperature sufficient to melt the reactants. As a result of the reaction, polybenzothiazole is formed that is then modified by reaction with 4-aminophthalimide. The amounts of mixed toluidines and elemental sulfur that are reacted are generally in the respective ranges of about 2 to 4 mols and 8 to 15 gram-atoms for each mol of 4-aminophthalimide.

A modified polybenzothiazole prepared by reacting 3 mols of mixed toluidines, 13 gram-atoms of sulfur and 1 mol of 4-aminophthalimide has been found to be particularly effective. In a preferred procedure, a mixture of the mixed toluidines and sulfur are heated under an inert atmosphere at a temperature in the range of 250° to 450°F for a period of 12 to 20 hours.

Examples of gases that can be used to provide an inert atmosphere include nitrogen, helium, argon, hydrogen sulfide, and the like. During the reaction period the 4-aminophthalimide is added to the reaction mixture and reacts with the polybenzothiazole that has formed. It is often preferred to add the 4-aminophthalimide in increments subsequent to commencement of the reaction, e.g., from 1 to 5 hours after the reaction mixture reaches the temperature of reaction.

The addition of the 4-aminophthalimide is generally completed in 2 to 5 hours. While the procedure described has been found to be the most desirable one, it is to be understood that the 4-aminophthalimide can be added continuously or at one time during the reaction period. In any event the addition should be completed from about 2 to 5 hours prior to the end of the 12 to 20-hour reaction period.

Example 1: A series of laminates was prepared in which plies of glass cloth were bonded with the adhesive composition of this process. The modified polybenzothiazole contained in the composition was prepared by charging 150 g (1.4 mol) of mixed toluidines (60% ortho, 37% para and 3% meta on a weight basis) and 195 g (6.1 g-atoms) of sulfur to a two-liter flask. After flushing with nitrogen, the flask containing the reaction mixture was heated to 350°F. The reaction temperature was increased from 350° to 420°F over a period of 13 hours.

During the reaction period a total of 79.6 g (0.49 mol) of 4-aminophthalimide was added in five equal portions at one-half-hour intervals. The reaction mixture was maintained under an atmosphere of nitrogen during the reaction period. A dark brown resin having a melting point of 275°F was recovered in the amount of 328 g. Utilizing the modified polymer prepared as described above, a formulation having the following composition was prepared:

Component	Parts by Weight
Modified polymer	100.0
Zinc oxide	15.6
Dimethylacetamide	150.0

The formulation was prepared by dissolving the modified polymer in the dimethylacetamide and then adding the zinc oxide. The solution containing zinc oxide was then ball-milled in order to break the insoluble material into fine particles. A solution of modified polymer with finely divided zinc oxide particles in suspension was thereby obtained.

Glass cloth was impregnated with the solution containing zinc oxide by a conventional dip tank technique. Following impregnation the prepreg was dried in a circulating air oven for 30 minutes at 150°F. After the drying step, the prepreg was tack-free. The dried prepreg was then precured in a forced-draft oven for 15 minutes at each of the following temperatures: 200°, 258°, 280° and 310°F.

Plies were cut to size from the B-staged cloth, and stacks of twelve plies each were formed. Each stack was placed in a press preheated to 500°F after which a pressure of 200 psi was applied. During the initial stages of the cure, the pressure was released for 20 seconds out of every minute to facilitate the evolution of volatiles. The platen temperature was raised to 600°F and held at that temperature for one hour. Thereafter, each laminate was cooled to room temperature while under pressure.

After each laminate was cured, its edges were trimmed. Each laminate was then postcured by heating in an oven in accordance with the following schedule: 22 hours from 350° to 550°F, 24 hours from 550° to 700°F, 3 hours from 700° to 750°F and 3 hours from 750° to 850°F.

During the postcure the laminates were wrapped in aluminum foil and covered with lampblack to prevent contact with air. The laminates each consisted of 21.5 weight percent modified polymer and 4.0 weight percent zinc oxide and had a specific gravity of 2.0.

Laminates were heat-aged for various time intervals at temperatures ranging from 600° to 800°F. After each specimen was heat-aged, it was tested to determine certain of its physical properties. The tests were performed in accordance with Federal Test Method Standard No. 406, "Plastics: Method of Testing," Method 1031. The results of the tests are shown in the following table.

Test conditions, ° F.	Flexural strength, p.s.i.	Modulus of elasticity, p.s.i. $\times 10^{-6}$	Weight loss, percent
Room temperature (RT)	60,000	3.65	
RT after 2 hours boiling in water	57,100	3.55	
600° after ½ hour at 600°	52,000	3.10	0.3
600° after 192 hours at 600°	42,000	3.00	1.0
600° after 300 hours at 600°	32,500	2.75	2.0
700° after ½ hour at 700°	50,000	3.00	0.5
700° after 48 hours at 700°	28,300	2.51	2.4
800° after 6 hours at 800°	30,700	2.71	4.1

From the data in the foregoing table, it is seen that the laminates exhibited excellent retention of physical properties after aging for 300 hours at 600°F. Furthermore, an outstanding retention of strength properties was obtained after the 800°F exposure. These results are particularly significant since they show that the laminates bonded with the adhesive composition of this process do not undergo the drastic loss of properties after short-time exposures to elevated temperatures as is characteristic of polyimide-based adhesives.

Example 2: A series of runs was conducted in which stainless steel panels were bonded to one another with the adhesive composition of this process. Control runs were also carried out in which arsenic thioarsenate, a conventional antioxidant, was included in the composition of the process and in which commercial adhesives were used to bond panels. Employing the modified polybenzothiazole prepared as described in Example 1, formulations having the following compositions were prepared.

Component	Parts by weight of— A	B
Modified polymer	100	100
Zinc oxide	17.6	17.6
Aluminum	100	100
$AsAsS_4$	0	10
Dimethylacetamide	100	100

The formulations were prepared by dissolving the modified polymer in the dimethylacetamide. After being pulverized, if required, the other components were added separately to the solution. After each addition the mixture was homogenized in a high-speed blender.

Glass fabrics were impregnated wtih each of the adhesive formulations, using a spatula to ensure good adhesive-fiber contact. After heating the impregnated fabrics to remove solvent, narrow strips of tape were cut from the fabrics and suspended in an air-circulating oven. B-staging was accomplished by exposing the strips of tape to the following temperature-time cycle: 30 minutes at 150°F, 15 minutes at 200°F, 15 minutes at 250°F, 15 minutes at 280°F and 15 minutes at 310°F. Portions of the B-staged tape exhibiting uniformity of color and cut to a width of 11/16 inch were selected for subsequent use in the metal bonding operation.

Prior to being bonded, each panel of stainless steel was thoroughly washed in an alkaline solution and rinsed in water. The panels were then degreased by condensing trichloroethylene vapor for a period of 10 minutes. After rinsing with water and drying, the panels were immersed for 4 minutes in an acid bath consisting of 87 ml of orthophosphoric acid (85%), 841 ml of hydrochloric acid (38%) and 56 ml of hydrofluoric acid (48%). The etched panels were then rinsed with water and dried with a hot air gun.

The etched panels were primed using the adhesive formulations as a paste. Equal amounts of the formulations were spread over the etched portion of the panels (about one-inch) using a spatula. After the panels were dried to remove solvent, the resultant primed surfaces were soft, glossy and quite uniform in appearance.

In the bonding operation, a primed panel was placed in a bonding jig. Adhesive tape, prepared as described above, was then applied over the primed area after which the tape was covered wtih a second panel so as to provide an overlap joint. The entire assembly was then placed between the platens of a hydraulically operated press. Before insertion the platens had been preheated to 500°F.

The temperature controls for the platens were then shifted to 600°F so as to provide a gradual increase in temperature. Contact pressure was alternately applied and released at 20-second intervals until the temperature reached 600°F. When this temperature was reached, a pressure of 55 psi was applied and held for 45 minutes. The platens were cooled to about 300°F prior to removal of the bonding jig.

The bonded panels were next heated according to the following time-temperature cycle in order to postcure the adhesive compositions: 4 hours to 400°F, 4 hours to 600°F and 10 hours to 825°F. The postcuring operation was carried out under a constant flow of nitrogen.

The bonded panels were tested at various temperature-time conditions. Aging of the specimens was carried out in a circulating air oven which maintained the temperature within plus or minus 5°F. Long-time-aging specimens were aged to within an hour or less of the total time required and then transferred to the test furnace for completion of aging prior to testing. In the testing, the tensile load was applied at a rate of 1,200 to 1,400 $lb/in^2/min$ as prescribed in Federal Specification MMM-A-132.

The test conditions and the results of the tests are shown in the following table. Also included in the table are the results of tests made with stainless steel panels bonded with two commercial adhesives in accordance with the recommendations of the suppliers. The figures shown in the table are the averages of the values obtained in testing four different specimens under each of the indicated conditions.

Exposure time and test temperature	Tensile shear strengths, p.s.i.			
	Composition [1] A	Composition [2] B	PBI [3] adhesive	PI [4] adhesive
Room temperature	1,980	1,320	3,670	3,680
½ hour at 700° F	1,860	1,230	2,600	1,090
75 hours at 700° F	1,400	675	0	1,750
½ hour at 800° F	1,840	1,210	1,900	500
4 hours at 800° F	1,335	1,430	0	----
½ hour at 900° F	1,760	1,235	1,425	200
1 hour at 900°	1,590	1,260	0	----

[1] Adhesive composition of this process.
[2] Adhesive composition containing $AsAsS_4$ in addition to other components of Composition A.
[3] Commercial polybenzimidazole adhesive.
[4] Commercial polyimide adhesive.

As seen from the data in the foregoing table, the stainless steel panels bonded with the adhesive composition of this process exhibited outstanding retention of tensile lap shear strengths at elevated temperatures. For example, at 800°F a 70% retention of room temperature properties was obtained. On the other hand the commercial adhesives tested provided bonds which had low strengths at elevated temperatures or failed altogether. The data also show that arsenic thioarsenate, a conventional antioxidant, has an adverse effect upon the strength property of the bonded panels.

OTHER APPLICATIONS

Abietate-Terminated Polysulfide Sealants

Curable liquid polysulfide polymer-based sealant and caulking compositions have long been known in the art. They are based on curable, liquid, mercaptan-terminated polysulfide polymers. The structure and preparation of these polythiopolymercaptan polymers are disclosed in U.S. Patent 2,466,963. These liquid, mercaptan-terminated polymers are used extensively in sealant and caulking compositions, especially in the building, automotive, and marine industries; for castings, e.g., solid rocket propellants; for leather and textile impregnating agents; for adhesives, coatings, etc.

Because of the wide range of applications in which these polysulfide polymer-based sealing and caulking compositions are used, they must be capable of bonding to various substrates, e.g., metal, aluminum, glass, concrete, wood, etc. However, the adhesive qualities of polysulfide polymers are such as to usually require the use of an adhesive additive in the polysulfide polymer-based sealant composition in order to insure adequate bonding of the cured sealant to the substrate.

Adhesive additives of the prior art such as those of a phenolic nature provided only a polar, rather than a chemical, type bonding of the cured sealant to the substrate. Thus, sealant compositions containing phenolic-type adhesive additives were found to be susceptible to attack by polar solvents such as water. The polar solvent tended to break down the polar bond thereby impairing the utility of the sealant composition.

More recently, as disclosed in U.S. Patents 3,297,473; 3,312,669 and 3,328,451, certain organo-silanes, e.g., (mercapto alkyl amino alkyl alkoxyl silanes), have been proposed to overcome these problems.

However, while the organo-silane adhesive additives have been found to protect the sealant composition bonds from attack by polar solvents, they do have certain disadvantages. Thus, the organo-silane adhesive additives are economically costly, somewhat deficient in storage stability, and exhibit erratic adhesion results with certain substrates, such as concrete.

The object of the process, therefore, is to provide a liquid polysulfide polymer-based caulking or sealant composition which will allow for the attainment of a strong and lasting bonding of the sealant to the substrates to be treated therewith.

J.L. Villa; U.S. Patent 3,838,078; September 24, 1974; assigned to Thiokol Corporation describes a liquid polysulfide polymer-based caulking or sealant composition containing adhesive additives which are less costly, exhibit improved storage stability, and which exhibit more reproducible and uniform adhesion characteristics to a wider range of substrates.

The adhesive additives of the process are the abietate-terminated polysulfide polymers, the abietate-terminated polyester polymers, and the epoxy-terminated polysulfide polymers. Of these adhesive additives, the abietate-terminated polysulfide polymers are preferred because they exhibit more uniform and reproducible adhesion results to various substrates when incorporated into a liquid polysulfide polymer-based sealant composition.

The abietate-terminated polyester polymer and epoxy-terminated polysulfide polymer adhesive additives show increased adhesion to certain substrates. However, it has been found that larger quantities of these latter adhesive additives are required to produce increased adhesion results. The use of these adhesive additives in the required quantities often results in some retardation of the cure rate of the sealant composition. Moreover, adhesion results on the abietate-terminated polyester polymer and epoxy-terminated polysulfide polymer adhesive additives have not proven to be as uniform nor as reproducible as the preferred abietate-terminated polysulfide polymer adhesive additive.

For that reason, it is recommended that the abietate-terminated polyester polymer and epoxy-terminated polysulfide polymer adhesive additives be used at not more than 5 parts by weight of additive per 100 parts by weight of liquid polysulfide polymer based on the total weight of the sealant composition. Larger quantities of these adhesive additives may require an adjustment in the amount of curing agents employed.

The abietate-terminated polysulfide polymer adhesive additive may be prepared by reacting a relatively low molecular weight, e.g., 500 to 4,000, preferably 500 to 1,500 SH-terminated liquid polysulfide polymer (referred to as a liquid polythiol polymer) with abietic acid at temperatures of from 150° to 200°C. The reaction is preferably carried out in a solvent medium and under an inert atmosphere.

Example 1: Preparation of Abietate-Terminated Liquid Polysulfide Polymer Resin — A 4,000-ml resin kettle equipped with a Barret trap, a thermometer and adapter, an electric stirrer, a gas inlet tube and a Fredricks condenser was charged with 2,000 grams (2 mols) of a liquid polythiol polymer of 1,000 mol weight having no crosslinking, 604.88 grams (2 mols) of abietic acid, and 150 ml of xylene.

The mixture was heated under a nitrogen atmosphere for approximately 1 hour 15 minutes at which time the temperature had reached 190°C and 10 ml of H_2O had collected in the Barret trap. The mixture was then heated for approximately 2½ hours at 190°C at which time 34 ml of H_2O had collected in the trap. The mixture was heated for an additional 10 minutes and the heat then removed. A sample of resinous product was removed from the resin kettle and analyzed for SH percentage.

SH percent was found to be 2.7%. The next morning, heat was again applied to the mixture for approximately 2 hours at 165°C. An additional 6½ ml of H_2O collected in the trap for a total of 50½ ml of H_2O collected and removed during the reaction. Heat was removed at this time and a sample of resinous product analyzed for SH percentage.

Analysis indicated that SH percentage was 0.5%. A vacuum was hooked up to the reaction apparatus and the reaction was heated to 70°C until all of the xylene had been evaporated. The resinous abietate-terminated polysulfide polymer was collected and the acid number and hydroxyl number determined. Results—Adhesive (1) Acid No., 29.85; OH No., 129.89. Adhesive (2) Acid No., 29.14; OH No., 129.49. About 5.5 lb of abietate-terminated polysulfide polymer resin was prepared by the above procedure and used as an adhesive additive in the examples which follow.

Examples 2 through 4: Evaluation of The Abietate-Terminated Polysulfide Polymer Resin of Example 1 as an Adhesive Additive in Liquid Polysulfide Polymer-Based Sealant Compositions — In order to determine the effectiveness of the abietate-terminated polysulfide polymer resin of Example 1 as an adhesive additive, a liquid polysulfide polymer-based sealant masterbatch of the following formulation was prepared using mixing procedures well known in the sealant art:

Ingredients	Parts by Weight
LP-32 Polysulfide Polymer*	100.0
Multiflex MM (Calcium Carbonate)	25.0
Icecap R (Anhydrous Clay)	30.0
Titanox RA 50 (Titanium Dioxide)	10.0
Arochlor 1254 (Chlorinated Diphenyl)	35.00
	200.0

* LP-32 polysulfide polymer has essentially the structure $HS(C_2H_4-O-CH_2-O-C_2H_4-SS)_{28}-C_2H_4-O-CH_2-O-C_2H_4-SH$ with about 0.5% crosslinking and a molecular weight of about 4,000.

A control formulation (Example 2) for this evaluation was prepared by combining the above sealant masterbatch with a lead peroxide curing paste (described hereinafter). One test formulation (Example 3) was prepared by combining the sealant masterbatch with the abietate-terminated polysulfide polymer resin and a lead peroxide curing paste.

An additional test formulation (Example 4) was prepared by combining the sealant masterbatch with the abietate-terminated polysulfide polymer resin and a zinc peroxide curing paste. The control and test formulations were as shown on the following page.

	Parts by weight		
Example number	2	3	4
Formulation:			
Polysulfide sealant masterbatch	200.0	200.0	200.0
Abietate-terminated polysulfide polymer adhesive additive		1.0	1.0
Lead peroxide curing paste	15.0	15.0	
Zinc peroxide curing paste			22.0

The lead peroxide curing paste shown above is a mixture of lead peroxide and arochlor 1254 containing 50% PbO_2. Arochlor 1254 is a chlorinated diphenyl plasticizer. The zinc peroxide curing paste shown above is a mixture of zinc peroxide, arochlor 1254, and amax containing 45.5% ZnO_2, 45.5% arochlor 1254, and 9% amax. Amax is N-oxy-diethylene benzothiazole-2-sulfamide and is used as a cure accelerator.

Peel adhesion strength specimens bonded to glass and aluminum substrates were prepared from the above formulations. These specimens were exposed to 7 days in air at room temperature, 7 days in an oven at 158°F, and 7 days' immersion in room temperature water. Peel adhesion valves were obtained upon removal of the specimens from water immersion while the samples were still wet using an Instron tester. The results of the peel adhesion tests are shown in the following table.

Adhesion Peel Strength Valves (psi)

	Substrate			
	Glass		Aluminum	
Example	Test 1	Test 2	Test 1	Test 2
2 (control)	[1] NA	NA	NA	NA
3	52	53	28	28
4	46	53	30	31

[1] NA=No adhesion.

The control specimens showed complete adhesion failure to both the glass and aluminum substrates.

Adhesive Liner for Flare

A process described by *G.C. Shaw and R. Reed, Jr.; U.S. Patent 3,895,578; July 22, 1975; assigned to Thiokol Corporation* is particularly concerned with colored signal flares, where the combustible composition is diluted with coloring materials, and hence, the combustion temperature is generally lower than that of illuminating flares.

The particular type of flare case with which the process is concerned comprises an aluminum tube lined with a kraft paper tube in which the adhesive liner is applied. In colored signal flares, a persistent problem has been the formation of chimney-like residues around the burning illuminant candles, as a result of failure to burn the case and its adhesive liner completely. This residue tends to obscure the light output of the flare, effectively limiting the practical length thereof and, hence, the duration of its operational usefulness.

The liner is customarily an elastomeric hydrocarbon bonding material that is primarily used to achieve a good bond between the illuminant composition and the case of the flare, so that combustion will be confined to the exposed end of the flare candle and cannot progress down the sides thereof, which occurs when the liner is absent or too thermally conductive. The conventional liner composition, being primarily an adhesive and a thermal insulation, is ordinarily deficient in oxidizers, so that its combustion uses oxidizer from the flare candle, thereby further decreasing the light output of the flare.

This process overcomes these disadvantages in prior art flares, through the use of an oxygenated adhesive liner composition for flare cases. This improved liner not only contains sufficient oxidizer to promote its own complete combustion and that of the case, so that light output of the flare is not obscured; but also enough to contribute to the brilliance of the flare.

The basic binder is an oxidizer-rich polymer, typically an elastomer similar to the binder of the illuminant composition. The added oxidizing coloring agent is chosen to be compatible with the color of the flare. For example, sodium nitrate is used for yellow flares, strontium chloride for red, and barium nitrate or chloride for green.

Appropriate curing agents are used for polymerizing the binder. Additional oxidizer, such as potassium perchlorate, may be included if the remainder of the liner composition is deficient in oxidizers. A preferred formulation for an adhesive liner useful for bonding red-flare, illuminant candles to their cases is, in approximate parts by weight:

		Practical Ranges
Formrez F-17-80 (carboxyl-terminated polyester)	42.0	15.0 - 50.0
Triglycidyl ether of trimethylolpropane	15.7	5.0 - 8.0
Chromium octoate	0.3	0.02 - 1.0
Strontium nitrate (50 micron diameter)	28.0	20.0 - 50.0
Strontium chloride	14.0	10.0 - 30.0

A preferred formulation for an adhesive liner to be installed in yellow flares is, in approximate parts by weight:

		Practical Ranges
Formrez F-17-80 (carboxyl-terminated polyester)	72.7	25.0 - 74.0
ERLA-0510 (triglycidyl ether of p-aminophenol)	15.0	5.0 - 18.0
Chromium octoate	0.3	0.02 - 1.0
Potassium perchlorate or sodium nitrate	12.0	8.0 - 70.0

Example 1: A liner composition was prepared by mixing 72.7 parts by weight of carboxyl-terminated diethylene glycol succinate prepolymer with 15 parts of p-N,N-dipropylene oxide phenyl glycidyl ether and 0.3 part of chromium octoate. This mixture was stirred for about 15 minutes, after which 12 parts of 6- to 20-micron potassium perchlorate was gradually added as stirring continued. After this composition was thoroughly mixed, it was applied by brush to a kraft paper flare case to a 30-mil thickness. It was then placed in an oven and cured at 135°F for 3 days. The pyrotechnic composition for a red illuminant candle was then cast in the tube and cured at elevated temperature for an appropriate length of time. When ignited, the liner and flare cases were observed to burn

at approximately the same rate as the flare candle, with no observable obscuration of the light output or color.

Example 2: A mixture was made of 64 parts by weight of carboxyl-terminated diethylene glycol succinate prepolymer, 21 parts of triglycidyl ether of trimethylolpropane, one part of iron linoleate, and four parts of fused silicon dioxide of submicron particle size. When these ingredients were mixed thoroughly, 12 parts of 6 to 30-micron-diameter potassium perchlorate was added gradually as stirring continued. The viscosity of the composition may be adjusted within limits by varying the quantity of silicon dioxide, to permit control of the tendency of the material to slump after it is applied to a surface.

This composition was then diluted to a sprayable consistency by adding methylene chloride. Any of several other volatile solvents may be used for this purpose. It was then applied to a flare case to an 0.60 inch thickness. The solvent evaporated rapidly and the remaining liner was cured at 135°F for 4 days.

Water-Resistant Labels for Glass Bottles

Bottled beverages such as soft drinks, beer and the like are often chilled by placing the bottles, prior to serving, in a tub or similar container, filled with ice and water. It is important that the labels which identify the contents should not peel off or become detached from the bottles during the submersion in the ice water.

A number of labeling adhesives based on casein have been described as resistant to the action of ice water. In particular U.S. Patent 2,276,148 describes suitable casein formulations containing certain water-soluble gums. However, casein is sometimes in short supply because of its increasing use in foodstuffs and it has become of interest to use compositions based on starch.

Conventional "jelly gum" adhesive, based on starch, urea and sodium hydroxide, have been used as bottle-label adhesives. They can be made with viscosity and tack properties that are quite well suited for applicability on labeling machinery, but they are deficient in water resistance, the labels becoming detached after only 24 hours' immersion in ice water. Manufacturers such as brewers now require labels to remain intact over a period of at least three days' immersion. The test can be made more severe by using water at room temperature (20° to 30°C) instead of ice water.

According to a process described by *R.S. Shenfeld and O.E. Brown; U.S. Patent 3,857,803; December 31, 1974; assigned to Borden, Inc.* a starch-based adhesive can be prepared which is capable of maintaining labels on bottles intact after exposure to room-temperature water for at least three days.

The adhesive composition comprises at least 40% by weight water, from 10 to 40% starch, from 5 to 20% nitrogeneous substance selected from the group consisting of urea, dicyandiamide and their mixtures and an ammoniated hydrolysis product of a styrene/maleic anhydride copolymer in amount corresponding to from 0.6 to 33 parts by weight of styrene/maleic anhydride copolymer for each 100 parts of combined weight of starch, urea and dicyandiamide, with sufficient ammonia to effect a pH in the adhesive in the range of 7.5 to 9.2.

The composition is made by a procedure which comprises the addition of a hydrolyzed styrene/maleic anhydride copolymer, in the presence of aqueous ammonia, to a conventional starch paste containing more urea and/or dicyandiamide. Various alternative procedures can be used to bring the components together. Preferably the starch is first dispersed by mixing in water at room temperature and after augmentation with the urea and/or dicyandiamide, the mixture is heated with stirring to about 85°C for fifteen minutes, then cooled to 60° to 65°C. Sufficient ammonia is then added to bring the pH to a value in the range of 7.5 to 9.2.

In a separate tank, a styrene/maleic anhydride copolymer, or a corresponding styrene/maleic acid copolymer, or an ammonium salt thereof, is added to water at 50° to 60°C and stirred with addition of sufficient ammonia to bring the pH to a value in the range of 7.5 to 9.2. When this mixture has become uniform, it is added to the augmented starch paste at 60° to 65°C and the paste is stirred until uniform.

At any stage during or after the pasting or mixing, conventional defoamers can be added as is customary to avoid formation of foam. Conventional preservatives such as phenol and various commercial phenol derivatives can also be added, as well as colorants which are customarily used in the preparation of label adhesives.

The adhesives can be employed with a wide variety of label types, being particularly suitable for labels comprising paper as the main substrate. Such paper can be prepared by any conventional paper process utilizing the conventional paper fibers and conventional sizing agents. The front label surface, reverse from the back surface to be glued to the glass vessel, can be provided with any kind of suitable message or design, applied by conventional means such as printing with suitable inks either directly on the paper itself or on a layer of fiber or foil laminated thereon, exemplarily aluminum foil. The printed side of the label can be protected by a coating of lacquer. The process is illustrated by the following examples, proportions being expressed as parts by weight except where specifically stated to the contrary.

Example 1: A solution was prepared containing 10.0 parts by weight of urea and 1.8 parts of a 29.4% aqueous ammonia solution (26°Be) in 67.8 parts of water. 25.0 parts of waxy cornstarch were then dispersed in the solution and a paste was made by heating this dispersion to 85°C for a period of fifteen minutes. The paste was allowed to cool to 65°C and there was then added an ammoniacal solution (pH 8.7) containing 24.87 parts water and 3.90 parts of a styrene/maleic anhydride copolymer having average molecular weight equal to 20,000.

The mixture was stirred until homogeneous. There were then added 0.20 part of a conventional defoamer and 0.20 part of a phenolic preservative available commercially as Dowicide A. The resultant product had pH equal to 8.8 and its viscosity was 65,000 cp as measured at 25°C on an HAF Brookfield viscometer with Spindle No. 3 at 1 rpm speed.

A five-gallon sample of the adhesive prepared as above was submitted to a test on a Biner Ellison bottle-labeling machine. The product had the requisite high tack for efficient labeling and exhibited clean machining during the entire test which continued through an eight-hour shift. Bottles from this test were stored at room temperature and three days later they were tested for resistance to submersion in water at room temperature. The labels were found to have at least three days' resistance to this submersion, the test being terminated after three days. The bottles were in close contact with each other during this test and although they were moved around at frequent intervals the labels showed no tendency to loosen.

In a second test made at the bottling establishment of a brewery, an adhesive made in the above manner showed desirable good quick tack in application of the labels and all the bottles of the test passed on the brewery's four-day water immersion test.

Example 2: Two adhesives were prepared following procedures similar to that described in Example 1 and having the following formulations in parts by weight.

	A	B
Waxy corn starch	25.00	25.00
Wheat flour	0.25	0.25
Urea	10.00	10.00
Styrene/maleic anhydride av. molecular wt. 20,000	0.00	3.90
Aqueous ammonia 26° Be	1.40	1.40
Water	63.35	59.45
Viscosity, cps.	360,000	64,000
pH	9.2	8.9

The adhesive A which did not contain the styrene/maleic component of this process failed the 3-day cold-water resistance test whereas all the labels fixed with adhesive B of this process passed the 3-day test.

Example 3: Two adhesives were prepared as in Example 2 but having the following formulations, with water to make 100 parts by weight.

	A	B
Waxy corn starch	25.00	25.00
Wheat flour	0.50	0.50
Urea	10.00	10.00
Styrene/maleic(av.M.W. 10,000	3.90	—
anhydride (av.M.W. 50,000	—	3.90
Aq. ammonia 26° Be	2.00	1.40
Viscosity, cps.	130,800	69,000
pH	8.2	8.8

All bottles treated with these adhesives passed the 3-day water resistance test.

Oil Well Sealer

According to a process described by *L.H. Eilers and C.F. Parks; U.S. Patents 3,839,260 and 3,839,262; October 1, 1974; both assigned to The Dow Chemical Company* the sealing of void spaces, e.g., as in geological formations and/or between metal shapes is accomplished with a liquid slurry of a particulate, water-soluble organic polymer in an organo solvent. The mixture of the polymer and solvent has a controlled set time to allow the emplacement of the slurry as a liquid. The slurry then sets to form a sealing cementitious material.

Organo solvents useful in the process fall into three general classes. Each class is characterized by unique gel-forming characteristics and in the properties of the final hardened gel. As used herein the term "organo solvent" refers to a homogeneous liquid composed substantially of an organic material, that is, it contains no more than 50% by weight of water, which organo solvent is absorbed by the particulate polymer to form a continuous phase.

The first class of organo solvents useful in the process, termed Group I solvents, are the liquid organic solvents which solvate or plasticize solid, water-soluble polymers. That is, the polymers, if they are not totally miscible with the solvent, absorb a significant proportion of the same.

Solvents of this nature, suitable for use with a wide variety of water-soluble polymers, include for example ethylene glycol, ethylenediamine, glycerol, dioxolane, formamide, pentaerythritol and acetic acid. Some materials may not function as solvents at normal room temperatures but can be activated as solvents for most water-soluble polymers at elevated temperatures. Such materials include diethylene glycol, dipropylene glycol, diethanolamine, monoethanolamine, propylene glycol, triethylene glycol, diethylenetriamine and triethanolamine.

A second class of useful organo solvents, termed Group II solvents, includes a large number of organic materials which are not solvents for the polymers (nonsolvents) i.e., they do not solvate or plasticize the water-soluble polymers to form a continuous phase, but which, when mixed with a small amount of water, produce organo solvents useful for the purpose of the process. Manifestly, such materials must be miscible with water. It has been found that as little as 1% by weight water dissolved in an organic liquid, which is not a solvent for the polymer, often renders the resulting mixture a suitable organo solvent for the purposes of the process.

Sometimes the organic material may normally exist as a solid. In such instances, enough water, e.g., up to 50% by weight of the total solvent composition may be used to provide a liquid solution of the organic material. The resulting aqueous solution is an organo solvent suitable for the purposes of the process. Since the set time of a formulation is decreased by increases in the water content of the organo solvent used, the amount of water is usually maintained at less than 25% by weight of the total solvent composition.

Materials which are not normally solvents for the water-soluble polymers, but which on addition of water become such solvents at room temperature, e.g., 75°F, include diethylene glycol, dipropylene glycol, diethylene glycol monomethyl ether, ethylene glycol monoethylether, dioxane, ethylene carbonate, ethanolamine, triethylene glycol, propylene glycol, sucrose, urea, dextrin, diethanol amine, triethanol amine, and diethylene triamine.

A most useful class of organic materials for formulating Group II solvents includes the liquid alkylene oxide polymers, which are not solvents in themselves for the water-soluble polymers. Such alkylene oxide polymers include, in addition to those mentioned above, higher polymers of ethylene oxide having molecular weights up to as much as 600 or so. Liquid propylene oxide polymers include tripropylene glycol, tetrapropylene glycol and higher propylene oxide polymers having molecular weights up to as much as 1,200 or so. In addition, water-soluble alkyl and aryl monoethers of the alkylene oxide polymers can form useful solvents when mixed with water. Exemplary alkyl and aryl etherifying groups include methyl, ethyl, propyl, butyl, dodecyl, phenyl, butylphenyl and the like groups.

A third class or organo solvents, termed Group III solvents, is composed of mutual solutions of an organic material, which is not solvent for the water-soluble polymers, and an organic material, which is a solvent for the polymer, i.e., a Group I solvent. In such mixtures, at least 0.05 part of the polymer solvent is employed for each part by weight of the nonsolvent. Such organo solvents allow convenient adjustment of the set times in nonaqueous cementitious formulations by increasing or decreasing the amount of polymer solvent relative to the nonsolvent.

Example 1: This example illustrates formulating cementitious gels according to the process using Group I and Group II organo solvents. To 100 ml of ethylene glycol at a temperature of 75°F was added 75 grams of a particulate carbamoyl polymer. The polymer was a polyacrylamide in which 7% of the initially available carboxamide groups had been hydrolyzed to sodium carboxylate groups. It was characterized by a molecular weight of at least one million. The polymer was stirred into the ethylene glycol to produce a uniform blend at 75°F (approximate room temperature). The composition set to a shape-retaining rubbery gel in 45 minutes.

The set gel is useful as a cement in a variety of oil well-treating applications. In a specific use, a slurry prepared as described above is placed in the annulus defined by two concentric oil well casings. When set, the composition provides a highly adherent and persistent annular seal.

To illustrate set time control with Group III organo solvents, a second formulation was prepared with a 50/50 mixture of ethylene glycol and diethylene glycol, which are a solvent and nonsolvent for the polymer respectively. To 100 ml of this mixture at 75°F was added 75 grams of the above polymer. After 4 hours, the composition achieved an initial set to produce a solid, rubbery cement. By way of comparison, 75 grams of the polymer in 100 ml of diethylene glycol remained fluid through 48 hours, after which the test was terminated.

Through the conjoint use of the nonsolvent and solvent to prepare the organo solvent, the set time of the composition was extented by over 3 hours. The longer set time permits batchwise preparation of the formulation and otherwise more extensive handling prior to use.

Example 2: To exemplify the metal surface adhesion of cementitious compositions prepared in accordance with the process a quantity of the formulation in the above example prepared with the Group III solvent was poured into a vertically aligned pipe, the lower end of which was plugged with a rubber stopper. The composition was allowed to cure for about 20 hours at 75°F.

In a second run, the temperature was increased to 150°C. In this manner, plug seals 12 inches long were produced in each pipe. The set time of the composition at these temperatures was determined as the length of time required for the organo solvent-polymer slurry to form a shape-retaining mass.

To measure seal quality, the upper end of the pipe was fitted with a high pressure cap and the rubber stopper removed. Nitrogen gas was applied gradually to the seal until it failed or withstood the maximum pressure for 10 minutes. The maximum pressure applied with this technique was 1,000 psi.

In still further runs similar to the above, a series of compositions was prepared in which portions of the polymer were replaced with corresponding amounts of magnesium hydroxide. The curing and testing conditions were as discussed above. The compositions, set times for cures at 75°F and 150°F, and adhesion strength properties for the above operations are included in the following table:

	Formulation				Properties		
Run	Carbamoyl Polymer, g	Ethylene Glycol, ml	Diethylene Glycol, ml	$Mg(OH)_2$, g	Yield Strength, psi	Set Time (hr) 75°F	Set Time (hr) 150°F
1	75	100	----	0	650	0.75	0.05
2	75	50	50	0	650	4	0.2
3	74	50	50	1	400	4	0.2
4	73	50	50	2	500	4	0.2
5	72	50	50	3	1,000	4	0.2
6	70	50	50	5	1,000	4	0.2
7	68	50	50	7	1,000	4.5	0.25

Example 3: Gel times for polymer-solvent mixtures of the process were ascertained according to the procedure set forth in Section 9, Schedule 6, of API-RP 10B, 12th Edition, March 1963. Basically, the procedure involved using the thickening time tester as therein described. The test device was charged with a polymer-solvent system to be tested. In these operations, the formulations were heated to a maximum of 144°F in 36 minutes during which time continuous viscosity measurements were made.

Thickening or gellation set times were measured as the elapsed time between starting the heating of the formulation and the occurrence of a slurry consistency of 100 poises. The thickening times of several formulations are set forth in the following table.

The formulations tested include a polyacrylamide having a molecular weight of at least about one million mixed with each ethylene glycol and diethylene glycol, and mixtures of the two glycols. Sixty grams of the polymer were slurried in 100 ml of the particular organo solvent used. The results of these tests are set forth in the following table.

Diethylene Glycol (vol %)	Ethylene Glycol (vol %)	Set Time (min)	Temperatures at Time of Set (°F)
0	100	15	106
50	50	27	128
80	20	49	144
100	0	80+	144

Of the above runs, Run 3 is a preferred formulation for employment in a well with a bottom hole temperature of 144°F. The set time of 49 minutes allows sufficient opportunity for placing the composition in the well prior to its setting and thus becoming unworkable.

Example 4: A series of formulations was prepared in accordance with the process using Group II solvents. The series illustrates the effect of increasing amounts of water on the set time of the formulation. To 100 ml of a particular organo solvent was added 60 grams of the polyacrylamide used in Example 1. The aliquots of each formulation were maintained at different temperatures ranging from 80° to 200°F. The elapsed time from formulation of the polymer-solvent slurry until a hardened gel formed was measured and is reported in the following table as the gel time.

Organo Solvent	Wt % Water Added	---Gel Time (hr) at Specified Temperature (°F)---					
		80	100	120	150	175	200
Diethylene glycol	0	–	–	48	14	3.5	1.5
Diethylene glycol	2	–	–	20	8	2	0.7
Diethylene glycol	5	–	18	5	1.4	0.5	–
Diethylene glycol	10	4	2	1	0.5	–	–
Diethylene glycol	16.6	0.7	0.5	0.3	–	–	–
Triethylene glycol	0	–	–	–	–	–	48
Triethylene glycol	5	–	–	–	–	48	15
Triethylene glycol	10	–	32	16	6	–	–
Propylene glycol	0	–	–	32	8	2.5	1
Propylene glycol	2	–	30	14	3	0.7	–
Propylene glycol	5	16	8	3	0.7	–	–
Propylene glycol	10	3	1.3	0.5	–	–	–
Propylene glycol	16.6	0.7	0.4	–	–	–	–
Dipropylene glycol	0	–	–	–	–	–	48
Dipropylene glycol	10	–	–	–	48	30	–
Diethanol amine	0	–	–	–	–	–	48
Diethanol amine	10	–	–	–	48	16	6

Phenolic Resins and Silanes for Foundry Sands

C.-D. Seiler, H.-J. Vahlensieck and H. Junger; U.S. Patent 3,847,860; Nov 12, 1974; assigned to Dynamit Nobel AG, West Germany describe mixtures of a plurality of silanes all of the formula:

$$[(R'O)_3\text{—Si—R—(NH—R)}_n]_m\text{—NH}_{3-m}$$

where R is an arylene or alkylene, R' is an aliphatic radical of 1 to 3 carbon atoms, n is 0 or 1 and m is 1, 2 or 3. Such mixtures, with each component present in a proportion of at least 5 weight percent, are used as additives for cold-setting and thermosetting bonding agents for inorganic materials to produce bonded products having significantly higher bending and compression strengths.

(1) Triethoxysilylpropylamine plus bis(triethoxysilylpropylamine) and tris(triethoxysilylpropylamine).
(2) Triethoxysilylmethylamine plus bis(triethoxysilylmethylamine) and tris(triethoxysilylmethylamine).
(3) Triethoxysilylphenylamine plus bis(triethoxysilylphenylamine) and tris(triethoxysilylphenylamine).

(4) $H_2N—CH_2—CH_2—NH—CH_2—CH_2—Si(OR_3)$ plus
$NH_2—CH_2—CH_2—N—[CH_2—CH_2—CH_2—Si(OR)_3]_2$.

It has been found that adding such silane mixtures, in quantities on the order of 0.02 to 3% of the weight of the resin, to cold-setting or thermosetting resin systems brings about a significant increase of the bond strength in comparison to products made with equal percentages of only one of these compounds in pure form instead of mixtures.

Thus increases in the improvement of bending strength were obtained of between 25 and 250% in the case of thermosetting resins and between 60 and 210% in the case of cold-setting resins. In the case of cold-setting resins the use of such silane mixtures was additionally observed to result in an 80% reduction of the setting time required for the achievement of equal final bending strengths, in comparison with resins prepared with one of the amino-silanes in pure form.

Resin 1: Cold-Setting Multiple-Component Phenolic Resin — 660 g of phenol (90%), 3,100 g of formaldehyde (37% aqueous solution) and 740 g of urea are heated at 100°C for 20 minutes in a glass flask provided with stirrer and reflux condenser. After this period, 1,540 g of phenol (90%), 2,070 g of formaldehyde (37% aqueous solution) and 150 g of technical furfuryl alcohol are added in that order. After the content of the flask is heated to 100°C, 12.7 g of NaOH (100%) is added drop by drop in the form of a 30 to 40% aqueous solution. After another 55 minutes at 100°C the condensation has ended. Water is then removed by vacuum distillation until a viscosity of 2,500 to 3,000 cp is reached.

Resin 2: Cold-Setting Phenol-Formaldehyde Resin — A mixture of 9 kg of phenol (90%) and 12.7 kg formaldehyde (37% aqueous solution) is heated at 100°C. After the addition of 120 g NaOH the temperature of the mixture is held at 100°C for 25 to 35 minutes. Then water is removed by vacuum distillation until a viscosity of 1,500 cp is reached. An additional 5.5 wt percent of phenol (90%) is added to the resin thus obtained, at a temperature between 25° and 40°C.

Example 1: 2,000 g of Haltener sand H 32 is mixed with 50 g of Resin 1, into which 0.2 g of silane additive had been admixed, and 7.5 g of a hardener consisting of an aqueous solution of 25 weight percent ammonium nitrate and 40% weight urea. The composition was mixed and blended in an appropriate, conventional positive-action mixer. The moist mixture is then put into a molding machine to prepare the test rods. The mixture is molded at a pressure of about 7 atmospheres gauge into a mold heated to 225°C. After various holding periods the mold opens automaticallly. The rods are removed and after cooling are tested for bending strength in a conventional +GF+ strength testing apparatus. The table on the following page lists the values which were found.

Silane A = γ-Aminopropyltriethoxysilane.
Silane B = γ-Aminopropyltriethoxysilane plus 40% bis-(triethoxysilylpropylamine).
Silane C = γ-Aminopropyltriethoxysilane plus 60% bis-(triethoxysilylpropylamine).
Silane D = 60% γ-Aminopropyltriethoxysilane, 30 to 35% bis-(triethoxysilylpropylamine), 5 to 10% tris-(triethoxysilylpropylamine).

	Bending strength in kp./m.² after hardening period of —			
	5 sec.	10 sec.	15 sec.	30 sec.
Without silane	64.0	76.0	81.0	72.0
With silane:				
A	67.0	76.0	81.0	90.3
B	67.0	83.0	91.0	95.0
C	74.0	90.0	90.0	----------
D	61.0	82.0	92.0	86.0

Example 2: 100 weight-parts of Haltener sand H 32 were intimately mixed with 2.5 weight-parts of Resin 2 containing 0.2 weight percent of silane, and with 16% by volume (with respect to Resin 2) of a hardener consisting of a 60% aqueous solution of p-toluenesulfonic acid.

The mixture is poured into +GF+ test rod molds, and rammed four times with the +GF+ rammer. The rods thus formed were allowed to set at room temperature. At different intervals of time the bending strength was measured by means of the +GF+ strength testing apparatus. The findings were as follows:

	Bending strength in kp./cm.² after setting at room temperature for—				
	1 hr.	3 hr.	5 hr.	8 hr.	24 hr.
Without silane	4.5	18.0	19.0	18.5	20.5
With silane:					
A	6.0	21.5	22.0	25.0	27.0
B	7.0	25.0	27.0	29.0	33.5
C	8.0	25.0	27.5	30.5	31.0
D	7.0	21.0	26.0	27.5	32.5

The silanes are as previously defined.

Acetoxysilicon Adhesion Promoter for Silicon Elastomers

Adhering silicone elastomers to various substrates is a continuing problem. There is an ever present need for better adhesion. With the changing substrates and silicone elastomers, the search for new methods of adhering the silicone rubber to substrates and for new materials to provide adhesion continues. One particular area wherein adhesion is required, is the application where room-temperature-vulcanizable silicone elastomer is to be adhered to substrates, such as metals.

Many of the prior art methods of adhering silicone rubber to substrates are not suitable for today's materials, provide an insufficient bond strength or are too expensive because of the complex method or materials.

U.S. Patent 2,751,314 shows that silicone rubber can be adhered to the surface of a solid by depositing a continuous coating having 50 to 100% by weight of an organotitanate and 0 to 50 weight percent alkylpolysilicate, depositing over this coating another coating having 1 to 10% by weight alkylpolysilicate and 90 to 99.9% of a toluene-soluble organopolysiloxane, thereafter applying a silicone rubber such as a room-temperature-vulcanizable silicone rubber.

U.S. Patent 2,979,420 shows that a room-temperature-vulcanizing silicone rubber can be bonded to a surface by applying certain monoorganotriacyloxysilanes to the surface as a primer. U.S. Patent 3,498,824 shows that an alkoxyacyloxysilane can be used as a primer to adhere room-temperature-vulcanizing silicone rubber to metal substrates.

U.S. Patent 3,671,483 describes certain compositions which are useful as primers for adhering silicone rubber to epoxy resin and polyurethane substrates. The compositions described contain alkylsilicates, organotitanates, solvent and a compound of the formula $CF_3CH_2CH_2SiOSi(CH_3)_2X_3$ where X is $—CH_2CH_2Si(OOCCCH_3)_3$ or a hydrogen atom and at least one X is $—CH_2CH_2Si(OOCCH_3)_3$.

U.S. Patent 3,714,109 describes a mixture of an organic solvent, bis(acetylacetonyl)diisopropyltitanate, an alkylpolysilicate and a mixture containing an organosiloxane having dimethylsiloxane units and methylhydrogensiloxane units and trimethylsiloxane units or dimethylhydrogensiloxane units and a modified organosiloxane having in addition to the above siloxane units,

$$(CH_3COO)_3SiCH_2CH_2(CH_3)SiO$$

units and $(CH_3COO)_3SiCH_2CH_2(CH_3)_2—SiO_{0.5}$ units. All of these primer compositions are suitable for adhering silicone rubber to substrates in certain applications but have insufficient bond strength for many applications and additionally have a potential to form hydrogen gas because of the presence of silicon-bonded hydrogen atoms.

In recent work *C.-L. Lee and J.R. Schulz; U.S. Patent 3,873,334; March 25, 1975; assigned to Dow Corning Corporation* describe reaction products which are obtained by mixing under anhydrous conditions

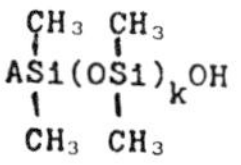

where k is 0 to 20 and A is methyl or vinyl and

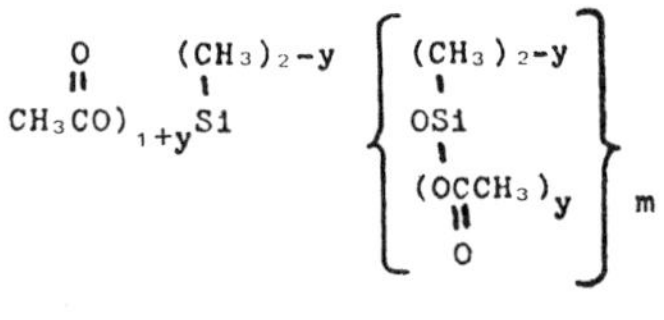

where m is 0 to 5 and y is 1 or 2. They are adhesion promoters and useful in primer compositions. The primer compositions are organic solvent solutions with 50 to 99.9 weight percent organic solvent, with the balance being made up of 5 to 100 weight percent of the adhesion promoter, 0 to 50 weight percent of an alkylsilicate and 0 to 50 weight percent of an organotitanate.

An example of a primer composition would be a solution of 76.9 weight percent inhibited 1,1,1-trichloroethane with 23.1 weight percent of equal weight amounts of an adhesion promoter where A is vinyl, k is 3 and m has an average value of about 0.7, ethylpolysilicate and bis(acetylacetonyl)diisopropyltitanate. The primer compositions are useful in adhering room-temperature-vulcanizable silicone elastomers to substrates such as aluminum.

Example 1: To a five-liter flask, 436 grams of acetonitrile and 43.6 grams of N,N-dimethylacetamide was added and 1,334 grams of hexamethylcyclotrisiloxane was dissolved therein. To this solution, 844 grams of dimethylvinylchorosilane was added and the mixture was stirred for 20 hours at room temperature. This mixture was stripped for three hours to a head temperature of 30°C at 2 mm of mercury and then fractionated on a vigreux column taking 40-ml cuts. The product was obtained at 65°C at 0.7 mm of mercury, and had a formula

```
          CH3  CH3
          |    |
    CH2=CHSi(OSi)3Cl
          |    |
          CH3  CH3
```

In a three-liter flask, suspended in an ice bath and fitted with a stirring bar, a mixture of 35.85 grams of concentrated ammonium hydroxide, 579 grams of diethylether and 409 grams of ice was stirred until the ice in the flask was melted. To this mixture, while rapidly stirring, 200 grams of

```
          CH3  CH3
          |    |
    CH2=CHSi(OSi)3Cl
          |    |
          CH3  CH3
```

was poured into the flask. The rapid stirring was continued for five minutes and the water was separated using a separatory funnel. The ether layer was washed with water until the water was neutral. The ether layer was then stripped under vacuum to a pot temperature of 50°C. The product was 100% pure as indicated by gas-liquid chromotography and infrared analyses, with a 94.9% yield of vinylsiloxane of the formula

```
          CH3  CH3
          |    |
    CH2=CHSi(OSi)3OH
          |    |
          CH3  CH3
```

In a three-liter flask, 1,838 grams of methyltriacetoxysilane was heated to 40°C. This silane was stirred rapidly while beginning to add 62.4 ml of water slowly from an addition funnel. After the first few milliliters, the temperature was reduced with an ice bath. The total addition time was about 1 hour. Volatile materials were removed under reduced pressure. The resulting partial hydrolyzate had a formula

```
     O      CH3  CH3   O
     ||     |    |     ||
   (CH3CO)2Si(OSi)mOCCH3
                 |
                 OCCH3
                 ||
                 O
```

where m had an average of about 0.7.

An adhesion promoter was prepared by slowly adding over a 1-hr period 287.5 g of

```
          CH3  CH3
          |    |
    CH2=CHSi(OSi)3OH
          |    |
          CH3  CH3
```

to 287.5 grams of the partial hydrolyzate prepared above. The product contained no hydroxyl radicals and was stored in a sealed polyethylene container.

Primer compositions were prepared by mixing under essentially anhydrous conditions in Chlorothene (for an inhibited 1,1,1-trichloroethane) the adhesion promoter prepared by Example 1, ethylpolysilicate and bis(acetylacetonyl)diisopropyltitanate. The amounts used are shown in the following table.

		Primer Composition		Composition of Reactive Ingredients in weight percent			Lap Shear Test	
Test No.	Number of Panels Tested	Weight percent Chlorothene	Weight percent Reactive Ingredients	Adhesion Promoter	Ethylpolysilicate	Titanate	Adhesive Strength, p.s.i.	% Cohesive Failure
1	27	none	none	none	none	none	176	0
2	5	99.1	0.9	33.3	33.3	33.3	854.4	73.4
3	8	94.7	5.7	33.3	33.3	33.3	827	95
4	8	89.3	10.7	33.3	33.3	33.3	823	85
5	8	87.0	13.0	33.3	33.3	33.3	767	93
6	39	76.9	23.1	33.3	33.3	33.3	804.1	85.7
7	10	81.6	18.4	44.5	44.5	11.0	929.7	100
8	5	81.6	18.4	44.5	11.0	44.5	669	37
9	5	74.1	25.9	28.6	42.8	28.6	873.6	88.6
10	5	76.9	23.1	33.3(1)	33.3	33.3	653.5	11

(1)Adhesion promoter prepared same as in Example 1, except $(CH_3)_3SiCl$ substituted for $(CH_2{=}CH)(CH_3)_2SiCl$ in the proper molar amount.

The primer compositions were wiped on aluminum test panels, then rubbed vigorously to remove any excess amounts, allowed to stand at least one hour before applying a silicone elastomer to make the adhesion test panels. The silicone elastomer was the same in each test and was basically a diorganopolysiloxane containing methyl and vinyl radicals, an organosilicon compound containing silicon-bonded hydrogen atoms and a platinum catalyst. The test panels were prepared and tested in accordance with the procedures set forth in ASTM-Designation: C273-61. The silicone elastomer was cured on the panel by heating for 1 hour at 150°C. The results for this lap shear adhesion test were reported in pounds per square inch and percent cohesive failure. The test results were as shown in the table above. The panels were tested the same day they were prepared.

Test panels were prepared and tested as described in Example 2 using the same materials. After the test panels were made, they were aged at room temperature for 1 day before the lap shear test was done. The results were as shown in the following table.

		Primer Composition		Composition of Reaction Ingredients in weight percent			Lap Shear Test	
Test No.	Number of Panels Tested	Weight percent Chlorothene	Weight percent Reactive Ingredients	Adhesion Promoter	Ethylpolysilicate	Titanate	Adhesive Strength, p.s.i.	% Cohesive Failure
1	33	none	none	none	none	none	139.3	0
2	31	76.9	23.1	33.3	33.3	33.3	902.7	99.2
3	10	80.0	20.0	40.0	20.0	40.0	835.4	75.5
4	4	81.6	18.4	44.5	33.3	22.2	779	80
5	5	78.4	21.6	36.4	36.4	27.2	683	41
6	5	80.0	20.0	40.0	40.0	20.0	769.2	67.6
7	5	83.3	16.7	50.0	50.0	0.0	965.4	98

POLYCHLOROPRENE-BASED ADHESIVES

Latex Mastic

Neoprene mastics are known to be useful in the adhesive art and are of two types, latex-based and solvent-based. Both types have outstanding resistance to degradation by weather, oxygen, ozone, solvents, and oils.

Latex mastics have been made by compounding neoprene latex with such components as fillers, adhesion-promoting resins, antioxidants, metal oxides, and curing agents, using conventional latex compounding techniques. The resultant compounded latex, which comprises a dispersion of neoprene and the other above-named components in a continuous water phase, is suitable without further processing for use in adhesive applications where a mastic is needed.

Solvent-based neoprene mastics comprise a solution of neoprene in one or more of its solvents, with the compounding components above-mentioned either dissolved or dispersed therein. Mastics of this type can be prepared by mixing dry, solid neoprene with the solvent and compounding components in a powerful internal mixer such as the sigma-arm type until the neoprene is completely dissolved or dispersed in the solvent. In solvent-based mastics, in contrast to the latex mastics, the solvent is the continous phase, and no water is present.

Solvent-based neoprene mastics are superior to neoprene latex mastics in that they have faster drying rates and develop higher adhesive bond strength. On the other hand, the latex mastics are easier to apply because they have better application rheology, that is, they have lower viscosity and nerve.

Moreover, the latex mastics are less costly to prepare because neoprene latexes are advantageously directly produced by polymerizing chloroprene monomer in aqueous emulsion and to make the latex mastics from such neoprene latexes requires no prior isolation of the polymer from its latex nor any costly internal mixing equipment.

According to a process described by *R.W. Keown; U.S. Patent 3,878,154; April 15, 1975; assigned to E.I. du Pont de Nemours and Company* mastics having excellent adhesive properties are produced by mixing a high-solids neoprene latex, directly prepared by polymerizing chloroprene in an aqueous emulsion, and an organic water-immiscible neoprene solvent containing an adhesion-promoting resin in the presence of an amphoteric or cationic surfactant.

The surfactant may be dispersed in the organic, water-immiscible neoprene solvent, or if it is amphoteric or a polyoxyethylated alkyl amine it may be dispersed directly in the latex. The mixing of the solvent and the latex in the presence of the surfactant is continued until a water-in-oil-type dispersion is obtained and the viscosity of the mixture has increased into the mastic range.

To operate this process, the expensive, high-power-consuming equipment of the prior art is not required. Prior isolation of the polymer is also not necessary; hence the need for water-separation can be dispensed with. In addition, the resulting mastic forms adhesive bonds with strengths and fast-drying rates approaching those of conventional solvent-based mastics.

Although prepared by latex compounding techniques, mastics prepared by the process differ from conventional latex mastics in that the neoprene solvent is the continuous phase in the composition, and the water is a dispersed phase. The fact that they contain a dispersed water phase similarly distinguishes them from conventional solvent-based mastics, which have a neoprene-organic solvent solution as the continuous phase but contain no dispersed aqueous phase.

The particular utility of the processes lies in the simplicity of their use. The initial mixing of the latex, neoprene solvent, and surfactant can be carried out in minimal mixing equipment; a simple kettle or drum with a paddle or propellor agitator will do.

During and immediately after this initial mixing step, the composition is a fluid of low viscosity. Its viscosity increases with time, however, and agitation is continued until the composition has taken on the pasty consistency characteristic of a mastic. This usually occurs between 30 seconds and 1 hour after mixing is initiated. The mastic remains workable for an ample period of time to permit its easy use for such purposes as adhesively setting tile and the like, that is, it does not gel prematurely, as do many otherwise useful mastics. All parts are by weight per 100 parts by weight of neoprene unless otherwise indicated.

Example 1: Rosin-modified phenolic resin (30 g) and a tert-butyl phenolic-formaldehyde resin (40.0 g) are dissolved in 236 g of a mixture of toluene and cyclohexane, of which 57.4% by volume is toluene. In a separate vessel, a mixture of 4.0 g of magnesium oxide, 5.0 g of zinc oxide, 140 g of kaolin, of which 78% has a particle size of less than 2 microns, and 2.0 g of 2,2'-methylene-bis(4-methyl-6-t-butylphenol) are wet with 236 g additional of toluene-cyclohexane mixture which has the same volume composition and has dispersed therein 3.0 g of surfactant (listed below).

The clay slurry and resin solution are combined, and the mixture is poured into 167 g of neoprene latex containing 60% solids. The latex is contained in a one-quart paint can and is stirred by hand with a spatula as the mixture is added. Immediately after addition of the mixture to the latex has been completed, the composition is a fluid of low viscosity much resembling the original latex. Viscosity increases sharply with passage of time beyond this point, and after approximately 2 to 6 minutes, it has reached that of the mastic range, where it levels off and becomes approximately constant with time.

Stirring of the composition is continued during this 2 to 6-minute time period following completion of addition. Mastic compositions thus obtained are ready for use as adhesives and sealants and exhibit exceptionally good application rheology and adhesive properties.

Surfactants that have given excellent mastics by this procedure are N-lauryl-beta-alanine, N-cetyl-N-ethylmorpholinium ethosulfate, C-cetylbetaine, C-decylbetaine, condensation product of coco amine (1 mol) with 2 mols of ethylene oxide, and lauryl ammonium sulfonic acid betaine.

Example 2: In 167 g of 60% neoprene latex contained in a one-quart paint can, 3.0 g of an amphoteric surfactant or a polyoxyethylated alkyl amine is dissolved. A slurry of clay, MgO, and ZnO in the solvent solution of resins and antioxidant is prepared according to the description in Example 1, except the surfactant is deleted. The slurry is added to the latex with hand-stirring of the latex. About 2 to 6 minutes after completion of mixing of the slurry with the latex, a mastic is obtained exhibiting excellent application rheology and adhesive properties.

Surfactants that have given excellent mastics by this procedure are N-lauryl-beta-alanine, C-cetylbetaine, C-decylbetaine, lauryl ammonium sulfonic acid betaine, and polyoxyethylated alkyl amine (Atlas G-3780-A). Mixtures of polyoxyethylated alkyl amine with polyoxyethylated sorbitol hexaoleate (Atlas G-1186) in various proportions including equal parts by weight, as well as a mixture of polyoxyethylated fatty amine and polyoxyethylated sorbitol oleate known as Atlas G-2090, have similarly given excellent mastics, in each case using 3 g total surfactant as above.

When using a polyoxyethylated fatty amine, alone or in admixture, it is advantageous to first disperse it in water to a concentration of about 25% by weight before adding it to the latex.

Latex with Phenol-Formaldehyde Resin

Phenolic resins contribute a variety of desirable properties to neoprene-based adhesives. Three significant factors in the utility of such adhesives are peel strength, the ability of the adhesive to not creep or flow under high-temperature conditions when being employed as an adhesive, (hot strength), and the ability of the adhesive to remain tacky for a sufficient length of time, (tack time) so as to be useable under normal application conditions.

Phenolic resins which are typically employed in such neoprene adhesive formulations are generally made by the reaction of formaldehyde with a substituted phenol particularly one which is substituted in the para position, such as para-tertiary butyl phenol and para-phenyl phenol.

Usually, in the manufacture of these resins one employs an excess of the formaldehyde so as to produce in the resin reactive methylol groups and methylene ether units which interbond phenol moieties therein. These resins are characterized in the art as oil-soluble, heat-reactive phenol-formaldehyde resin.

In response to ecological demands, various resins formulations which are used as coatings, adhesive, caulks and sealants, are being reformulated such that on setting or curing, a minimum amount of organic solvent or other organic volatile matter is eliminated. For example, resins which are used as particles suspended in water and containing a minimum of organic solvent, are favored.

Neoprene elastomers are available in latex form. They have been described as useful in making contact adhesives. Certain phenolic resins are described as modifiers for these neoprene latices. It is understood that such resins do not provide, in combination with the latex, an extremely effective contact adhesive.

For a complete discussion on neoprene latexes, see Carl, J.C., *Neoprene Latex, Principles of Compounding and Processing*, Elastomers Chemicals Department, E.I. du Pont de Nemours and Co. (Inc), Delaware (1962), 97 and U.S. Patent 3,425,978.

To avoid the gelation, which is considered to be the result of incompatibility, this process employs a specified sulfonated phenolic resin for incorporation and in combination with the neoprene emulsions. It was found that at least 10% sulfonate groups based on the total weight of the sulfonated phenolic resin is needed in order to achieve the level of compatibility which is necessary in order for effective incorporation in a neoprene latex.

B.P. Barth; U.S. Patent 3,887,539; June 3, 1975; assigned to Union Carbide Corporation has determined that a neoprene latex and an aqueous phenolic resin dispersion can be intermixed without the necessity of sulfonating the phenolic resin as described in U.S. Patent 3,425,978, with the consequent achievement of an adhesive composition which can be employed very effectively as a contact adhesive to provide adhesive joints which have excellent peel strength, good open tack times, good initial bond strength and excellent elevated temperature strengths, as determined by peel strength determinations.

The neoprene latexes suitable for use in this process are described the above-noted book by Carl, particularly those which are listed on the inside cover of the text. These latexes are typically made by emulsifying chloroprene in water, usually by means of a rosinate soap, and polymerized with the aid of a potassium persulfate catalyst.

Progress of the polymerization is followed by means of specific gravity changes and the polymerization may be terminated by the addition of a "short stop," if desired, or the polymerization is carried to full conversion. Usually a buffer is added to aid in maintaining proper alkalinity during storage and processing. Reference is made to Table 1 on page 2 of the text by Carl for physical properties of a variety of neoprene latex films.

The aqueous phenolic resin dispersion usable in the process is an alkaline-catalyzed reaction product of formaldehyde and a difunctional phenol compound, such as the para-substituted phenols. Illustrative of para-substituted phenol are para-alkyl phenol where the alkyl group contains 1 to 12, inclusive, carbon atoms; p-tolylphenol, p-phenyl phenol; and para-cycloalkyl phenol. Particularly desirable para-substituted phenols are para-n-propylphenol, para-isobutylphenol and para-tert-amylphenol, etc.

The condensation of a phenol and an aldehyde to form the resin is carried out in a reaction medium containing an alkaline catalyst. Illustrative of suitable catalysts are the following: alkali metal and alkaline earth metal hydroxides or oxides, as for example, sodium hydroxide, potassium hydroxide, calcium hydroxide, calcium oxide, barium hydroxide, barium oxide and the like, or an amine such as hexamethylenetetramine or ammonia.

The amount of catalyst used in producing the resole resins is 0.01 to 0.1 mol per mol of phenol, preferably 0.02 to 0.05 mol per mole of phenol. For a detailed discussion of condensates produced from a phenol and an aldehyde and methods for the production thereof, reference is made to the following: "Phenolic Resins" by W.A. Keutgen, *Encyclopedia of Polymer Science and Technology*, Volume 10, pages 1-73 Published by Interscience, John Wiley, 1969.

Example 1: To a still was added 40 parts of 37% formalin, 100 parts of para-tertiary-butylphenol and 8 parts 25% aqueous NaOH. Atmospheric reflux of the contents was carried out for 1¼ hours at which time 50 parts toluene was added to the still and the batch was cooled to 70°C by vacuum reflux. Then 2.5 parts of concentrated H_2SO_4 was added to 5 parts of water and the resulting solution to the still.

The batch was agitated 20 minutes at 70°C and then allowed to settle into two layers. The bottom water layer was removed and 30 parts of water was added to the remaining solvent layer, agitated for 20 minutes and again allowed to settle into two layers. The bottom water layer was again removed and the resin solution was stripped of solvent atmospherically to 110°C, when, at the end of the solvent stripping step, vacuum (27 to 28 inches Hg) was applied and the temperature was raised to 115°C. The resin was held at 115° to 120°C to a tripod flow of 110 to 120 seconds before discharging. The labile formaldehyde content of this resin was 0.

Example 2: A second resin synthesized by the procedure of Example 1 using, instead, 113 parts of 40% formalin/100 parts of p-t-butyl phenol. This resin had a labile formaldehyde content of 6.9%.

Example 3: The phenolic resin of Example 2 was dissolved in toluene to form a 65 weight percent resin solution. To 154 grams of this 65 weight percent solution was added 6 grams oleic acid to form the oil phase, or Part A, of the emulsion. Part B, the water phase, contained 4 grams of 25 weight percent aqueous NaOH, 30 grams of 10 weight percent ammoniated casein solution, 60 grams of 5 weight percent gum arabic solution, and 77 grams of water. The formula is summarized below.

		Parts by Weight
Part A	Phenolic Resin of Ex. 2	100
	Toluene	54
	Oleic acid	6
Part B	25% Aqueous NaOH	4
	10% Ammoniated casein*	30
	5% Gum arabic**	60
	Water	77

*To prepare this solution, 100 parts of ammoniated casein were added to 497 parts by weight of water, 3 parts by weight Dowicide G (Dow Chemical Co.), and 2 parts by weight ammonia. After stirring for 30 minutes at room temperature the mixture was heated to 80°C. and stirring continued until the dispersion was complete (1 hr.). It was then diluted to 10% solids by adding 400 parts by weight water.

**The gum arabic solution was prepared by adding one part gum arabic (Stein. Hull & Co., Inc., grade A-10) to 19 parts water and stirring at room temperature until solution was complete.

To prepare the emulsion, Part A was added to Part B with high-speed agitation using an Eppenbach mixer. Agitation was continued for at least 5 minutes.

Example 4: The adhesive consists of neoprene latex, the phenolic resin emulsion, zinc oxide dispersion, and an antioxidant dispersion. The adhesive is prepared by simple mixing of these ingredients. The formula is shown in the following table.

Adhesive Formulation

	Parts by Weight	
	Wet	Dry
Neoprene Latex 750*	200	100
Phenolic Resin of Ex. 3	167	50
Zinc Oxide	10	5
Neozone D*	4	2

*E. I. duPont de Nemours & Co. (Inc.)

The zinc oxide and Neozone D dispersions were prepared by ball-milling the following blends of ingredients for 24 hours.

	Zinc Oxide Dispersion	Neozone D Dispersion
Zinc oxide (Kadox 15, N.J. Zinc)	100	—
Neozone D	—	100
Daxad 11* solution (10%)	30	30
Ammonium caseinate (10%)	30	30
Sodium silicate solution (10%)	5	—
Water	35	40

*Dewey and Almy Chemical Division, W. R. Grace & Co.

Example 5: One significant difference between water and solvent-based neoprene contact adhesives is the level of pressure needed to obtain good contact between the mating adhesive-coated surfaces. In solvent systems, light pressure is usually sufficient to produce a satisfactory bond. With water-based adhesives, higher pressure must be applied.

Convenient methods for doing this is to use either a set of nip rolls or a Jay roller. When either of these assembly methods is used, good contactability is obtained with the adhesive of Example 4. Peel strength (pounds per square inch width) data of canvas to canvas bonds are shown in the following table.

Peel Strength* of Adhesive
(*Rate of peel in 2 in/min; 180° peel)

Open Time Hrs.	Assembly Method	Tested at 25°C. after 1 hr	1 day	1 week	3 weeks	At 70°C. after 1 week	3 weeks
¾	nip roller	8C	26C	22A	22A	15C	22C
	Jay roller	10C	22C	22C	23A	15C	22C
2½	nip roller	10C	23C	21A	22A	14C	19C

A = Adhesive failure
C = Cohesive failure

Repeating Examples 3, 4 and 5 but replacing the phenolic resin of Example 2 with a commercial p-t-butyl phenol resin that had a labile formaldehyde content of 15.1%, resulted in an adhesive that had no contact properties after ½ hour open time (following application of the final adhesive coat). Peel strength was essentially zero.

Example 6: The adhesive of Example 4 was compared with two commercial neoprene water-borne contact adhesives. Comparison of peel strengths, listed in the following table, show that the adhesive of Example 4 produces peel strengths in a commercially acceptable range with elevated temperature peel strength comparable to the best commercial adhesive but with superior green strength (initial peel strength following assembly). It was found that there was no sigificant change in peel strength within the first hour after assembly of the two adhesive-coated substrates.

	Peel Strength at 25°C.				at 70°C
Open Time = ¾ hr.					
Tested After	1 hr.	1 day	1 week	3 wks.	3 wks.
Adhesive of Example 4	13C	19C	32A	44A	25C
Commercial Adhesive A	2C	3C	52A/C	53A/C*	25C
Commercial Adhesive B	4C	21A	19A	23A	11A

*dry on edges (only ~ 30% contact)
A = Adhesive failure
C = Cohesive failure

Fire-Resistant Compositions for Building Construction

G.R. Perkins; U.S. Patent 3,843,592; October 22, 1974; assigned to Spotnails, Inc. describes a self-curing polychloroprene-base adhesive composition which exhibits superior resistance to fire and high-temperature conditions, shows reduced staining properties, exhibits good shelf life and is especially useful in the building construction field in lieu of traditional fasteners such as nails and screws.

The composition contains polychloroprene elastomer, asbestos, a combination of heat-advancing oil-soluble phenolic resins, nonstaining antioxidants and fillers. The properties of this composition are achieved by a process which includes forming a solvent solution of the phenolic resin which is admixed with magnesium or cadmium oxide. This mixture is held for a period of time for prereaction, preferably at least two weeks, or more, although shorter times will suffice. A separate mixture of neoprene, magnesium or cadmium oxide, zinc oxide, antioxidants and asbestos is formed and is thoroughly mixed in the dry state below 140°F.

The two mixtures are blended for a period of about eight hours or more to provide a high-viscosity composition and solvents are added to the mixed composition to reduce the viscosity to from 23,000 to 90,000 centipoises.

Example: An adhesive was prepared employing polychloroprene, phenolic resins, asbestos, antioxidants and fillers having the following composition in the weight percent portions listed below:

	Parts or Percent
Chloroprene (soft grain, duPont type ac neoprene)	44
Heat-reactive alkylphenolic resins	22
8% (CKR No. 1634)	
8% (SP-174)	
6% (SP-154)	
Hard clay	10
Hydrated calcium silicate	6
Asbestos (chopped fiber)	6
Magnesium oxide	5
Zinc oxide	3.5
Heat-resistant accelerator (NA 101)	2
Phenyl-alpha-naphthylamine	0.5
Nickel dibutyl dithiocarbamate	0.5
Dilauryl thiodipropionate	0.5

The foregoing materials are mixed in the following sequence procedure.

Step A — The phenolic resins are combined and mixed with an equal amount of hexane followed by addition of one-half of the magnesium oxide (or cadmium oxide, if employed). The composition is agitated for a period of 4 to 6 hours to assure full dispersal of the solid components. The mixture is preferably permitted to prereact for an additional period of up to 2 weeks or more for best results.

Step B — The neoprene elastomer is dry-milled for a period of 5 minutes on a rubber mill, after which period the remainder of the magnesium and the antioxidants are blended in, followed by another 1 or 2 minutes of mixing. Further additions of hydrated calcium silicate and asbestos are made to the mixture followed by further milling or mixing for a period of from 10 minutes to assure complete dispersion. In this specific embodiment a Banbury was employed. The rubber or neoprene material is then diced and reduced with toluene until the mixture is smooth. The accelerator is then added to the solvent-reduced elastomer and blended for a period of 1 hour. The zinc oxide is also added at this time, although it can be added to the rubber in the milling step.

Step C – The compositions resulting from steps A and B are blended and admixed for a period of 8 hours and then reduced to the desired viscosity (in this instance 52,000 centipoises) with a solvent admixture comprising: 40% hexane, 30% toluene, 20% methyl ethylketone and 10% acetone. The final product is packaged into cartridges.

The adhesives pass the Flame Penetration Tests as described in Forest Products Laboratory Fire Tests, Report No. 1443, for 20 x 20-inch plywood panels. These tests demonstrated that the adhesives were capable of withstanding the elevated temperatures of the direct flame impingement test. The test was terminated after 14 to 16 minutes due to plywood burning away prior to failure of the glue line. The test showed glue line temperatures (thermocouple-registered) of 430° to 510°F.

Single-Phase Polymerization Process

According to a process described by *P. Branlard and J. Modiano; U.S. Patent 3,872,043; March 18, 1975; assigned to Distugil, France* polychloroprene for use in adhesive compositions and having the advantage of remaining in a single phase is prepared by effecting polychloroprene polymerization in the presence of a binary emulsifying system of 1.8 to 3% of resinic derivative and 0.1 to 1% of a fatty acid or salt. Althernatively, a ternary emulsifying system may be used comprising 1 to 3% of the resinic derivative, 0.1 to 1.5% of the fatty acid or salt thereof and 0.1 to 2.5% of a sticking agent of a colophony derivative or a polydiene resin.

Use of the ternary or binary systems provides an effective operation in which the latex is stabilized sufficiently during fabrication and treatment so as to inhibit phase separation. Moreover, the presence of the fatty acid or salt permits the polymerization of the chloroprene to be effected to a conversion of 80%, in a commercially suitable time of from 8 to 10 hours.

Example 1: The following ingredients are loaded in a polymerization reactor:

	Parts by Weight
Chloroprene	100
Water	100
Resinic acid: disproportionated rosin (Phedre V)	2
Sodium oleate	0.4
Soda (100%)	0.46
Sodium salt of sulfonic methylene-bis-alkyl-naphthalene acids (Distabex LS)	0.875
n-Dodecylmercaptan	0.145

With an initial catalyst composed of:

		Parts by Weight
A)	Ferrous sulfate·$7H_2O$	0.00384
	Disodic salt of ethylenediaminetetraacetic acid (Masquol B)	0.00577
	Soda	0.00124
	Water	1.026
B)	Sodium dithionite	0.0294
	Water	0.588

Polymerization is carried out for 10 hours at +10°C by adjusting to a convenient rate the feed of an aqueous solution of ammonium persulfate. Polymerization is stopped at a conversion of 80% of chloroprene with the mixture:

	Parts by Weight
Chloroprene	0.8
Sodium lauryl sulfate	0.02
Phenothiazine	0.01
Water	2.93

Lastly, 0.4 part by weight of di-tert-butyl para-cresol in solution in latex chloroprene (2.93 parts by weight) is added. Vapor removal and latex insulation are conducted according to the methods commonly used for polychloroprene. Especially pure latex is obtained at pH 6.5 on a drum cooled at -20°C. The obtained film is washed and dried. A polychloroprene rubber containing 2.15% by weight of resinic acid is obtained.

Example 2: Loading in the polymerization reactor according to Example 1 is modified by replacing the disproportionated rosin with hydrogenated rosin, parts by weight per parts by weight and by using 0.6 part by weight of sodium oleate and 0.48 part by weight of soda (100%). Polymerization is carried out for 9 hours under Example 1 conditions. The obtained polychloroprene rubber contains 2.4% of resinic acid.

With polychloroprene rubbers obtained in the above examples, adhesive mixtures are prepared according to the following formula:

	Parts by Weight
Polychloroprene rubber	100
Magnesium oxide	8
Zinc oxide	5
Di-tert-butyl para-cresol	2
Phenolic resin (CKR 1634)	40
Methylethylketone	106
Ethyl acetate	106
Oil C*	85
Cyclohexane	169

*Oil C corresponds to petroleum cut SBP No. 2 (Standard boiling point)

After the adhesive compositions using the rubbers of Examples 1 and 2 are prepared, the possible appearance of phase separation phenomenon in the adhesive mixtures is studied. In all the cases, after 6 months of storage, no separation is observed, the mixtures always being homogeneous.

COMPANY INDEX

The company names listed below are given exactly as they appear in the patents, despite name changes, mergers and acquisitions which have, at times, resulted in the revision of a company name.

INVENTOR INDEX

U.S. PATENT NUMBER INDEX

NOTICE

PRESSURE SENSITIVE ADHESIVES 1974

Formulations and Technology

by Louis F. Martin

Chemical Technology Review No. 34

Pressure sensitive adhesives, made for instantaneous bonding with high cohesive strength, are based on film-forming elastomers such as styrene butadiene rubber, polychloroprene, polybutadiene and acrylics. Additionally, silicones and many fatty nitrogen compounds are used as release agents.

Pressure sensitive tapes, for instance, are made from the following materials: **1. A suitable tape material such as plastic, metal foil, paper or other nonwoven fabric. 2. A primer coating sometimes necessary on plastic, paper, or metal foil. 3. The adhesive itself. 4. A release coating which must be applied to the back side of the tape so that the user can unroll the tape.**

In view of the wide utility and consumer acceptance of these adhesives, it was considered necessary that this book describe not only the synthetic resin processes which are basic to this technology, but to provide hundreds of formulations for developing new and improved adhesives and to describe in some detail the many possible end uses. 220 different processes are described.

A partial and condensed table of contents is given here. Numbers in parentheses indicate a plurality of processes per topic. Chapter headings are given, followed by some important subtitles.

ISBN 0-8155-0540-X

308 pages

NONWOVEN TEXTILES 1975

by M. S. Casper

Chemical Technology Review No. 43

The continual growth of nonwoven textiles is based on their ability to outperform conventional woven fabrics at lower cost, wherever nonwovens can be used, and their application sphere increases steadily.

This large sphere of applications is best illustrated by the headings in the end use section of this book. It was therefore considered necessary to describe in detail all major routes to nonwovens, such as dry-laid webs, wet processes (modified papermaking techniques), spunbonding, and the manufacture of laminated materials.

Nonwovens can be engineered for specific end uses (such as form-fitting diapers or specially-shaped conveyor belts) easier than woven goods, and the most recent developments for doing this have been included. More than 400 processes based on recent patents are described.

A partial and condensed table of contents follows. Numbers in parentheses indicate a plurality of processes per topic. Chapter headings and some of the more important subheadings and subtitles are given here.

ISBN 0-8155-0561-2 **414 pages**

HOT MELT ADHESIVES 1974

Manufacture and Applications

by M. J. Satriana

Chemical Technology Review No. 33

Hot melt adhesives are solid at room temperature, but fluid when hot as specified by their melting range. Depending upon their formulation, they form a hard, more or less flexible bond instantly upon cooling. No curing time is required, as is the case with thermosetting adhesives. Because hot melts are 100% solid, no evaporation of solvents, and no drying equipment is required. There is a high antipollution value built in, since no flammable or toxic fumes are evolved from hot melts.

In addition to rapid setting time, safety and low cost, other advantages are resistance to moisture and freezing, ability to bond impervious surfaces, reduced machinery cleanup-time and stability during storage.

Hot melts have gained wide acceptance in packaging, shoemaking, and bookbinding. They are being used increasingly in product assemblies of all kinds.

This book gives considerably more information than our previous volume on this subject. It now describes 134 processes in 6 chapters. A partial and condensed table of contents follows here. Numbers in parentheses indicate the number of processes per topic. Chapter headings are given, followed by examples of important subtitles.

ISBN 0-8155-0538-8

301 pages

POLLUTION CONTROL IN THE PLASTICS AND RUBBER INDUSTRY 1975

by Marshall Sittig

Pollution Technology Review No. 18

Air pollution of the plastics industry are assuming greater importance as more becomes known about the toxicity of such monomers as vinyl chloride, vinylidene chloride and styrene. Stricter standards are being set by OSHA (Occupational Safety and Health Administration, U.S. Dept. of Labor). Design and control techniques are being improved to permit conformance to these standards.

Water pollution problems of the plastics industry are being given increased attention by EPA (Environmental Protection Agency). Waste effluents are being more strictly characterized and treatment methods adapted to meet zero discharge standards.

Solid waste problems of the plastics industry consist of the twin problems of reuse and ultimate disposal. Reuse of plastics waste is as yet of little commercial value, but deserves continuing research and economic evaluation. Incineration requires most careful manipulation and furnace design to avoid the generation of noxious and corrosive gases. Research thus far has not produced plastic packaging materials which are truly biodegradable. High pressure compaction of solid waste for sanitary landfills, however, represents a temporary working solution.

The solid waste problem of the rubber industry is truly monstrous. The problem of handling discarded tires has become worse with the introduction of the steel-belted tire, adding weight and impossibility of comminution. The most practical disposal of waste rubber appears to be pyrolysis or energy recovery, but such technology is unknown and unproven in the market place. In addition, a critical cost and logistics problem is the collection of tires to central processing locations. Fiscal subsidies are clearly indicated here.

In the meantime much can be done by industry during manufacture of raw materials, intermediates and end products. Therefore the present volume is not just a book on pollution control in the plastics and rubber industries, but also a comprehensive treatise on plastics and rubber manufacturing processes and product finishing. The point of view is an integrated one: New and timely reviews of industrial practice are given together with constant attention to air pollution, water pollution, and solid waste problems.

PLASTIC PRODUCTS INDUSTRY

Air Pollution

Water Pollution

Solid Waste Generation From Specific Operations
- Compounding
- Injection Molding
- Film & Sheet Extrusion
- Extrusion Coating & Coating of Wire & Cable
- Pipe, Rod, Tubing & Special Profile Extrusion
- Coextrusion
- Blow Molding
- Rotational Molding
- Compression & Transfer Molding
- Dip & Slush Molding
- Casting
- Calendering
- Cellular Plastics
- Thermoforming
- Laminating
- Coating
- Reprocessing

RUBBER PRODUCTS INDUSTRY

Air Pollution

Water Pollution
- Tire & Inner Tube Manufacture
- Molded Products
- Extruded Products
- Fabricated Products
- Reclaimed Rubber Production
- Latex-Based Products

Aqueous Waste Characterization
- Tire & Inner Tube Area
- Molded, Extruded & Fabricated Area
- Reclaimed Rubber Area
- Latex-Based Products Area

Water Pollution Control & Treatment Methods
- Tire & Inner Tube Area
- Fabricated & Reclaimed Products Area

Water Pollution Control Economics
- Tire & Inner Tube Area
- Molded, Extruded & Fabricated Area
- Reclaimed Rubber Area
- Latex-Based Products Area

Solid Waste Generation from Manufacture of Specific Products
- Tire Manufacturing
- Footwear Manufacturing
- Belting Manufacturing
- Hose Manufacturing
- Sponge & Foam Rubber Product Manufacture
- Mechanical Rubber Goods Manufacture
- Wire & Cable Manufacture

RECOVERY OF PLASTICS PRODUCTS FROM SOLID WASTE

Separation of Plastics from Mixed Waste
- Air Classification
- Liquid Media Separation
- Screening & Electrostatic Separation
- Solvent Extraction
- Proposed Combination Processes

Refabrication

Chemical Recovery by Pyrolysis
- Polyvinyl Chloride
- Polyvinylidene Chloride
- Polytetrafluoroethylene
- Polyurethanes
- Polyolefins
- Polystyrene
- Mixed Plastics

Recovery by Solvent Extraction

Manufacture of Composite Materials from Waste Plastics

RECOVERY OF RUBBER PRODUCTS FROM SOLID WASTES

DISPOSAL OF PLASTIC PRODUCTS

Intermediate Disposal

Collection

Treatment Processes
- Baling
- Pulverizing, Shredding & Milling
- Melting

Environmentally Disposable Plastic Products
- Biodegradable Plastic Products
- Water-Soluble Plastic Products
- Photodegradable Plastic Products

Incineration

Composting

Landfills

DISPOSAL OF RUBBER PRODUCTS

Incineration

Chemical Recovery by Pyrolysis

Other Waste Rubber Utilization Techniques
- Road Building
- Reef Building

FUTURE TRENDS

LITERATURE REFERENCES

ISBN 0-8155-0572-8 **306 pages**

POLLUTION CONTROL AND CHEMICAL RECOVERY IN THE PULP AND PAPER INDUSTRY 1973

by H. R. Jones

Pollution Technology Review No. 3

The pulp and paper industry has serious problems to solve, both in the air pollution and water pollution areas.

Many effluent wastes from paper mills are biodegradable, but treatment costs are increasing, effluent discharge requirements are becoming more stringent, and urbanization increasingly limits the availability of land. Economic control procedures applied by the industry, therefore, form an essential part of the pollution control procedures.

In this book are condensed vital data from government sources of information that are scattered and difficult to pull together. Important processes are interpreted and explained by examples from 54 U.S. patents. One should have to go no further than this condensed information to establish a sound background for action towards combating pollution in the pulp and paper industry.

A partial and condensed table of contents follows. Chapter headings are given, followed by examples of important subtitles.

ISBN 0-8155-0479-9 **337 pages**

PAPER MANUFACTURE 1975

by M. G. Halpern

Chemical Technology Review No. 47

This volume describes recently developed processes for converting fibrous pulp into paper. Since paper today is manufactured to possess specific properties for specific applications, it is necessary to add other substances to the cellulose fibres. The pulp is diluted with water and loaded with various additives, before it is transferred to the paper machine, where a wet web is laid from the pulp.

This wet end loading determines the properties of the finished product. It is at this stage that pigments, fillers, sizes, retention aids, compounds for improving strength and other physical characteristics are added.

Cost cutting, shortages of raw materials, and demand factors, such as light weight, but strong paper for periodicals because of high postage rates, have produced a most interesting and efficient technology described here in detail. This includes treatments for increasing the drainage of the white water from the furnish, forming machine variations, evaluation of fibers while processing, drying improvements, as well as computerized systems for controlling moisture, fiber content, base weight, etc.

A partial and condensed table of contents follows. Numbers in parenthesis indicate a plurality of processes per topic. Chapter headings and some of the more important subtitles are given here.

ISBN 0-8155-0576-0 **379 pages**